**Protozoology
2nd edition**

Protozoology

Klaus Hausmann
Norbert Hülsmann

with contributions by
Hans Machemer, Maria Mulisch,
and Günther Steinbrück

Foreword by John O. Corliss

772 illustrations
10 tables

2nd edition

1996
Georg Thieme Verlag
Stuttgart · New York

Thieme Medical Publishers, Inc.
New York

Prof. Dr. Klaus Hausmann
Freie Universität Berlin
Institut für Zoologie
AG Protozoologie
Königin-Luise-Straße 1–3
D-14195 Berlin
Germany

Dr. Norbert Hülsmann
Freie Universität Berlin
Institut für Zoologie
AG Protozoologie
Königin-Luise-Straße 1–3
D-14195 Berlin
Germany

Prof. Dr. Hans Machemer
Universität Bochum
Lehrstuhl für Allgemeine
Zoologie und Neurobiologie
AG Zelluläre Erregungsphysiologie
Universitätsstraße 150
D-44780 Bochum
Germany

Priv.-Doz. Dr. Maria Mulisch
Universität zu Köln
Zoologisches Institut
Experimentelle Morphologie
Weyertal 119
D-50931 Köln
Germany

Dr. Günther Steinbrück
Universität Tübingen
Zoologisches Institut
Abt. Zellbiologie
Auf der Morgenstelle 28
D-72076 Tübingen
Germany

Library of Congress Cataloging-in-Publication Data

Hausmann, Klaus.
  [Protozoologie. English]
  Protozoology / Klaus Hausmann, Norbert Hülsmann : with the cooperation of Hans Machemer Maria Mulisch, and Günther Steinbrück.
     p. cm.
  Includes bibliographical references (p.) and index.
  1. Protozoology. I. Hülsmann. N. II. Title.
QL366.H3613   1995
593.1--dc20   95-643
                                                  CIP

1st German edition 1985
1st Russian edition 1988
1st Japanese edition 1989

© 1996 Georg Thieme Verlag, Rüdigerstraße 14,
D-70469 Stuttgart, Germany

Thieme Medical Publishers, Inc.,
381 Park Avenue South,
New York, NY 10016

Typesetting by primustype Robert Hurler GmbH,
D-73274 Notzingen

Printed in Germany by
K. Grammlich, D-72124 Pliezhausen

ISBN 3-13-110301-9 (GTV, Stuttgart)
ISBN 0-86577-571-0 (TMP, New York)
                           1 2 3 4 5 6

This book is an authorized and revised translation of the 1st German edition published and copyrighted 1985 by Georg Thieme Verlag, Stuttgart, Germany. Title of the German edition: Protozoologie.

Some of the product names, patents and registered designs referred to in this book are in fact registered trademarks or proprietary names even though specific reference to this fact is not always made in the text. Therefore, the appearance of a name without designation as proprietary is not to be construed as a representation by the publisher that it is in the public domain.

Foreword

In this day and age of ever-increasing interest in the "lower" eukaryotes—the protists—there is a pressing need for a general textbook that will accomplish two principal goals. Eager students are demanding a clear exposition of the most modern advances in biochemical and molecular biological research on these fascinating microorganisms. At the same time, they want to know what the unicellular algae and protozoa actually look like under the microscope as whole organisms, and what roles they may play in the overall bionomics of life on Earth. Such young biologists are in luck, for the present volume appears to meet those two criteria admirably, and more besides!

Authors Hausmann and Hülsmann are abundantly well qualified to carry out the objectives mentioned above. Their own numerous cytological (including ultrastructural) investigations on a diversity of protists are highly respected by the protozoological community, and their recent forays into various physiological, ecological, systematic, and evolutionary areas have also resulted in significant contributions to the research literature. Their Leeuwenhoekian and Ehrenbergian love for their "wee beasties" is clearly evident throughout the book and is beautifully manifest in the several films that they have produced.

Readers will recall an earlier (1985) edition of Hausmann's Protozoologie and may wonder about differences between that and this newer version. In addition to the obvious fulfillment of "up-dating" in every area treated, there are several outstanding features that deserve brief mention. Most delightful to many budding protozoologists/protistologists around the world, it must be confessed, is the appearance of the new edition in the English language! There are more than 100 additional pages of text, and each page is larger in size. There are also over 100 additional plates of figures, giving the reader a better visual understanding of both whole organisms and of their numerous organelles as seen in sectioned material. Many of the figures are original, prepared purposely for use in this textbook. There are line drawings, life cycle outlines, etc. in addition to strikingly attractive light and electron micrographs.

The authors have been even bold enough to present a phylogenetically based classification of the protistan groups covered in this book. Undoubtedly, we shall be seeing revisions in this controversial area in the future, with availability of still newer molecular data, but it is salutary to note changes over the past and to see where we stand at the moment in this exciting field of evolutionary protistology.

Hopefully, this Foreword (if read!) will forewarn and foretell the student what to expect: so, go ahead, turn the page!

John O. Corliss
University of Maryland

Preface to the Second Edition

With the second edition of Protozoology, we are meeting an often-heard wish among students and teachers outside of Germany to publish this textbook in English as well, so that it can be used in protozoology courses worldwide. The first edition appeared 10 years ago in German, and was a great success.

The book is aimed at introducing students to the amazing and bewildering world of protozoa by giving basic information on the biology of these creatures. Since the publication of the first edition in 1985, there has been tremendous progress in biological research, especially in the fields of cell and molecular biology, as well as phylogeny and ecology. These new developments have been accounted for in this second edition. This is especially evident in our treatment of such topics as phylogeny and taxonomy. We discuss—probably for the first time in a protozoological textbook—a radically different, but accessible approach to the phylogenetic relationships of the eukaryotic organisms. We are aware that some of our colleagues will passionately disagree with these new ideas, but we are absolutely sure that others will agree with them. Moreover, they may use this compilation of information as a basis for further considerations and deductions. We acknowledge that we are still a long way from a complete understanding of protozoan and eukaryotic phylogeny.

We also try to present an up-to-date synthesis of the many other facets of the biology of protozoan organisms. We have therefore made extensive use of illustrative material.

As for the first edition, we thank many colleagues for providing with light- and electron micrographs and diagrams that bring this book to life. We indicated the exact sources of the corresponding illustrations in the corresponding figure legends.

Of the numerous persons responsible for the appearance of this book, a few should be mentioned individually. First of all, we thank Bland J. Finlay, Institute of Freshwater Ecology, Windermere Laboratory, Ambleside, United Kingdom, who helped in making the English more readable. Secondly, Peter Adam, scientific illustrator of our institute, should be mentioned for his patience and painstaking endurance during the preparation of the numerous drawings and diagrams adapted or newly designed for this book. Finally, we warmly acknowledge the cooperation of our colleagues Hans Machemer, University of Bochum, Maria Mulisch, University of Cologne, and Günther Steinbrück, University of Tübingen, for contributing the chapters Behavior (HM), Nuclei and Sexual Reproduction (MM), Morphogenesis and Reproduction (MM), and Molecular Biology (GS). They contributed greatly to the book with their specialized knowledge.

Furthermore we wish to thank Margrit Hauff-Tischendorf, Rainer Zepf, and Lieselotte Brlečić at Georg Thieme Verlag, Stuttgart, for their patience and cooperation during the final preparation and eventual realization of this book.

As usual: last, but not least and privately, our families should not be forgotten for being neglected to a remarkable degree for a long time in favor of the beloved protozoa during the preparation of this new edition.

Berlin, September 1995   Klaus Hausmann
Norbert Hülsmann

Contents

System of Protozoa

Part I: Introduction and Overview

Definitions and History of Nomenclature

Fig. 1 Compilation of the most important nomenclatural designations for protozoa (graphic by M. Gradias, Wolfenbüttel).

This book deals with the organism we usually call protozoa. Just what are the protozoa? We start the book with one of the most difficult questions to answer. While the remainder of the book represents an assemblage of factual information concerning the protozoa, the fundamental issue of the definition of the group is difficult and rather controversial. This is partly due to the rapid expansion of our knowledge concerning these organisms, and especially to the relatively recent rejection of the idea that protozoa represent a monophyletic taxonomic unit. Thus, a phylogenetic–systematic definition of the protozoa cannot be given. We must consider the protozoa as a paraphyletic (or even polyphyletic) assemblage of small organisms of mostly microscopical dimension, which do not constitute a natural group.

Controversial definitions of the protozoa have been a characteristic of three centuries of research. In 1818, the term "protozoa" was introduced by the German zoologist and paleontologist Georg August Goldfuß (1782–1848). Only the Greek prefix "proto-" (= first) and the Greek suffix "-zoa" (= living creatures, animals), or the etymological realities were clear (as in the German designation "Ur-Thiere"). We now know that the protozoa were not the first creatures on earth: the pioneer role was taken by the prokaryotic-organized unicells, possibly the fermenting bacteria.

The approach of a modern understanding should take account of the term "zoon." The use of "zoon" in terms such as zoology, zoogeography or zoophysiology led to the assumption that protozoa also are exclusively animal-like creatures, or animals that have to be carefully separated from their green, plant-like relatives, the "protophytes" (Ernst Haeckel 1866). This separation has grown historically, become established in tradition, and has been partly maintained with some fanatical devotion. However, the differentiation of our animalcules into plant-like/phototrophic and beast-like/heterotrophic organisms bears problems. These become obvious when monophyletic taxa, such as the euglenids or dinoflagellates, are grouped depending on the presence or on the primary or secondary absence of plastids, as plant-like or as animal-like creatures within the corresponding zoological or botanical systems. In the middle of the last century, this unsatisfactory situation led to efforts to establish, alongside the systems of botany and zoology, an independent system which accommodated, more accurately and formally, the "exceptional position" of the lower eukaryotes.

The frequently used circumscription of protozoa as unicellular organisms or unicells avoids such difficulties, but it is still inadequate. Of course, most protozoa are organized as cells, but not all are fixed in their structural organization to the status of a unicell. There exist, for instance, cases where many similar conspecific individuals are able to gather together to form temporary feeding-communities (in heliozoans) or to form longer-lasting syncytia (in vampyrellids). In other cases, the daughter cells do not separate after karyokinesis, at least not completely; they form colonies, as in the Volvocida, or they grow to the multinucleated plasmodia of acellular slime molds, as in the Myxogastra. Also, multicellular creatures occur that have a complex architecture and differentiate into specialized cell types, as in the Myxozoa.

The problem of characterizing the protozoa with the necessary brevity, however, is not the result of etymological restraints, but of the fantastic diversity of organisms that we now know and must be embraced by the term. In contrast to the scientific designations for species, genera, and families, the suprafamilial categories of ranking (and therefore also the term "protozoa") are not subject to the restrictive rules of nomenclature. They are maintained as terms for the mostly great taxa in which organismic composition is often unknown, and neither the name of the author nor the year of publication are given. As in other disciplines, it is useful to give a historical overview of the conceptual framework.

In the following, the most important terms are listed in chronological order. They give some information about the criteria used for distinguishing the protozoa from other groups of organisms (Fig. **1**).

Animalcula (Antoni van Leeuwenhoek 1676). The term is a diminutive of "animalia" and was used in the sense of "water insects" or "small animals." The collective term embraced all microscopic creatures collectable from standing rain waters, springs, lakes, rivers, ect., but also from body fluids of higher animals. Even sperm cells ("spermatozoa") were included.

Monads (Gottfried Wilhelm Leibniz 1714). The term was adapted from mathematical and philosophical theories of the Greek and Roman classical eras and came into fashion again after the first observations made using microscopy. Primarily used in a metaphysical sense, the word served to designate the indivisible, unbegetable and everlasting smallest units, which were thought to be the elements and sources of all creatures. Because the monads (according to Leibniz, with an origin in the unification of soul and matter) could later be demonstrated as visible realities, as in the example of a spermatozoon or a unicellular flagellate, the term was adapted by naturalists for nomenclatural purposes, mainly with reference to flagellates, e.g., *Monas* O.F. Müller, 1786; Cryptomonadina Ehrenberg, 1838; Chrysomonadida Engler, 1898; Diplomonadida Wenyon, 1926; Trichomonadida Kirby, 1947; Proteromonadida Grassé, 1952; Prasinomonadida Christensen, 1962. At present, the flagellated phase in life cycle of some algal groups, or the flagellate organization in general, is designated as "monadial" or "monadoid."

Infusoria or **Animalcula Infusoria** (Martin Frobenius Ledermüller, between 1760 and 1763). Aufgußtierchen or infusion animals: The term primarily embraces all organisms which are able to produce desiccation-resistant stages (as, for instance, also rotifers) and which can be reactivated by an infusion of water to hay or pepper contaminated with such resting stages. Jean Baptiste de Lamarck (1744–1829) established the zoological taxon "Infusoria" alongside other "invertebrates". Since the acceptance of the cell theory in the 19th century, and later on until the mid–20th century (and even later in the Russian literature), the term has been used exclusively as a synonym for the ciliates.

Urthiere (Lorenz von Oken 1805). This German term served as a synonym for Infusoria, but also for the separation of unicellular organisms from higher plants and metazoa. It is the philological basis for the term protozoa (= Urtiere), but the name was not used in this taxonomic sense.

Protozoa (Georg August Goldfuß 1818). The "Urtiere" embrace, besides the Infusoria in the sense of Ledermüller, also some Cnidaria, Spongia, and Bryozoa. The term was used after 1845, when Carl Theodor von Siebold formulated the definition that they represent animals that can be reduced to the status of a cell. At this point, the cell theory was incorporated into the new branch of systematic biology.

Animalia Microscopica (Jean Baptiste Bory de Saint-Vincent 1826). Synonym for "Infusoria." However, for the bell animalcule *Vorticella* and its relatives, a new kingdom ("regne psychodiaire") was erected (1822–1831).

Eithiere or Oozoa (Carl Gustav Carus 1832). Synonym for Ledermüller's Infusoria and Goldfuß's Protozoa. However, this therm did not incorporate the rather attractive idea of organisms retaining protozoan-like characters at the initial stages of their development.

Archaezoa (Maximilian Perty 1852). This etymologically beautiful expression, meaning "original creatures," was used initially as a synonym for Protozoa. This meaning has not survived. The term is presently used (spelled Archezoa) to designate original heterotrophic eukaryotes without mitochondria.

Microzoaires (Emile de Fromentel 1874). This term was used exclusively for various microscopic unicellular creatures and has fallen out of use.

Against this more or less zoologically oriented terminology (despite the compromises made by Peter Simon Pallas [1741–1881] and Felix Dujardin [1801–1860] in establishing the taxa Zoophyta or Zoophytes Infusoires), new ideas developed in the middle of the last century. They arose from multi-kingdom concepts based on genealogical principles. As a first step, Rudolf Leuckart (1822–1898) excluded the heterotrophic unicellular organisms from the animal kingdom (1848).

Acrita (Richard Owen 1861). This kingdom of "nondifferentiated cells", which contains not only the protozoa but also some smaller metazoans, all characterized by their morphologically indifferent architecture, was erected alongside the classical zoological kingdom ("Animalia") and the botanical kingdom ("Vegetabilia"). Today the term is no longer in use.

Protoctista (John Hogg 1861). With the "Protoktista" (Greek, = first creatures), a kingdom was erected which documents separation from the kingdoms of the animals, plants, and molds. Nowadays the term is used in combination with the five-kingdom concept of R. H. Whittacker (1959). The four eukaryotic kingdoms of Plantae, Animalia, Fungi, and Protoctista are held to be quite separate from the kingdom Prokaryota (or Monera). The term allows a precise but negative characterization: Protoctista are those microscopic and macroscopic eukaryotes that remain after exclusion of (1) all animals developing from a blastula, (2) all plants developing from embryonic stages, and (3) all higher Fungi without a flagellate stage in their life cycles (Margulis et al., 1990). Thus the Protoctista embrace, besides protozoa, Phaeophyta, Chytridiomycota, Oomycota, Rhodophyta, and other taxa.

Protista (Ernst Haeckel 1866). According to the various definitions given by Haeckel himself, the taxon Protista contains many unicellular but not exclusively eukaryotic organisms. Therefore, it is different from the unequivocally phototrophic plants (including unicellular green algae such as *Closterium*) and the clear animals (metazoa and ciliates), according to the equation: protophyta + protozoa = protista. In the modern view, the protists are eukaryotic organisms of cellular organization. Therefore, the term embraces classical protozoa, unicellular autotrophic organisms such as diatoms, and also lower unicellular fungi.

As seen by this list of terms with their short definitions, biological concepts and interpretations are involved even in the nomenclature. From the list, only four designations have survived: Protozoa, Archezoa, Protoctista, and Protista.

We will continue to use the term protozoa. This is only partly due to the reason of priority; mostly, however, because the other terms are not better alternatives. It should be emphasized that with this term, no monophyletic or holophyletic group of organisms has been assumed. This is also likely to be true for the future because in the phylogenetic sense, the "Protozoa" do not represent an evolutionary lineage. We can, however, use the term for the designation of creatures based on their body plan and the plesiomorphic character of their eukaryotic cellular organization.

Historical Overview of Protozoological Research

ANTONIUS A LEEUWENHOEK.

Regiæ Societatis Londinensis membrúm.

J Verkolje pinx: A de Blois fec:

Fig. 2   Antoni van Leeuwenhoek, initiator of scientific microscopy.

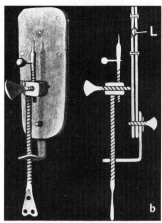

Fig. **3**   Antoni van Leeuwenhoek in 1670 (**a**). Leeuwenhoek's microscope has only one lens (L) with a magnification power of up to 250 times (**b**) (from: C. Dobell: Antony van Leeuwenhoek and his "little animals." Bale, Sons & Danielsson, London 1932).

Since the protozoa are almost always organisms of microscopic dimension, they could only be detected with certainty following the development of suitable magnifying instruments. Nevertheless, the compound microscope (composed of both an objective and an ocular lens) was invented long before unicellular organisms were discovered. Around 1590, the Dutch opticians Hans and Zacharias Janssen built the first of such microscopes. Initially, they were not accepted as bona fide scientific instruments but rather as upper class, adult toys. The starting point for scientific protozoology dates from about 80 years later.

The discoverer of unicellular organisms and hence the "father of protozoology" was the Dutchman Antoni van Leeuwenhoek (1632–1723) (Figs. **2** and **3a**). As a hobbyist, he used simple microscopes of his own design and manufacture. He cut and ground his own lenses and fabricated them into simple microscopes in a special support (Fig. **3b**). They were all single, high power magnifying lenses. It has been said that he built about 400 of these microscopes over the years and retained them for his own use.

After training to be a merchant in Amsterdam, van Leeuwenhoek moved back to his home town

of Delft, Netherlands, where he was the local draper and also acted as a member of the town council. As he was not a professional biologist who carried out his studies as a consequence of training or occupation, he must have been considered an unusual character who pursued his hobby with vigor motivated by his insatiable curiosity. He described his new findings in over 100 letters, which he sent to the Royal Society of London. His formal observations began in 1676, and van Leeuwenhoek was the first to observe, draw, and describe the protozoa (mainly ciliates). He called his little creatures the "animalcula." Two years later, in 1678, the famous Dutch physicist Christian Huygens (1629–1695) confirmed van Leeuwenhoek's findings. This independent confirmation spurred intense research by natural scientists.

Both van Leeuwenhoek and Huygens believed that the protozoa they observed in standing waters or in hay infusions originated from "air germs." The question of the origin of these protozoa would become the cause of animated disputes and discussion until the middle of the 19th century.

The next important date in protozoology and microscopy was 1718, when the French scientist

Louis Joblot published a book on the applications of the microscope. In addition to his consideration of different types of microscopes, Joblot illustrated a variety of protozoa. He included some of the first descriptions of subcellular structures. Nuclei, contractile vacuoles, ciliature, and "intestines" of ciliates were described in some detail. Naturally, it was not possible for Joblot to explain the significance of these structures, although he did reflect on the question of the origin of his protozoa. He concluded that they would contain eggs that develop through stages resembling embryos and fetuses to identical images of their parents. Such ideas about the ontogeny of protozoa were explainable at that time nothing was known about the existence of cells and their ability for reduplication. Nevertheless, he carried out experiments on the origin of the protozoa with boiled and unboiled hay infusions. As protozoa did not appear in boiled infusions, but did appear in natural infusions or those exposed to normal air, he concluded that the "eggs" must be found in the air, and if they drop into the water, the infusoria or animalcula develop. Thus the air-germ theory of van Leeuwenhoek and Huygens was reborn.

In 1727 an anonymous Parisian physician based a satire on these ideas. He claimed that the air is completely filled with "animalcula" and "homunculi" that cause different diseases. He gave the creatures different names such as bellyache-ists, diarrhea-ists, pestilence-ists, faint-ists, sensual-ists, and so on. In a second publication, he went on to describe the antagonists to these creatures, calling them anti-bellyache-ists, anti-sensual-ists, and so on. Further, he explained how they could be used to combat disease. If we regard the animalcules and homunculi as pathogenic agents, he was right in some respects even though he was actually trying to ridicule these ideas.

Of far more importance to the scientific community was the doctrine of abiogenesis ("generatio spontanea"), which was introduced in 1749 by the French zoologist George Buffon (1707–1788) and the English naturalist John Needham (1713–1781). According to Buffon, the doctrine means that plants and animals are composed of organic, living molecules that are ingested with food. At puberty, surplus molecules are deposited in spermatozoa. The infusoria are evidence of those living molecules, which are then liberated when plants and animals deteriorate. This theory was held to be valid until the 19th century, when the German natural philosopher Ludwig von Oken (1779–1851) wrote in 1805

that the biogenesis of spermatozoa (which were later identified by himself as his "Urthierchen") would be the result of a vital putrefaction within the testicles, a process continuing in the female after pregnancy and resulting in the formation of the fetus from decay products of blood.

Needham used a different approach when he stated that organic matter is able, due to a "principle of expansion," to create life under favorable circumstances. In some cases, however, it is more difficult than in others. The degree of difficulty is due to a "principle of resistance" that is inherent in all matter. The decay of organic matter is then the process whereby the principle of resistance is broken down. The final product of this decay is a gelatinous mass which he called "zoogloea," from which new life can eventually arise.

In support of their postulates, both Buffon and Needham carried out experiments with boiled and unboiled, covered and uncovered infusions. According to their reports, infusoria appeared in all cases without exception. The doctrine of spontaneous generation was enthusiastically accepted by the scientific community.

After a few years, however, the Italian scientist Lazzaro Spallanzani (1729–1799) argued vehemently against this dogma. He built his argument on a foundation of experimental evidence and differentiated between more or less heat-resistant organisms. In spite of his investigations and insight, which were in accordance with the modern approach of science, he was not recognized for finally disproving spontaneous generation. Buffon, Needham, Oken, and Lamarck remained unconvinced until their deaths. Not until the experimental work of Louis Pasteur (1822–1895) and Robert Koch (1843–1910), was the hypothesis refuted definitively.

The philosophizing about general human questions, such as the origin of life and the finality of death, was accompanied by more pragmatic attempts to collect and order the living world visible under the microscope. Following Joblot, numerous articles were published that dealt with the protozoa.

As examples, we might mention Henry Baker and his 1754 book "Beiträge zu nützlichem und vergnügendem Gebrauch und Verbesserung des Microscopii" (= Contributions for the useful and amusing employment and improvement of the microscope) and August Johann Rösel von Rosenhof, the discoverer of amoebae (1755). In 1769 Nicolas Theodore de Saussure was the first to observe transversal division of the infusoria (= ciliates); he was also the first to establish a clonal

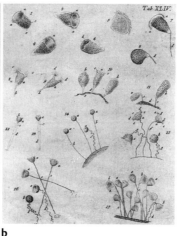

Fig. **4**   Title page of O. F. Müller's work (**a**) and one of the tables in the book (**b**).

a   b

culture, that is, a culture started with a single organism. Around 1775, the process of encystment was first observed, and it was about this time that experimental research began. New observations, including the description of newly discovered organisms, have continued until the present day.

In 1768 the first systematic plan was established for the Infusoria by the Danish naturalist Otho Fridericus Müller (1730–1784; Fig. **4**). His systematics included rotifers, in addition to planktonic metazoa and those organisms we presently consider to be protozoa. As his book appeared after Carl von Linné (1707–1778) and therefore after the starting point of the scientific biological nomenclature, many valid genera and species are combined with his surname; e.g., *Monas, Ceratium hirundinella, Bursaria truncatella, Euplotes patella, Lacrymaria olor, Stylonychia mytilus.*

The early 19th century was a very exciting and interesting era for protozoology. On the one hand, there were the most important developments regarding terminology, on the other, the initiation of detailed taxonomic research. The scientific activity of the French micropaleontologist Alcide D'Orbigny (1802–1857) may be considered the starting point of this era. He is responsible for the taxon "Foraminifera" (= organisms with shell chambers connected by holes).

In contrast to D'Orbigny, who focused his interests exclusively on the inanimate shells, his compatriot Felix Dujardin (1801–1860) also investigated the living foraminifers. He detected the living matter inside the shells and their ability to produce pseudopods and undergo active movements. This living substance, which he described also from other unicellular organisms and which he considered as a general (homologous) character, was named "sarcode" (according to the Greek term for "meat-like"). His 1835 thesis, the so-called "sarcode doctrine," states that the "living substance" is a motile fluid with vacuoles and granular inclusions. However, the "living substance," which is synonymous with "protoplasm," fell into disuse and survived only in the designation for the taxa "Sarcodina" or "Sarcomastigophora", which are no longer in use.

After the emergence of the protoplasm theory and cell doctrine, originated in the mid–19th century by the Czech Jan Evangelista Purkinje (1787–1869) and the Germans Theodor Schwann (1810–1882) and Matthias Jacob Schleiden (1804–1881), new horizons opened for protozoologists. Could the results of work on cells of higher animals and plants also apply to protozoa? At first, the complicated architecture of ciliates did not seem to fit with the cell concept.

The special status of the ciliates led two Germans, Ernst Haeckel (1834–1919) and Christian Gottfried Ehrenberg (1795–1876), to neglect their unicellular nature. Haeckel hesitated until 1873 to group the ciliates within his Protista, and Ehrenberg asserted that the unicellular organisms are "perfect" miniature animals that reflect the macroscopic, visible fauna. This is summarized in the so-called "polygastric" theory. From his point of view, it should not be surprising to find that he described stomach and intestines (= food vacuoles), vascular systems, salivary glands (= special vacuoles or probably zoochlorellae), testicles (= micronuclei) with

DIE

INFUSIONSTHIERCHEN

ALS

VOLLKOMMENE ORGANISMEN.

EIN BLICK IN DAS TIEFERE ORGANISCHE LEBEN
DER NATUR.

VON

D. CHRISTIAN GOTTFRIED EHRENBERG
ZU BERLIN.

NEBST EINEN ATLAS VON VIERUNDSECHSZIG COLORIRTEN KUPFERTAFELN,
GEZEICHNET VOM VERFASSER.

LEIPZIG,
VERLAG VON LEOPOLD VOSS,

1838.

a   b

Fig. **5** Title page of Christian Gottfried Ehrenberg's most famous book (**a**) and of the accompanying pictorial atlas (**b**).

seminal vesicles (= contractile vacuoles), and ovaries (probably macronuclei) in his marvellous 1838 book "Die Infusionsthierchen als vollkommene Organismen" (= The infusion animalcula as perfect organisms) (Fig. **5**). However, even with these pardonable misinterpretations, Ehrenberg is one of the protagonists of protozoology – so important are his merits in the fields of systematic zoology and micropaleontology. He showed that chalk cliffs are composed of microscopic organisms (mostly diatoms), and he described many free-living protozoa. The names of numerous well-known genera are a tribute to his productivity: *Actinophrys, Amoeba, Arcella, Bodo, Carchesium, Chlamydomonas, Cryptomonas, Dinobryon, Euglena, Euplotes, Loxodes, Nassula, Peridinium, Prorodon, Spirostomum, Synura,* and so on.

Around 1840, Ehrenberg's views were violently attacked by Dujardin and the German zoologist Carl Theodor von Siebold (1804–1885). For both it was not too hard to imagine that an animal could consist of only a single cell. Siebold tried to lay the dispute to rest by defining the protozoa as "animals, in which the different organ systems are not sharply defined and which consist of a single cell." In time his concept was accepted and it was he who was the first to separate the protozoa from the multicellular metazoa. It is also interesting to learn that the first definition of the cell was elaborated by studying living foraminifers. The German anatomist Max Schultze (1825–1874), who introduced the use of

osmium tetroxide for fixation purposes, gave, in 1861, the classical description of the cell as "a little clump of protoplasm with a centrally located nucleus."

While it is not our intention to review all of the protozoologists who made a mark for themselves, perhaps the following scientists should be mentioned to complete the historical perspective: the German botanist Heinrich Anton de Bary (1831–1888), who investigated slime molds; the Italian physician Alphonse Laveran (1845–1922), who detected the malaria parasites; the Swiss zoologist Édouard Claparède (1832–1871), who, together with his co-worker Johannes Lachmann, produced an impressive monograph on the infusoria and rhizopods; the British zoologist W. Saville Kent (1845–1908) with a famous "Manual of the Infusoria"; the German zoologist Richard Hertwig (1850–1937), who cleared up the karyological events during the conjugation of ciliates; the Russian zoologist Vladimir T. Schewiakoff (1859–1930), who studied acantharian sarcodines; the German botanist Georg Klebs (1857–1918) with investigations of the life cycles of flagellates; the German zoologist and parasitologist Rudolf Leuckart (1822–1898), who created the taxon "Sporozoa"; the German heliozoan and coccidian specialist Fritz Richard Schaudinn (1871–1906); the German zoologist Franz Eilhard Schulze (1840–1917) with his investigations of rhizopods; the expert on radiolarians Haeckel, who described over 4000 species and presented them in elegant il-

Table **1**   Important dates of protozoology

Epoch 1:	**Use of microscopes for detection of small "invisible" organisms or cells** A. van Leeuwenhoek (1675)
Epoch 2:	**Systematic research** G. A. Goldfuß (1818): Protozoa A. D'Orbigny (1826): Foraminifera Chr. G. Ehrenberg (1838): Free-living and fossil protozoans
Epoch 3:	**Protozoa recognized as unicellular organisms** C. T. von Siebold (1845): "Animals, in which the different systems or organs are not clearly differentiated and those with irregular shape and simple organization can be reduced to only one single cell.—The protozoans can be divided into the rhizopods and infusorians." Sarcode doctrine and protoplasm theory: F. Dujardin (1835), H. von Mohl (1846) Establishment of the cell doctrine: M. Schultze (1861, 1863): "A cell is a little clump of protoplasm with a nucleus in the centre" Studies on protozoa are combined with cell biology and embryology
Epoch 4:	**Combination of protozoology with the emerging disciplines of microbiology and parasitology** Protozoa as pathogens:

Under Epoch 4, the following two-column sub-listing:

E. Gruby (1843):	*Trypanosoma* (in frogs)
C. W. von Naegeli (1857):	pébrine (*Nosema bombycis*)
F. Lösch (1875):	amoebic diarrhea (*Entamoeba histolytica*)
R. Leuckart (1879):	Sporozoa
A. Laveran (1880):	Malaria and its agent (*Plasmodium*)
O. Bütschli (1881):	Myxosporidia (= Myxozoa)
G. Balbiani (1882):	Microsporidia (= Microspora), Sarcosporidia
J. E. Dutton (1902):	*Tryponosoma gambiense*—causative agent of sleeping sickness

Epoch 5:	**Institutionalization of the discipline of protozoology** **Establishment of zoological institutes by protozoologists** O. Bütschli, University of Heidelberg (1878) F. E. Schulze, University of Berlin (1884) R. Hertwig, University of Munich (1885) F. R. Schaudinn (1902): Founder and editor of the first protozoological journal (Archiv für Protistenkunde) F. R. Schaudinn (1904): Founding of protozoological labs at the Reichsgesundheitsamt in Berlin-Lichterfelde, and 1906 founding of the department for protozoological research at the Institute of Nautical and Tropical Diseases (presently: Bernhard-Nocht-Institute) in Hamburg Establishment of a school of protozoology in St. Petersburg (F. Dogiel) M. Hartmann (1914): Founding of the Department for Protistology at the Kaiser-Wilhelm-Institute for Biology in Berlin-Dahlem **Institutionalization of protozoology at the international level** 1947: Founding of the Society of Protozoologists in USA, presently an international society with several national sections; since 1954 editing the Journal of Protozoology (renamed 1993 Journal of Eukaryotic Microbiology) 1963: Founding of the Polish journal Acta Protozoologica 1968: Founding of the French journal Protistologica (1987 continued as the German journal European Journal of Protistology) 1961, Prague: First International Congress of Protozoology, followed by congresses at four yearly intervals: 1965 in London, 1969 in Leningrad (St. Petersburg), 1973 in Clermont-Ferrand, 1977 in New York, 1981 in Warsaw, 1985 in Nairobi, 1989 in Tsukuba, 1993 in Berlin

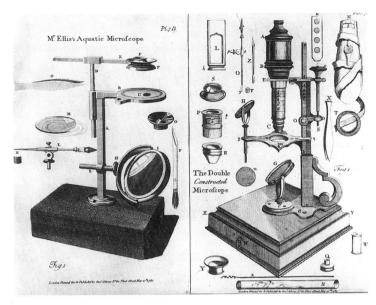

Fig. **6** Anno 1787: two pages from the catalogue of a British microscope manufacturer.

lustrations; the French zoologist Édouard Gerard Balbiani (1823–1899) as the first protozoan geneticist; and the German zoologist Otto Bütschli (1848–1920), who was — as "the Architect of Protozoology" (according to Clifford Dobell, 1951) — the first to write a comprehensive three-volume textbook on the protozoa.

While most early protozoological studies were carried out in Europe, by the end of the 19th century the protozoa were being studied all over the world. In 1879 Joseph Leidy and in 1888 Alfred C. Stokes wrote monographs on the protozoa of North America. Many of the protozoa already described from Europe were rediscovered by these pioneers, and many new species were described. By the turn of the 19th century, a solid base of knowledge had been accumulated, and to a large extent the parasitic and/or pathogenic forms were beginning to be known. In some cases partial knowledge was known of the complex life cycles of some of these forms.

By 1900, a sufficient body of knowledge had been accumulated to require that new textbooks be published, and Gary Nathan Calkins (1901 and 1933) and Franz Doflein (1901, 1909, 1911, 1916, continued by Eduard Reichenow in 1929 and 1953) responded with texts that provided not only the necessary descriptive information but also added to the general biology of the protista. It was then that protozoology gained sufficient respect to join other organism-oriented biological disciplines (Table **1**).

Although the description of new species continues, the protozoa have also been found to be ideal models for the investigation of general principles in biology. In particular, the discipline of cell biology has benefited from the special characteristics of the unicellular organisms. Notable characteristics are the ease of cultivation of these organisms, which typically have rapid rates of growth, and correspondingly short generation times. Since we have learned how to establish mass cultures of protozoa, these unicellular organisms have become popular objects for investigations in molecular biology and biochemistry. There are, however, certain unique attributes of these organisms that have not yet been adequately explored or exploited, and we will try to point out some possibilities here as the opportunity arises.

We would like to conclude this historical overview with a short survey of the development of what was the most important instrument for the development of the field, the microscope. Van Leeuwenhoek's microscope was fairly simple. It was basically just a magnifying glass (see Fig. **3b**). Some of the instruments of his day had a rather bizarre form as judged by today's microscopes (Fig. **6**). By the end of the 19th century, microscopes had evolved to a form similar to those used today. Those who are experienced realize, however, that it is not the appearance of the microscope but its optical components that are important in the formation of the sharp and informative image that can be produced. Great progress had been made in the intervening years; by about 1830 optics had improved to the point where resolution of approximately 1 µm was possible by a skilled microscopist. It was no longer necessary to have the skilled (and perhaps special) eye of a van Leeuwenhoek. In fact, the advances in lens design were responsible for the observation of the cell nucleus, and in 1830s Europe the cell and subcellular structures became known due to observations made by using these improved instruments. About 1880, immersion objectives came into generel use. They guaranteed general availability of higher magnifications at improved brightness and higher resolution, and around 1885, improvement of resolution advanced near the present objectives following the correction for chromatic aberration by the development of apochromatic objectives with suitable compensating oculars. Later, around 1932, the first phase contrast microscopes were constructed. With this instrument it was possible to observe faint details in living cells. This was a major breakthrough for protistologists and cell biologists alike, honored by the Nobel prize given to the Dutch inventor Frits Zernike. Today, there are diverse microscopical instruments available for our use. Brigthfield, darkfield, phase contrast, fluorescence, and differential interference contrast microscopy are instruments used in the routine investigation of the unicells.

The electron microscopes are of special importance in present times. The transmission electron microscope became a useful tool for the protistologist starting in the mid-1950s when techniques of specimen preparation improved to the point that meaningful data could be acquired. This invention was also honored by a Nobel prize (1986 to the German physicist Ernst Ruska). The development of the scanning electron microscope in the mid-1960s by the German physicist Manfred von Ardenne provided graphic information on the general features of the protistan surface and overall morphology. These electron microscopes provided a new dimension to the study of protozoa.

Cellular Organization of Protozoa

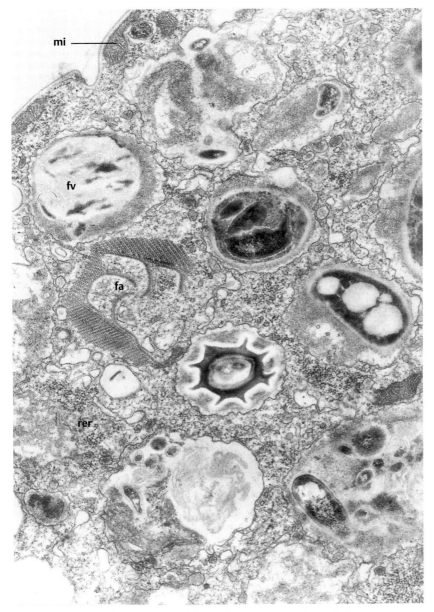

Fig. **7** Ultrastructural organization of the euglenozoon *Entosiphon sulcatum*.
fa = feeding apparatus, mi = mitochondrion, fv = food vacuole,
rer = rough endoplasmic reticulum. Magn.: 20 000×.

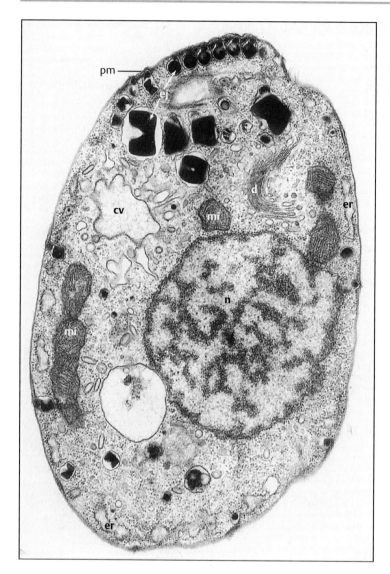

Fig. **8**   Section through the cryptomonad *Cyathomonas*. The essential elements of a protozoan cell are present (apart from the flagellum): d = dictyosome, ej = ejectisome, er = endoplasmic reticulum, n = nucleus, cv = contractile vacuole, mi = mitochondrion, pm = plasma membrane. Magn.: 10 000×.

It is generally the case that unicellular protozoan organisms have the same construction as eukaryotic cells. Using electron microscopy, it is possible to find some differences in the structure and number of organelles found in the different protozoa, but one would also find such differences between the cells of a multicellular eukaryote. There exist few if any organelles that are known to occur exclusively in protozoa.

Membranes and Compartments

A basic characteristic of eukaryotic cells is the compartmentilization of the cytoplasm by membranes (Figs. **7, 8**). As with all living cells, the protozoan is separated from its environment by a cell membrane (Fig. **9**). As seen in the electron microscope, this membrane has the typical trilamellar appearance of biological membranes. Aggregates of regularly arranged intramembranous particles have been detected in the

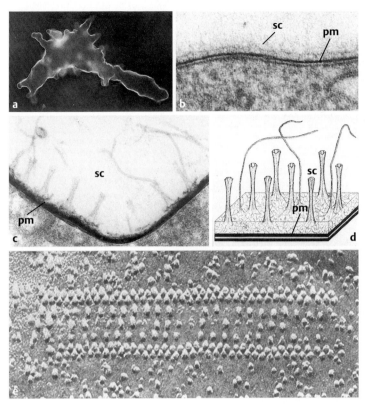

Fig. **9** Plasma membranes (pm) of various protozoa; glycocalyx of *Amoeba proteus* stained with FITC-Con A (**a**); the amoebae *Vampyrella lateritia* (**b**) and *Vannella simplex* (**c**) with surface coat (sc). Schematic presentation of the highly structured glycocalyx of *Vannella* (**d**). Regular arrangement of intramembraneous particles in *Cyclidium* (Ciliophora) (**e**). (a, courtesy of R. Radek, Berlin; c+d, courtesy of E. Hausmann, Berlin; e, courtesy of C. F. Bardele, Tübingen).
Magn.: a 150×, b 185000×, c 150000×, e 100000×.

plasma membrane, for instance, of some ciliates (Fig. **9e**). The function of these aggregates is at present unknown.

Externally, the plasma membrane is usually covered by a mucoid layer, the glycocalyx (Fig. **9b, c**). This layer is considered to be composed of the oligosaccharide chains of the membrane glycoproteins (= intramembraneous particles, IMP) and glycolipids plus absorbed secreted glycoproteins and proteoglycans consisting of glycosaminoglycans attached to a protein core. The glycocalyx may have a regular structure, but when such structures occur they do not necessarily follow the pattern of the IMP. The glycocalyx is involved in the information system of the cell, and it is in this layer that receptor molecules reside. Further, the glycocalyx allows the cell to selectively absorb solutes from the surrounding medium. Absorbed molecules can be transported into the cell by a variety of mechanisms which are presently under investigation.

The perilemma is a membrane-like structure known from some hypotrich and tintinnid ciliates (comp. Fig. **24f**). While this structure has the appearance of the plasma membrane, it apparently represents an outer envelope that is an extracellular structure of unknown function.

In addition to the glycocalyx, some protozoa possess extracellular scales (Fig. **10**), fibrillar systems or even cell wall-like structures. Extracellular loricae may be produced, and in some cases these can become elaborate structures.

Intracellular membrane systems, such as the endoplasmic reticulum, lysosomes (Fig. **11**), peroxisomes (Fig. **11**), Golgi apparatus (Fig. **12**), mitochondria (Fig. **13**), and plastids (Figs. **14, 15**) appear similar to those in metaphytan or metazoan cells. However, it is interesting to note that members of the protozoa contain both the fewest and the largest number of Golgi cisternae known (one to two in *Tetrahymena* during vegetative growth, to as many as 30 or more in *Trichonympha*).

Two special types of organelles are represented by the glycosomes and the hydrogenosomes. Glycosomes are vesicles which are found as 0.2–0.3 µm diameter spheres in African trypanosomatids. These organelles, which do not show any special morphological features, might be present as hundreds in one cell. In the glyco-

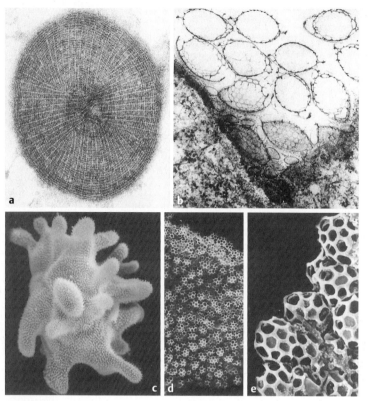

Fig. **10** Scales of unicellular organisms. Scale of the prymnesiomonad *Pleurochrysis* (**a**) and of the amoeba *Cochliopodium* (**b**); *Dactylamoeba* completely covered with scales (**c**); scales of the ciliate *Spetazoon* (**d**, **e**) (a, courtesy of W. Herth, Heidelberg, c–e, courtesy of W. Foissner, Salzburg). Magn.: a 55 000×, b 20 000×, c 1000×, d 18 000×, e 80 000×.

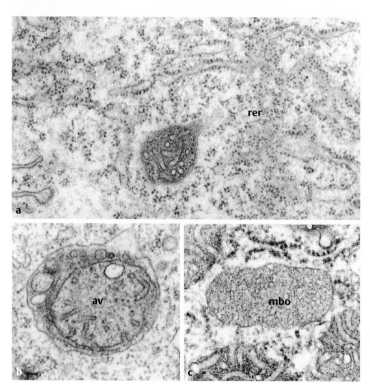

Fig. **11** Membrane systems in the ciliate *Paramecium*: rough endoplasmic reticulum (rer) (**a**); autophagic vacuole (av) with mitochondrion during degradation (**b**), microbody (mbo) (**c**). Magn.: a 25 000×, b 35 000×, c 50 000×.

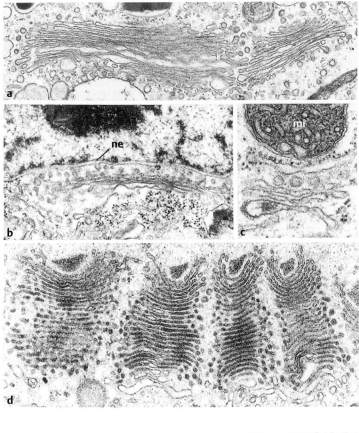

Fig. **12** Organizational types of dictyosomes in the cryptomonad *Rhodomonas* (**a**), in an unidentified amoeba (**b**), in the ciliate *Pseudomicrothorax* (**c**), and in the hypermastigid flagellate *Joenia* (**d**). mi = mitochondrion, ne = nuclear envelope (d, courtesy of R. Radek, Berlin).
Magn.: a 15 000×, b 25 000×, c 50 000×, d 30 000×.

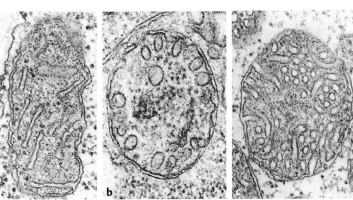

Fig. **13** Organizational types of mitochondria in the cryptomonad flagellate *Cyathomonas* (**a**), in the filopodial amoeba *Vampyrella lateritia* (**b**) and in the ciliate *Paramecium caudatum* (**c**).
Magn.: a 55 000×, b 40 000×, c 60 000×.

somes, glycolysis takes place more efficiently than in all other eukaryotic cells, where it takes place, probably exclusively, in the cytosol.

Hydrogenosomes (Fig. **16**) are the organelles that act like substitute mitochondria in some protozoa living in anaerobic environments. They metabolize pyruvate derived from glycolysis into acetate, CO_2, and H_2. They might contain crystalline inclusions. However, their contribution to ATP production appears to be substantially less than that of mitochondria and they may have other functions in addition.

Extrusomes are basically exocytotic vesicles. In their complexity, extrusive organelles are largely specific to protozoa, although related structures may exist in some lower metazoa,

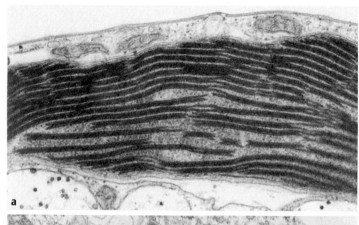

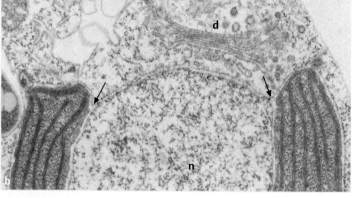

Fig. **14** Partial view of the plastid of the cryptomonad *Rhodomonas* with double stacks of thylakoids (**a**). Continuity of nuclear envelope and exterior plastidial envelope (arrows) in the chrysomonad *Ochromonas* (**b**). d = dictyosome, n = nucleus. Magn.: a 20 000×, b 18 000×.

e.g., rhabdites in flatworms. Extrusomes are membrane-bounded organelles which are usually located in the cortical cytoplasm of these cells, although immature forms arise in the cytoplasm. While they are known to have different functions, depending upon their type, they all exhibit one general characteristic: They are readily discharged when subjected to a wide range of stimuli, i.e., mechanical, electrical, and chemical. During the transition from the resting state to the ejected form, the organelles undergo characteristic morphological changes. The best known type of extrusome is the trichocyst of *Paramecium* (Fig. **17**). At present about ten different types of extrusomes are known.

In the protozoa, the contractile vacuole is a characteristic membranous structure. The contractile vacuole seen in the light microscope is associated with other structural elements visible only by electron microscopy. Therefore, the term "contractile vacuole complex" is presently used to identify those structures responsible for osmoregulation.

Some organelles, e.g., mitochondria and plastids, exhibit morphological variations that may be significant for our understanding of their evolutionary history, and researcher working on higher level systematics are interested in these variations. In particular, comparative data on the associations of ER-cisternae with plastids and the arrangement of thylakoid membranes within plastids of diverse groups of photosynthetic flagellates (see Fig. **1**) have generated considerable interest.

All protozoa possess at least one nucleus. Not infrequently, and especially in larger protozoa, multiple identical nuclei may be found within a single cell. Some foraminiferans and the ciliates have two types of nuclei; a generative micronucleus and a somatic macronucleus (Fig. **18**), a phenomenon that is called nuclear dimorphism or nuclear dualism. Depending upon the individual species, the nuclei might be diploid or haploid. The mode of nuclear division, the karyokinesis, is by mitosis. Nuclear division is usually followed by cell division, also known as cytokinesis. There are several morphological alternatives for karyokinesis, e.g., external or internal spindle, nuclear envelope intact or fragmented (see Fig. **255**).

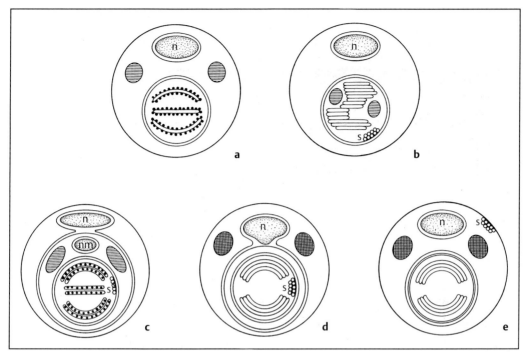

Fig. **15** Arrangement of membranes associated with the chloroplast, the type and location of stored carbohydrate, and the location of the stigma (s) found in five algal groups: Biliphyta (**a**), Chlorophyta (**b**), Cryptomonada (**c**), Chrysomonadea (**d**), and Euglenoidea (**e**). Starch deposits are shaded with horizontal lines and deposits of leucosin or paramylon with cross-hatched shading. Where there are four chloroplast membranes, the outermost of these may be continuous with the outer membrane of the nucleus (n), and in cryptomonads, the periplastidial space contains both a nucleomorph (nm) and food storage deposits (from Sleigh: Protozoa and other protists. Arnold, London 1989).

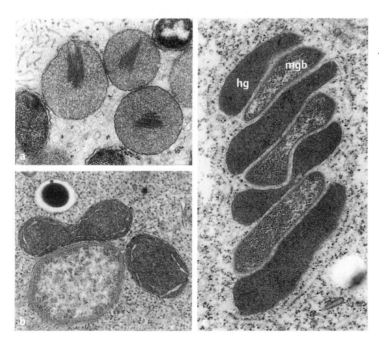

Fig. **16** Hydrogenosomes in the hypermastigid flagellate *Joenia annectens* (**a**) and in the ciliates *Metopus contortus* (**b**) and *Plagiopyla frontata* (**c**). hg = hydrogenosomes, mgb = methanogenic bacteria (a, courtesy of R. Radek, Berlin; b and c, from Fenchel and Finlay: Europ. J. Protistol. 26 [1991] 201). Magn.: a 25 000×, b 35 000×, 20 000×.

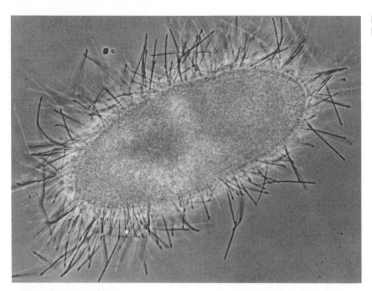

Fig. **17** Trichocysts extruded by *Paramecium*. Magn.: 500×.

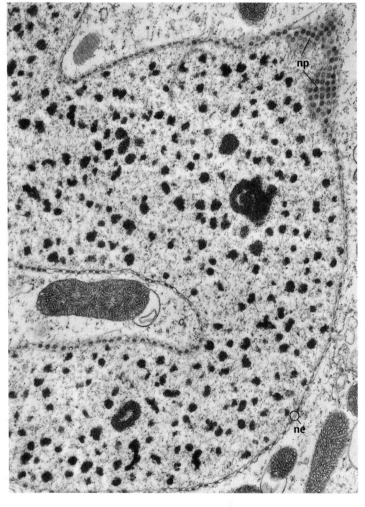

Fig. **18** Partial view of the macronucleus in *Paramecium*. ne = nuclear envelope, np = nuclear pores. Magn.: 15 000×.

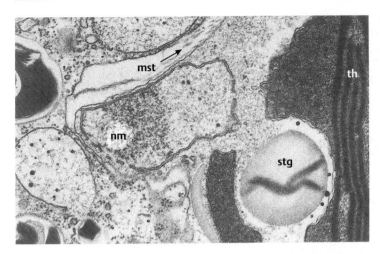

Fig. **19** Nucleomorph (nm) in the cryptomonad *Crypto-monas ovata*. mst = mastigonemes, stg = starch granule, th = thylakoid. Magn.: 35 000×.

Sexual processes involving fusion of two gametes following meiotic division of the nuclei exist in numerous protozoa. The details of the process may vary considerably, but these organisms follow the general rules known for higher organisms. In many protozoa only asexual reproduction is known. This prevents us from making generalized statements concerning sexuality among the protozoa.

The nucleomorph is a DNA-containing compartment, found especially in cryptomonads (Fig. **19**). It is covered by a double membrane interspersed by pore complexes, thus resembling a nuclear envelope. This organelle is a remnant of a eukaryotic endocytobiont.

Microfilaments and Microtubules

Filaments with a diameter of 4–10 nm are very common in protozoa, and they often combine to form thicker bundles and fibrils. In certain amoebae it has been shown that these fibrils represent actomyosin complexes that are responsible for cellular locomotion (Fig. **20a**). Such microfilaments have also been shown to be involved in cell division. In the ciliates they form a contractile ring in the plane of cytokinesis, and they are involved in the constriction, which ultimately results in the separation of the daughter cells.

Intermediate filaments have also been detected in protozoa. So far their biological function is virtually unknown, but they seem to play a cytoskeletal role in shaping the cell.

Other types of filaments appear to have a different chemical nature, but they may still be involved in contractile activity. For instance, the protein spasmin, which is found in the stalks of peritrich ciliates, is involved in the extremely fast contraction of their stalks. Other types of filaments such as those of *Stentor* and other heterotrich ciliates are involved in contractions of the cell body. At present, these filament systems are poorly understood, but it does seem clear that their activity is different from the actomyosin system.

Microtubules play an important role in the protozoan cell. They often act as supporting cytoskeletal elements in giving rigidity to the cortical system (Fig. **21**), by the stabilization of certain types of pseudopodia (Fig. **22**) in concert with actin filaments (Fig. **23**), and shaping the oral apparatus of flagellates and ciliates. Another function might be to hold organelles in specific positions as is, for instance, the case with the contractile vacuole complex of *Paramecium*.

Furthermore, they are involved in many dynamic processes. It has been shown that vesicles may be transported along bundles (Fig. **23**) or ribbons of microtubules to areas of the cell where the vesicles are needed. Other, larger organelles, such as mitochondria, are also known to be transported along microtubular networks. Microtubules are also involved in the transport of chromosomes to daughter cells during mitosis; the fine details of this process are still being determined.

It is not clear yet whether microtubules carry out fast movements by themselves. The available

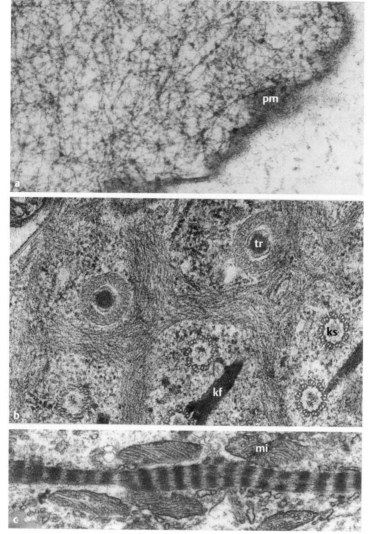

Fig. **20**   Filament systems in *Amoeba proteus* (**a**; pm = plasma membrane), in the ciliate *Paramecium caudatum* (**b**; ks = kinetosome, kf = kinetodesmal fiber, tr = trichocyst tip), and in the ciliate *Loxophyllum meleagris* (c; mi = mitochondrion) (a, courtesy of M. Hauser, Bochum). Magn.: a 120 000×, b 60 000×, c 20 000×.

information shows that microtubules, microtubule associated proteins, and microfilaments are functionally integrated in movement. Of special interest are motor proteins such as dynein and kinesin, which are involved in interactions with microtubules or compartments to create movement. The best known example is the axonemal arrangement found in cilia and flagella, the 9×2+2 pattern. It is also possible that slower movements are created by the assembly and disassembly of microtubules, e.g., during mitosis.

The protozoa provide striking examples of the motility of cilia and flagella. While their ultrastructure is identical, their types of movement can differ. Some may be involved in food uptake, others in locomotion, and still others may be capable of both types of activity. These motile organelles occur in group-specific orientations and the number of cilia and flagella is also group specific (comp. Fig. **42**). In some protozoa the presence of these organelles is restricted to specific stages of the life cycle.

Although the 9×2+2 pattern is a conservative feature of eukaryotic cells, not all flagella or cilia of a single cell are necessarily identical. It has been shown that the ciliate *Tetrahymena* has at least four ultrastructurally distinct types of cilia, and many flagellates possess flagella of different, but characteristic lengths. Flagella of some metakaryotic organisms may be covered by hairlike appendages, called mastigonemes (Fig. **24 a**) or scales (comp. Fig. **145**).

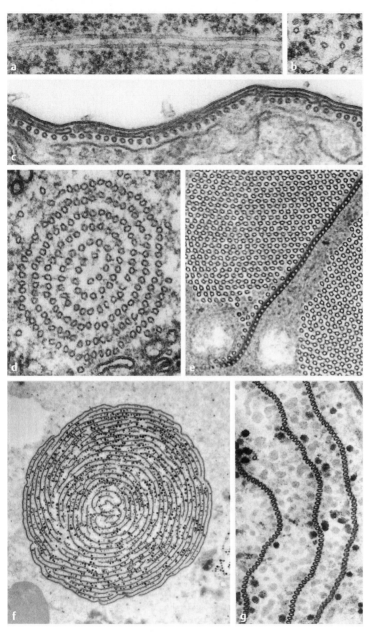

Fig. **21**   Microtubules. Longitudinal (**a**) and transverse sections (**b**) of cytoplasmic microtubules in the ciliate *Paramecium*. Subpellicular microtubules in the ciliate *Euplotes* (**c**). Axopodial microtubules in the heliozoon *Actinophrys* (**d**). Nematodesmal microtubules in the ciliate *Nassula* (**e**). Microtubules of the axostyle in the flagellate *Joenia* (**f, g**) (f and g, courtesy of R. Radek, Berlin). Magn.: a 40 000×, b 40 000×, c 45 000×, d 80 000×, e 35 000×, f 8000×, g 32 000×.

Both cilia and flagella are covered by the plasma membrane. In the cytoplasm the axoneme is anchored by a basal body, the kinetosome. In flagellates and ciliates there may be additional root structures that originate from basal bodies. These structures are so characteristic that their arrangement has been shown to be of taxonomic value (Fig. **25**).

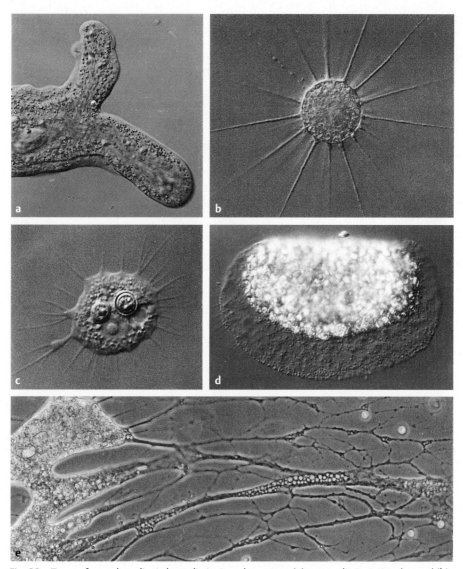

Fig. **22**   Types of pseudopodia: Lobopodia in *Amoeba proteus* (**a**), axopodia in *Actinophrys sol* (**b**), filopodia in *Nuclearia* (**c**), lamellipodium in *Hyalodiscus pedatus* (**d**), reticulopodia in *Reticulomyxa filosa* (**e**), (c, courtesy of D. J. Patterson, Sydney). Magn.: a 500×, b 280×, c 500×, d 750×, e 150×.

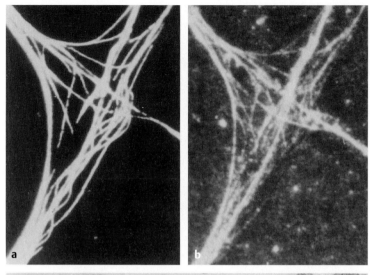

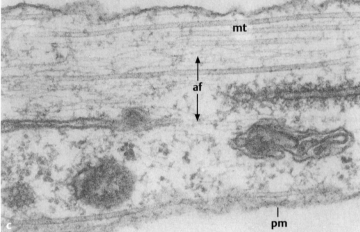

Fig. **23** Cytoskeletal elements in *Reticulomyxa filosa* (Granuloreticulosa). Immunocytochemical visualization of tubulin (**a**) and actin (**b**) in the same part of the organism. Association of actin filaments (af) and microtubules (mt) (**c**); pm = plasma membrane (courtesy of M. Schliwa, Munich). Magn.: a and b 1700×, c 70 000×.

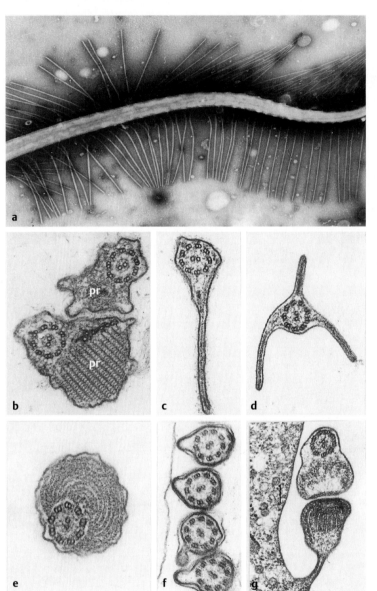

Fig. **24**   Additional structures of flagella and cilia. Mastigonemes in chrysomonads (**a**). Paraxial rod (pr) in the euglenid *Entosiphon* (**b**). Lateral flagellar fin-like extensions in the bodonid flagellate *Colponema loxodes* (**c**) and in the flagellate *Retortamonas* (**d**, compare Fig. **43 b**). Paraxonemal concentric layers in a flagellum of the trichomonad *Foaina* (**e**). Perilemma around cilia of the ciliate *Stylonychia* (**f**). Undulating membrane in the flagellate *Tritrichomonas angusta* (**g**). (b, courtesy of D. J. Patterson, Sydney, c, courtesy of J. P. Mignot, Clermont-Ferrand, d and g, courtesy of G. Brugerolle, Clermont-Ferrand, e, courtesy of R. Radek, Berlin). Magn.: a 20 000×, b 45 000×, c 40 000×, d 40 000×, e 50 000×, f 40 000×, g 35 000×.

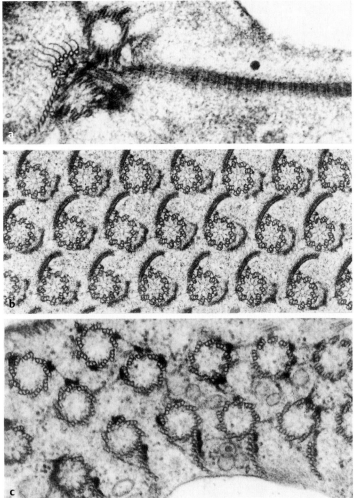

Fig. **25** Flagellar roots and kinetosome-associated structures in the cryptomonad *Chilomonas* (**a**), hypermastigid flagellate *Joenia* (**b**) and the ciliate *Eufolliculina* (**c**) (b, courtesy of R. Radek, Berlin). Magn.: a 60 000×, b 45 000×, c 60 000×.

Shape and Size of Protozoa

The diverse groups of protozoa have traditionally been identified by their shapes, although sometimes rather confusing forms can be observed (Fig. **26**). Characteristic cell outlines are produced by the intra- and extracellular skeletal elements.

The dimensions of protozoan cells cover wide ranges (Fig. **27**). The smallest uninucleate forms are only a few micrometers long, whereas others measure several hundreds of micrometers. Multinucleated forms can be several centimeters long; in extreme cases they may reach even meters (Fig. **28**).

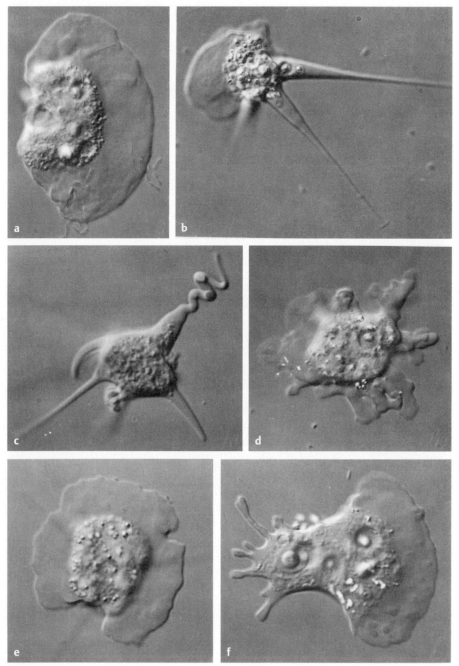

Fig. 26   Shape transformations in the amoeba *Vannella simplex*. Motile (**a**) and various radiating forms (radiosaforms) (**b** and **c**). **d–f** reversion to normal locomotory form. Magn.: 800×.

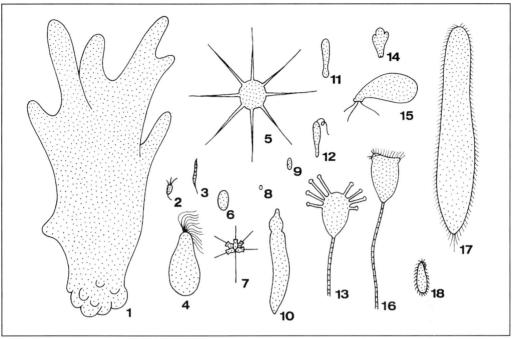

Fig. **27** Variability of dimensions in protozoans (all organisms drawn at same scale). 1 *Amoeba*, 2 *Trichomonas*, 3 *Trypanosoma*, 4 *Joenia*, 5 *Actinophrys*, 6 *Eimeria*, 7 *Codonosiga*, 8 Microspora, 9 Myxozoa, 10 *Gregarina*, 11 *Saccamoeba*, 12 *Euglena*, 13 *Discophrya*, 14 *Entamoeba*, 15 *Trinema*, 16 *Vorticella*, 17 *Paramecium*, 18 *Tetrahymena*.

Fig. **28** Gigantic specimen of the slime mold *Physarum polycephalum* covering an area of 5.54 m² with a maximum depth of 1 mm (courtesy of F. Achenbach, Bonn).

Part II: Evolution and Taxonomy

Evolution of Unicellular Prokaryotes and Eukaryotes

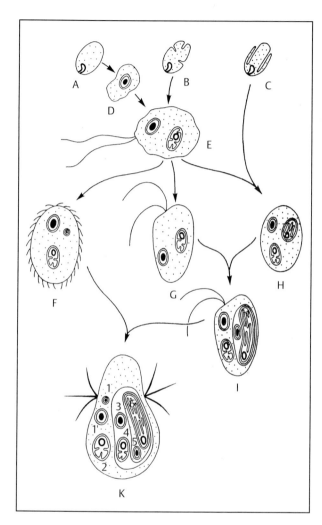

Fig. **29** Theory of the evolution of eukaryotic cells from prokaryotic organisms (**A–C**) and evolutionary development of more complex Eukaryota from various, simply organized eucytes (**D–K**). (**A**) prokaryote I (later to become the nucleated and flagellated host cell); (**B**) prokaryote II (later transformated to a mitochondrion); (**C**) prokaryote III (later transformed to a chloroplast); (**D**) first eukaryotic cell; (**E**) eukaryotic cell with paired flagella and mitochondrion (= metakaryote, = eucyte); (**F**) ciliate with macro- and micronucleus and mitochondrion; (**G**) heterotrophic flagellate with mastigonemes (e.g., euglenoid or cryptomonad flagellate or dinoflagellate); (**H**) unicellular organism with mitochondrion and chloroplasts after secondary loss of flagella (e.g., rhodophyte); (**I**) import of photosynthetically active rhodophyte by heterotrophic flagellate (evolution of secondarily autotrophic flagellates as in many cryptomonads, chrysomonads or dinoflagellates); (**K**) the ciliate *Mesodinium rubrum* with six heterogeneous DNA-containing compartments after internalization of a cryptomonad: 1, micronucleus and 1', macronucleus of the ciliate; 2, ciliate mitochondrion; 3, cryptomonad nucleus; 4, cryptomonad mitochondrion; 5, nuclear remnant of the rhodophyte (= nucleomorph); 6, rhodophyte chloroplast; the seventh compartment, the rhodophyte mitochondrion, is completely reduced (after several authors).

According to the generally favored theories, the historical rise of the antecedents of modern protozoa followed almost the same route as the evolution of the other recent prokaryotic and eukaryotic organisms. Approximately 2500–1500 million years ago their pathways were separating, leading to eukaryotic plants, molds, and animals on the one hand and the modern prokaryotes on the other. The early stages of organism evolution will be described only briefly.

Assuming a total age of the globe of about 4550 million years, the first microbial ecosystems possibly arose 4000 million years ago. Their sedimented biomass is fossilized in the so-called stromatolith limes. The $^{13}C/^{12}C$ ratio within these biosediments reflects the relationship between inorganic carbonates and organic carbon and serves as an indicator of the presence of autotrophic carbon fixation processes. It is virtually certain that during the photosynthetic primary production of organic matter, even at an early stage, water (H_2O) served as reductant, and that considerable quantities of oxygen were produced. However, the actual oxygen concentration in the oceans and the O_2 partial pressure in the atmosphere must have been very low because of widespread inorganic oxygen consuming processes (such as oxidation of ferrous (II) to ferric (III) iron and the weathering of the continental crust).

Reconstructing the hydrosphere at that time as the only region supporting the existence and evolution of life, the following abiotic parameters were probably dominant:

- Partial oxygen saturation occurred only close to photosynthetically active cells (possibly mostly antecedents of recent cyanobacteria).
- CO_2 of inorganic (volcanic) origin and potassium were present in abundance (soda model of ancient oceans).
- The Ca^{2+} content of the sea was extremely low (10^{-7} M in contrast to the present 10^{-2} M), but allowed some biogenic sedimentation of lime.
- The pH of the oceans was about 10 (presently between 7 and 8).
- Divalent soluble ferrous compounds changed slowly after oxygen enrichment of the atmosphere into insoluble Fe(III) compounds (ironstone).
- The temperatures are believed to have been nearly the same as at present, whereby the elevated carbon dioxide of the ancient atmosphere with its superior hothouse effect would compensate for the lower intensity of solar radiation.

Many extant recent prokaryotes are known to live under such extreme conditions. From this it is argued that some of their antecedents were associated with the origin of the more advanced ancient cells. As important as their structural organization or body plan, however, is the fact that none of the ancient organisms were successful in the evolution of more complex life forms. This did not occur until the combination of simple unicells into eucytic two-cells or three-cells; a process embraced within the so-called serial endosymbiosis theory (Fig. **29**).

In reconstructing the phylogeny of all animal and plant organisms, one obviously has to proceed from the premise that all eukaryotic beings are reducible to only one ancient species and that they consistently form the monophylum Eukaryota. One of the consequences of this postulate is the realization that the recent unicellular eukaryotes correspond only in a plesiomorphic character, unicellularity or absence of cellular somatic differentiation, and that they do not form a monophyletic taxon.

According to the dates of micropaleontology, the basic form of all eukaryotic unicellular and multicellular tissue–organized creatures belonged to those unicellular organisms that already existed about 2000 million years ago. Between this Proterozoic Era and the first certain appearance of metazoa and higher plants in the Paleozoic (700 million years ago), lies an epoch characterized by the evolution of the typical morphological organization of recent unicellular and multicellular organisms. Because most of the fossils of the Proterozoic are only vaguely interpretable, the reconstruction of phylogeny has to be done by comparing all available morphological, biochemical, and molecular biological data of recent organisms and by searching for homologies. With this aim, the dates of geobiochemistry are very important, for they allow the reconstruction of the ecological conditions during the Paleozoic Era.

The evolution of all recent and extinct phyla has been accompanied by radical alterations in the atmosphere, hydrosphere, and geosphere. These variations have mostly been caused by the activity of prokaryotes, especially free-living and endosymbiotic cyanobacteria. Due to the elaboration and activity of photosynthetic apparatus and the pathway of H_2O photolysis, the oxygen content of the atmosphere increased from nearly 0% to the present value of 21%, which was stabilized about 400 million years ago. By the same photosynthetic pathway, the carbon dioxide content of the oceans was

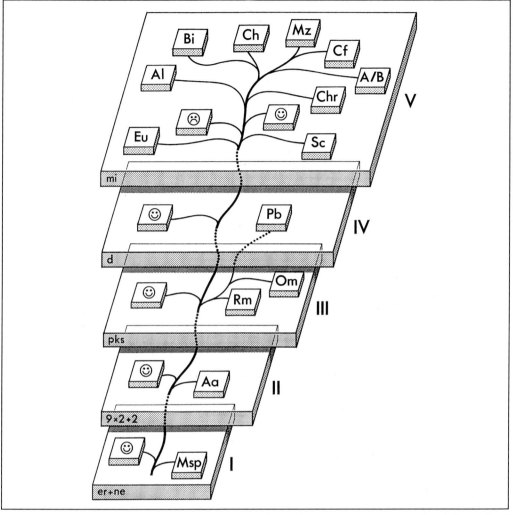

Fig. **30** Simplified and tentative phylogenetic system of eukaryotic organisms. – I. First eukaryotic level with endoplasmic reticulum (er) and nuclear envelope (ne) and sexuality as apomorphic characters; branching taxon: Microspora (Msp) – II. Level of monomastigote-flagellated eukaryotes: evolution of flagellum (9×2+2) and kinetosome; branching taxon. Archamoebaea (Aa). – III. Level of all dimastigote-organized Eukaryota: kinetosomes are paired (pks), branching taxa: the primarily tetramastigote flagellates Retortamonadea (Rm), Oxymonadea (Om) [and Parabasalea (Pb)]. – IV. Level of convergently evolving endomembranous systems, especially of classical and atypical dictyosomes (d). – V. Level of Metakaryota: evolution of eucytic cells (= metakaryotes) by endosymbiosis with prokaryotic organisms which later on evolve into mitochondria and chloroplasts. Typical taxa appear: Euglenozoa (Eu), Alveolata (Al), Biliphyta (Bi), Chlorophyta (Ch), Metazoa (Mz), Choanoflagellata (Cf), Ascomycetes (A) and Basidiomycetes (B), Chromobionta (Chr), Schizopyrenida (Sc). Boxes with ☺ are reserved for taxa which are extinct or might be defined in the future (adapted from Leipe and Hausmann).

diminished, leading to the formation of extensive silicious and calcareous deposits.

An important consideration in reconstructing the path of evolution in that era is that free oxygen acts as a poison. Recent cells are known to deal with this poisoning by oxygen-consuming and detoxifying processes. This ability is associated with the presence of distinct compartments, namely the peroxisomes (or glyoxysomes), hydrogenosomes, and mitochondria and respiring endosymbiotic bacteria lying freely in the cytoplasm or sheltered within vacuoles.

The evolution of mitochondria is obviously the result of a symbiosis between an eukaryotic and a prokaryotic cell. The overall scheme of this process is described by the so-called serial endosymbiosis theory and the idea is now favored that the evolution of mitochondria from such an endosymbiotic association was realized only once. The question of whether hydrogenosomes, which occur only in mitochondria-less cells, result from a conversion of already established mitochondria, from endosymbionts similar to mitochondria or from the stepwise differentiation of rather simple endomembrane systems, cannot presently be answered with certainty. It is noteworthy that anaerobic ciliates with hydrogenosomes often have close relatives with mitochondria, and it is now clear that anaerobic ciliates are embedded in the phylogeny of aerobic ciliates. Thus, it seems likely that anaerobic free-living ciliates and their hydrogenosomes are derived from aerobic ciliates and mitochondria respectively. However, there are reasons to believe that the hydrogenosomes of rumen ciliates have derived independently from hydrogenosomes in termite flagellates (Parabasalia). In every case, the possession of mitochondria or hydrogenosomes serves as a derived character useful for the evaluation of monophyly (Parabasalia, Metakaryota). In contrast, the presence of prokaryotic endosymbionts in combination with the absence of mitochondria or hydrogenosomes represents a plesiomorphic character (Fig. **30**).

The endosymbiosis theory is also taken as a model for the acquisition of plastids. As expected, such considerations are restricted to cells equipped with preestablished respiratory mitochondria, as found only within the Metakaryota. Plastids are genetically related to recent cyanobacteria. Apparently two different types of cyanobacteria have evolved independently into true plastids with a double membrane (Fig. **31**): one within the Chlorophyta and another within the Rhodophyta. They differ in the principal and accessory pigments; Chlorophyta and their descendants have plastids with chlorophylls a and b, Rhodophyta those with chlorophylls a and c. On the other hand, both eukaryotic groups contain unicellular representatives able to act as plastid-like endosymbionts in many other eukaryotic hosts: for instance, green algae within ciliates, hydrozoans, and turbellarians, and red algae within cryptomonads. In other cases these relations are stabilized, as in green euglenozoa or in more or less reddish dinoflagellates or chrysomonads.

The possession of typical flagella and their attachment structures is believed to play an important role in the reconstruction of cell evolution. Even if the historical process of the origin of this motile apparatus is controversial (anagenetic proceeding from preexisting simple microtubular associations as used for karyokinesis, versus (questionable) endosymbiotic acquisition from former spirochetes), the flagellation itself or their different insertion types serve as derived characters and can be used for constructing monophyletic taxa. As a consequence, primitively devoid of flagella (or cilia) means a plesiomorph character if combined with other primitive signs (absence of dictyosomes or mitochondria). Secondary absence of flagella, however, can serve as an apomorphic character, especially when other derived attributes are present (e.g., possession of mitochondria or even plastids). As an example, the secondary loss of flagella is a character of numerous groups with amoeboid motility.

Similar methods are used to evaluate presence or absence of dictyosomes. It is a surprising fact that this association complex of cisternae with the endomembrane system can be found only in cells with mitochondria or hydrogenosomes. It seems reasonable that functional changes during the evolution of the Golgi apparatus have been conditioned by a change in the mode of nutrition: from osmotrophy and feeding on bacteria to a carnivorous behavior among the unicellular eukaryotes. The Golgi-mediated synthesis of cystic or shell elements on the one hand or of extrusomes on the other hand within the same system supports the argument that

Fig. **31** This suggestion for the evolutionary lineages within the photoautotrophic eukaryotes, which result, from molecular and morphological data, resembles a three-dimensional network of relationships. The individual genomes of the host cell and the endosymbiont are symbolized by lines: the pathway of autotrophic lineages is drawn in green and that of the heterotrophic lineages in blue (prokaryotes) or red (eukaryotes). – HM = Level of heterotrophic metakaryotes with two genomes (2) from the starting point of endobiosis between one heterotrophic eukaryote (HE) and one mitochondrial precur-

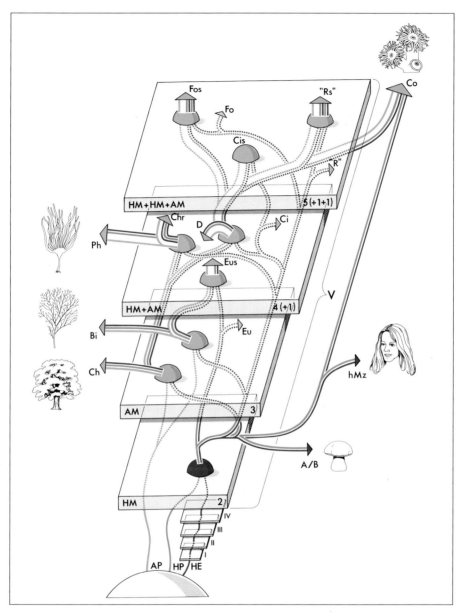

sor (heterotrophic prokaryote, HP). – AM = Level of the autotrophic metakaryotes in which, after endosymbiotic incorporation of an autotrophic prokaryote (AP), three genomes are present (3). – HM + AM = Level of endosymbiosis between one heterotrophic metakaryote of the level HM and one of the autotrophic partners of the level AM representing the evolution of autotrophic organisms such as dinoflagellates (D), Chromista (Chr), euglenids without (E) and with symbionts (Es), and other taxa; primarily, the number of genomes is 4 (+ 1 reduced mitochondrial genome). – HM + HM + AM = level of the most complex associations: unicellular organisms of the level HM + AM live as endosymbionts in heterotrophic metakaryotes such as ciliates without (Ci) or with endosymbionts (Cis), foraminifers without (Fo) or with endosymbionts (Fos), "radiolarians" without ("R"), or with endosymbionts ("Rs"), or metazoans (Co = corals). In such cases, the number of genomes might reach up to seven, but mostly only five genomes are present. Within the three lower levels, the organization principle of unicellularity is evolved independently into multicellularity which is symbolized by left- or right-handed arrows: multicellular organisms originate with the Biliphyta (Bi), Chlorobionta (Ch), Phaeophyta (Ph), higher fungi (A/B = Ascomycetes/Basidiomycetes), and higher Metazoa (hMz). I–V: compare Fig. 30 (after Hülsmann and Hausmann).

cells equipped with protective devices produced from organelles could develop better strategies of survival. However, the morphological simplicity of dictyosomes in metakaryotes and the surprising complexity of dictyosomes in Parabasalia leads to the assumption that such organelles have developed independently from primitive preexisting structures, at least in these two taxa.

In combination with the available biochemical and molecular biological dates, the comparison and evaluation of primitive and derived morphological characters allows the construction of a phylogenetic tree that is still hypothetical in several respects (see Fig. **30**). The starting point for speciation is an organism characterized by cells with the following basic functional–morphological inventory: endomembrane system with nuclear envelope, rough, smooth ER and lysosomes, chromosomes, mitotic and meiotic apparatus made of microtubules, microfilament systems (actomyosin) for static and motile purposes and for cytokinesis, and 70 S ribosomes. Primitive means the absence of flagella, all kinds of respiratory organelles, and dictyosomes.

Development of Classification Systems

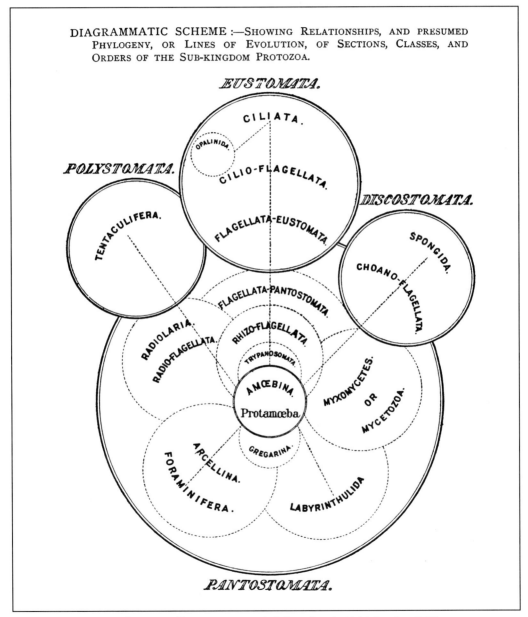

Fig. **32**   Kent's system of protozoa (from Kent: Manual of the Infusoria, Vol. I, London 1882).

The primary goals of systematic zoology are to establish the relationships between taxa (Fig. **32**), model the historical phases of the evolution of organisms (phylogenetics), and rationalize the fundamental processes and causes of the phylogenetic development itself (evolutionary biology).

From the time of van Leeuwenhoek up to the 20th century, systematically oriented research on protozoa was driven mostly without genealogical components: The only available criterion for the assessment of higher taxa was the morphologically-based similarity of organisms. As an example, the small dimensions of bacteria, eukaryotic unicells, and several plants and metazoa led to the establishment of the Animalcula or Infusoria. The study of microscopical anatomy and the consideration of historical dimensions started with Ehrenberg and Dujardin. Initiated by the theory of the origin of species (Charles Darwin, since 1842), and tackled at the protistan level by Haeckel, the first genealogical trees were developed.

At that time several taxonomic designations were created which are partly in use at present: Polycystinea Ehrenberg, 1838; Rhizopoda von Siebold, 1845; Ciliata Perty, 1852; Flagellata Cohn, 1853; Radiolaria J. Müller, 1858; Suctoria Claparède and Lachmann, 1858/59; Mycetozoa de Bary, 1859; Heliozoa Haeckel, 1866; Mastigophora Diesing, 1866; Sarcodina Schmarda, 1871; Filosea Leidy, 1879; Sporozoa Leuckart, 1879; Acantharea Haeckel, 1881; Myxosporidia Bütschli, 1881. Bütschli (1881) was the first to propose a general scheme on the system of Protozoa:

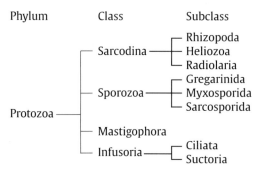

Phylum	Class	Subclass
		Rhizopoda
	Sarcodina	Heliozoa
		Radiolaria
		Gregarinida
	Sporozoa	Myxosporida
Protozoa		Sarcosporida
	Mastigophora	
	Infusoria	Ciliata
		Suctoria

This scheme, adopted for its didactic simplicity, is presently still in use, mostly in general zoological textbooks. However, new knowledge, including new evaluations of characters, led Honigberg and ten co-workers to establish a more modern system in 1964. The scheme they proposed is presented here:

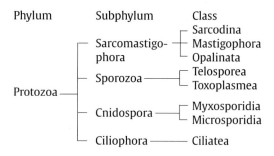

Phylum	Subphylum	Class
		Sarcodina
	Sarcomastigo-phora	Mastigophora
		Opalinata
	Sporozoa	Telosporea
Protozoa		Toxoplasmea
	Cnidospora	Myxosporidia
		Microsporidia
	Ciliophora	Ciliatea

The differences between this scheme and that of Bütschli are mostly based on the liquidation of the old taxon Sporozoa into the Cnidospora (spores with filament) and into the "new Sporozoa" (spores without filament), and by combining the Mastigophora with the Sarcodina in a new subphylum Sarcomastigophora.

During the following decades, new ultrastructural work allowed new insights. As a consequence, the realization of great diversity within the unicells led to the supposition of independent genealogical lineages. In 1980, a classification committee (Levine and 15 co-authors) proposed a system in which the postulated heterogeneity is acknowledged by the introduction of seven phyla instead of the one old phylum Protozoa:

Subkingdom: Protozoa
1. Phylum Sarcomastigophora Honigberg & Balamuth, 1963
2. Phylum Labyrinthomorpha Page, 1980
3. Phylum Apicomplexa Levine, 1970
4. Phylum Microspora Sprague, 1977
5. Phylum Ascetospora Sprague, 1978
6. Phylum Myxozoa Grassé, 1970
7. Phylum Ciliophora Doflein, 1901

The illustrated guide to the Protozoa, edited in 1985 by the Society of Protozoologists, contains – under the subkingdom Protozoa – only six phyla with 28 classes. In comparison with the system of Levine et al., additional taxa are marked by round brackets ( ) and classes with new scientific names are symbolized by square brackets [].

Subkingdom Protozoa (Goldfuß, 1818) von Siebold, 1846
   Phylum Sarcomastigophora Honigberg & Balamuth, 1963
      Subphylum Mastigophora Diesing, 1866
         Class Phytomastigophorea Calkins, 1909
         Class Zoomastigophorea Calkins, 1909
      Subphylum Opalinata Corliss & Balamuth, 1963
         Class Opalinatea Wenyon, 1926

Subphylum Sarcodina Schmarda, 1871
Superclass Rhizopodea von Siebold, 1845
    Class Lobosea Carpenter, 1861
    Class Acarpomyxea Page, 1976
    (Class Acrasea Schröter, 1886)
    Class Mycetozoea de Bary, 1859
    (Class Plasmodiophorea Cook, 1928)
    Class Filosea Leidy, 1879
    Class Granuloreticulosea de Saedeler, 1934
    (Class Xenophyophorea Schulze, 1904)
Superclass Actinopodea Calkins, 1902
    Class Acantharea Haeckel, 1881
    Class Polycystina (Ehrenberg, 1838)
      Riedel, 1967
    Class Phaeodarea Haeckel, 1879
    Class Heliozoea Haeckel, 1866
Phylum Labyrinthomorpha Page, 1980
    (Class Labyrinthulea Levine & Corliss, 1963)
Phylum Apicomplexa Levine, 1970
    Class Perkinasida Levine, 1978
    Class Sporozoasida Leuckart, 1879
Phylum Microspora Sprague, 1969
    Class Metchnikovellidea Weiser, 1977
    [Class Rudimicrosporea Sprague, 1977]
    Class Microsporididea Corliss & Levine, 1963
    [Class Microsporea Delphy, 1963]
Phylum Myxozoa Grassé, 1970
    Class Myxosporea Bütschli, 1881
    Class Actinosporea Noble, 1980
Phylum Ciliophora Doflein, 1901
    Subphylum Postciliodesmatophora Gerassimova & Seravin, 1976
    Class Karyorelictea Corliss, 1974
    Class Spirotrichea Bütschli, 1889
    Subphylum Rhabdophora Small, 1976
    Class Prostomatea Schewiakoff, 1896
    Class Litostomatea Small & Lynn, 1981
    Subphylum Cyrtophora Small, 1976
    Class Phyllopharyngea de Puytorac et al., 1974
    Class Nassophorea Small & Lynn, 1981
    Class Oligohymenophorea de Puytorac et al., 1974
    Class Colpodea de Puytorac et al., 1974

The Handbook of Protoctista, edited in 1990 by Margulis, Corliss, Melkonian and Chapman, enumerates within the protoctista 35 more or less "true" as well as "two" questionable phyla. According to the definitions given in Part I, they are recruited from additional as well as from fragmented phyla of protozoans and from classical algal subkingdoms or classes (Rhodophyta, Phaeophyta, Chlorophyta) and some "Fungi" (Hyphochytriomycota, Plasmodiophoromycota, Chytridiomycota and Oomycota). This (only informal) "system" already contains parts of a phylogenetically oriented classification, but it is based simply on the possession of flagella and on the presence or absence of complex sexual cycles.

Kingdom Protoctista (Hogg, 1861) Margulis & Schwartz, 1988

Phylum	Class
Section I:	Phyla without flagellated stages and without complex sexual cycles
1. Rhizopoda	Lobosea
	Filosea
2. Haplosporidia	Haplosporea
3. Paramyxea	Paramyxidea
	Marteiliidea
4. Myxozoa	Myxosporea
	Actinosporea
5. Microspora	Rudimicrosporea
	Microsporea
Section II:	Phyla without flagellated stages, but with complex sexual cycles
6. Acrasea	Acrasida
7. Dictyostelida	Dictyostelida
8. Rhodophyta	Rhodophyceae
9. Conjugatophyta	Conjugatophyceae
Section III:	Phyla with flagellated stages, but without complex sexual cycles
10. Xenophyophora	Psamminida
	Stanomida
11. Cryptophyta	Cryptophyceae
12. Glaucocystophyta	Glaucocystophycea
13. Karyoblastea	Karyoblastea
14. Zoomastigina	Amoebomastigota
	Bicoecids
	Choanomastigota
	Diplomonadida
	Pseudociliata
	Kinetoplastida
	Opalinata
	Proteromonadida
	Parabasalia
	Retortamonadida
	Pyrsonymphida
15. Euglenida	Euglenophyceae
16. Chlorarachnida	Chlorarachniophyceae

17. Prymnesiophyta	Prymnesiophyceae
18. Raphidophyta	Raphidophyceae
19. Eustigmatophyta	Eustigmatophyceae
20. Actinopoda	Polycystina
	Phaeodaria
	Heliozoa
	Acantharia
21. Hyphochytriomy-cota	Hyphochytrida
22. Labyrinthulomy-cota	Labyrinthulida
	Thraustochytrida
23. Plasmodiophoro-mycota	Plasmodiophorida

Section IV:	Phyla with flagellated stages and complex sexual cycles
24. Dinoflagellata	Dinophyceae
	Syndiniophyceae
25. Chrysophyta	Chrysophyceae
	Pedinellophyceae
	Dictyochophyceae (= Silicoflagellata)
26. Chytridiomycota	Chytridiomycetes
27. Plasmodial slime molds	Protostelida
	Myxomycotina
28. Ciliophora	Karyorelictea
	Spirotrichea
	Prostomatea
	Litostomatea
	Phyllopharyngea
	Nassophorea
	Oligohymenophorea
	Colpodea
29. Granuloreticulosa	Athalamea
	Foraminifera
30. Apicomplexa	Gregarinia
	Coccidia
	Hematozoa
31. Bacillariophyta	Coscinodiscophyceae
	Fragilariophyceae
	Bacillariophyceae
32. Chlorophyta	Prasinophyceae
	Chlorophyceae
	Ulvophyceae
	Charophyceae
	(and 4 other groups at the rank of order)
33. Oomycota	Saprolegniomycetidae
	Peronosporomycetidae
34. Xanthophyta	Xanthophyceae
35. Phaeophyta	Phaeophyceae
36. Incertae Sedis:	
(a)	Ellobiopsida
(b)	Ebridians

In this system, it becomes apparent that the heterogeneity of unicellular creatures is honored by a surprisingly large number of taxa. There are also some drastic reorganizations within well-known taxa. This is especially true for the Rhizopoda of the older systems, which are divided here into nine different phyla. On the other side, it becomes apparent that some taxa (for instance the Ciliophora or the successors of the former "Sporozoa") show relative stability.

It must be stated that at present only preliminary systems can be constructed in the field of protozoan and prot(oct)istan systematics. These will undergo changes when apomorphic characters become recognized, convergences are exposed (Fig. **33**), and molecular chronometers become established (Fig. **34**). In this context, the interim utilitarian "user friendly" hierarchical classification and characterization of the protists, proposed by Corliss in 1994, can only be regarded as transient because it does not sufficiently take into account, for instance, the actual, continuously growing knowledge inferred from small subunit rRNA sequence data (Fig. **34**). This approach is especially promising because the sequence diversity of the relevant molecules in protozoa by far exceeds that of any prokaryote or multicellular taxon. The inferred trees result from computer calculations applying distance matrix-, maximum parsimony-, and maximum likehood-methods, using additionally computations such as bootstrapping and analysis of signature sites.

It is unreasonable to again establish old designations such as Protoctista or Protista and to use them as names for distinct monophyletic taxa. A more fundamental problem is to solve the independent nomenclatural rules of botanists, microbiologists and zoologists that will lead to a renaming of many well-known genera and species. The main disadvantage, however, is not based on the terminology itself, but on the fact that all terms represent paraphyletic (or even polyphyletic) groups of organisms that are inappropriate for the elucidation of historical and genealogical connections. They can be used, as the term protozoa is in this book, only for informal, not for systematic purposes.

Even subordinated taxa must be revised. The main criterion used for the erection of the taxa "Flagellata Cohn, 1853" and "Mastigophora Diesing 1866" was the equipment of unicellular organisms with persistent flagella. In the classic systematic systems based on Bütschli, the additional requirement was that such flagella must be present, for correct classification, at least in

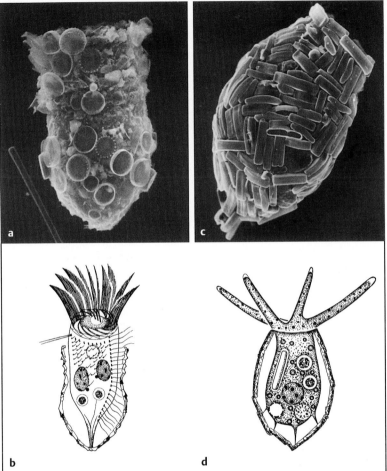

Fig. **33** Convergently evolved shells in nonrelated taxa. **a** and **b** *Codonella cratera* (ciliate), **c** and **d** *Difflugia* (testate amoeba) (from Foissner & Hausmann: Mikrokosmos 76 [1987] 258). Magn.: a 1500×, b 700×, c 450×, d 220×.

the main phases of the life cycles, i.e., during feeding and growth. By this restriction, a separation became possible between "true flagellates" and those that only occasionally exhibit a flagellate (swarmer) stage, such as gametes or zoospores. However, even Bütschli was conscious that such a characterization would represent only a pragmatic acknowledgement of vague boundaries between the unicellular eukaryotes. He tried, until his death, to prove a genealogical relationship between rhizopods and flagellates, and he emphasized the homology of flagella and slowly swinging pseudopods of distinct amoebae. In 1910 he coined the term "Undulipodia" to emphasize the evolutionary origin of flagella and pseudopodia from a common ancestral structure. Later on, this term was transferred also to the flagella of prokaryotic organisms. With the growing knowledge of ultra-structural research, the morphological differences between these two types of flagella became obvious, and in Europe the term has fallen out of use (in contrast to some authors in America).

The fundamental uniformity of the axonemal $9 \times 2 + 2$ structure within the flagella and cilia of all phyla is seen as an apomorphic character and as an argument for justification of the taxon Mastigota. However, the attempt to establish the monophyletic status of the "Mastigophora" or "Flagellata" using an original (plesiomorphic) character was recognized as rather problematic even during Bütschli's era. The proposition is undermined by the postulate of a common origin of metazoa, molds, metaphytes, and mastigotes out of one common flagellate ancestor (see Fig. **30**). Considering the convergent acquisition of photosynthetically active endosymbionts,

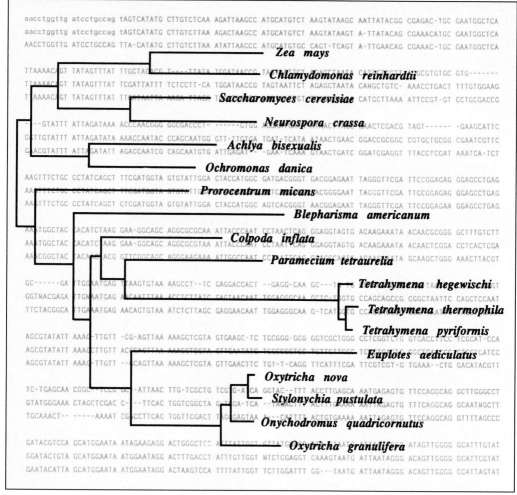

Fig. 34   Distance matrix tree inferred from complete small subunit rRNA sequences superimposed on a portion of the sequence data of three different ciliates (from Schlegel: Europ. J. Protistol. 27 [1991] 207).

even the taxon "Phytomastigophora" cannot be considered a monophylum. The taxon "Zoomastigophora" is a paraphyletic group: the term is therefore substituted with the more descriptive designation "heterotrophic flagellates."

In the following chapter, the traditional taxa "Sarcodina" and "Rhizopoda" are no longer used. This is because the enigmatic homology between the different modes of organization of pseudopods (lobopodia, filopodia, reticulopodia, axopodia) cannot be demonstrated and should be neglected. The characteristic construction

principles are based on only two elements of the cytoskeleton, namely actomyosin and tubulin systems, which are present in all eukaryotic organisms. From this it is not surprising that rhizopodial organization types could emerge from almost all protozoan taxa. However, for didactic and less formal reasons, the use of terms such as flagellates, amoebae, rhizopods, heterotrophic flagellates, and phototrophic flagellates should be seen as pragmatic designations of distinct organization types. Again, these terms are not used to indicate any taxonomic group.

System of Protozoa

Fig. **35** Antique cabinet for microscope slides (courtesy of G. Teichert, Berlin).

As explained in the foregoing chapters, taxa such as "Zoomastigophora," "Phytomastigophora," or "Sarcodina" and some of the subordinated well-known subgroups (e.g., "Radiolaria") are para- or polyphyletic and should no longer be used. At the same time, some less well-known taxa achieve the rank of an important branch such as the Microspora as the sister group of all other eukaryotes. However, it is presently not possible to provide a complete cladistic system, because some groups sharing common characters such as those known under the designation "Rhizopoda" must be separated as "Eukaryota incertae sedis" (Fig. **35**).

Empire	**EUKARYOTA**
	Chatton, 1925

The organization scheme of all eukaryotes embraces as its most relevant character, the compartimentalization of the cell into distinct reaction rooms. From the morphological point of view, the most important compartments are created by the endoplasmic reticulum, which separates, e.g., the cytoplasm from the karyoplasm. The characteristic architecture includes the cytoskeleton with filamentous systems (actin and others) and tubular systems (microtubules) which are responsible not only for maintaining shape and locomotion, but also for intracellular transport phenomena, such as the separation of chromatids during mitosis. Because of the wide variation in eukaryotic taxa, it is reasonable that sexual processes, such as meiosis and karyogamy, were present at the earliest stages of cell evolution.

Kingdom	**MICROSPORA**
	Hülsmann & Hausmann, 1994

The organisms grouped within this taxon are characterized by the absence of both flagella and highly evolved endomembrane systems. They also exhibit some aspects typical of primitive eukaryotes such as smaller ribosomal subunits. The apomorphic characters are the same as for the taxon Microspora, the only presently known taxon.

Phylum	**Microspora**
	Sprague, 1982

The regulary nonflagellate cells are mostly very small. Their spore diameters range from 1 μm to a little more than 20 μm. Without exception they are intracellular parasites, living sometimes within a parasitophore vacuole but mostly free in the cytoplasm of the host cell. The host spectrum of the 900 known species ranges from uni-cells (Apicomplexa, Myxozoa, Ciliophora) to metazoans (coelenterates, platyhelminths, nematodes, bryozoans, annelids, molluscs, arthropods and vertebrates). The most widespread distribution is in the arthropods and bony fishes. Within the mammals, they occur mostly in rodents and carnivores, but probably also in primates. Surprisingly, plants are not infected.

As primitive (plesiomorphic) characters, the Microspora have 70s ribosomes (with 16s and 23s rRNA subunits) and some other features which they share with prokaryotes. Flagella, centrioles, as well as all kinds of respiratory organelles and true dictyosomes (with more than three cisternae) are also missing. The spore walls are made of chitin. To the general apomorphic characters belong diplokaryotic (paired) nuclei and a very conspicuous extrusion apparatus (Fig. **36**).

In resting spores, this apparatus has a coiled and tubular polar filament thickened by proteinaceous layers, and in the class Microsporea, an additional lamellated polaroplast made of densely packed or lamellar membrane staples. The whole system arises as an invagination of the cell membrane. The cell, with one or two nuclei (sometimes designated as amoeboid germ, amoebula or sporoplasm) also contains rough ER, free ribosomes, flat staples of membranes resembling dictyosomes and, in the posterior region, an empty body interpreted as a vacuole (posterosome).

The incompletely understood infection maneuver is visible and analyzable using cultivated cells as hosts. In vivo, it normally begins with spores acquired from food. In the intestine of hosts, the internal pressure of the spore increases, possibly by dilatation of the polaroplast vesicles and the posterosome. The pressure causes the polar filament to turn over (Fig. **37**). The kinetic energy of the injection-like extrusion process is strong enough to penetrate cell membranes, whole cells, and even cyst walls. Through the lumen of the long (up to several hundreds μm) polar filament, the amoebula is injected into the host cell (usually an epithelial cell). Inside the cell, the parasite changes into a simply organized "meront" and usually undergoes several asexual multiplicative stages (called schizogony or merogony). It is an interesting phenomenon that some afflicted host cells are able to adapt their metabolic rates and to increase the number of nucleoli required to fulfill the protein

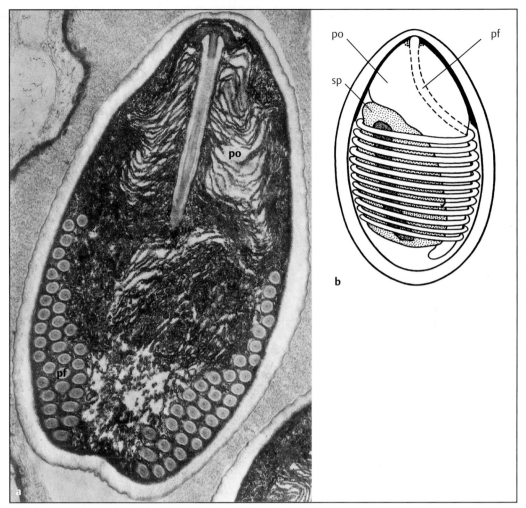

Fig. **36** Microspora: spore of *Plistophora hyphessobry-conis* with coiled pole filament (pf) and lamellated polaroplast (po) (**a**); scheme of a spore of *Thelohania california*. sp = sporoplasm, pf = pole filament, po = polaroplast (**b**) (a from Lom & Corliss: J. Protozool. 14 [1967] 141; b from Grell: Unterreich Protozoa, Einzeller oder Urtiere. In: Lehrbuch der Speziellen Zoologie, 4th edition, ed. H.-E. Gruner. Gustav Fischer Verlag, Stuttgart 1980). Magn.: 25 000×.

requirements of the meronts. In other cases, they feed on neighboring uninfected tissue cells causing tumor-like bodies several millimeters in diameter.

In the meronts, nuclear division takes place as intranuclear mitosis, as in most lower eukaryotes. During sporogony, so-called sporoblasts appear showing a higher degree of differentiation; in some species meiotic phenomena (formation of synaptonemal complexes) are also detected. The sporoblasts ultimately form infective spores that are released by defecation or by putrefaction of the host. In more complex life cycles, het-eromorphic stages are present with one or more host changes (heteroxeny). The method of dispersal within an infected host is not known.

The practical significance of the Microspora results from their parasitic way of life. They play an important economic role as causative agents of pébrine (*Nosema bombycis*), foul brood in bees (*Nosema apis*), and some fish infections (several *Glugea* species). The bulk appearance of Microsporea (*Encephalitozoon* ssp). in AIDS-affected humans is possibly opportunistic. On the other hand, there have also been several attempts to utilize microsporeans in biological pest control

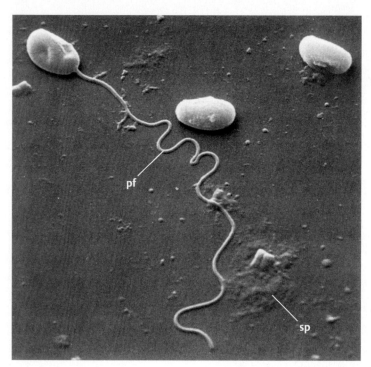

Fig. **37** Microspora: extruded pole filament (pf) and freely moving amoeboid sporoplasm (sp) of *Nosema tractabile* (from Larsson: Protistologica 17 [1981] 407). Magn.: 4000×.

as hyperparasites or as parasites of pests: *Vairimorpha necatrix* against caterpillars or *Nosema locustae* against tropical grasshoppers.

The systematic division of Microspora follows the mode of differentiation of the extrusion apparatus.

Class   **Rudimicrosporea**
                Sprague, 1977

The representatives have a rudimentary (secondarily simplified) extrusion apparatus without polaroplast and without posterosome. Their distribution is restricted to gregarines (Apicomplexa) parasitizing annelids.

Examples: *Amphiacantha, Desportesia, Metchnikovella*

Class   **Microsporea**
                Delphy, 1963

The Microspora *sensu stricto* mostly have a complex extrusion apparatus (with polaroplast). The multilayered spore wall contains chitin and other compounds.

Examples: *Amblyostoma, Encephalitozoon, Glugea, Nosema, Pleistophora, Vairimorpha*

Kingdom   **MASTIGOTA**
                    Hülsmann & Hausmann, 1994

In the evolution of the flagellum from microtubular systems originally belonging to the distribution apparatus of the chromosomes, and at least morphologically and functionally correlated with the nucleus, a process took place which is characteristic of the majority of the eukaryota. Cells equipped with this motile apparatus have a greater repertoire of actions: They are able to swim or to collect food particles from greater distances. Because of hydrodynamic interactions with the fluid medium, it is necessary for optimum efficiency that the number of flagella and the size of the cell body are correlated; therefore it may be assumed that the first freely swimming flagellates with only one or a few flagella belonged to an extremely small-sized class (2–10 μm) corresponding to the recent representatives of the nanoplankton.

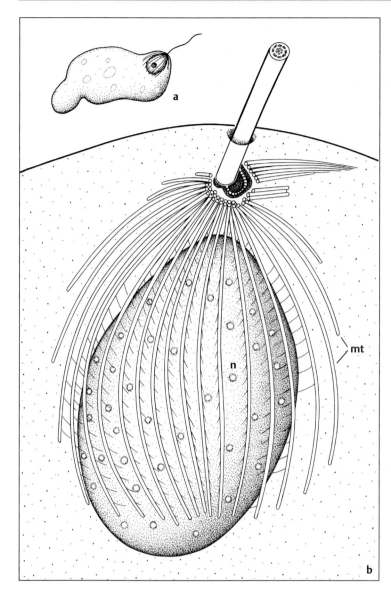

Fig. **38** Archamoebaea: karyomastigont system in *Mastigina* (**a**), functional and morphological relationships between kinetosome and nucleus (**b**). n = nucleus, mt = microtubules (after Brugerolle). Magn.: a 350×.

Subkingdom **ARCHAMOEBAEA**
Cavalier-Smith, 1991

The archamoebae branch from the stem line of the flagellated Eukaryota after evolution of the flagellum. They represent a primitive type, based on the presence of singular and always unpaired kinetosomes. The axonemes of the flagella are reduced, and the cells are also capable of amoeboid locomotion.

Phylum **Caryoblasta**
Margulis 1974

A typical characteristic is the flagellar-cytoskeletal complex (= karyomastigont) comprising the kinetosome of the flagellum, the microtubules deriving from a MTOC, and the nucleus. In typical cases, the nuclei are conically embraced by such systems (Fig. **38**). The flagellar apparatus itself (= mastigont) is present as one to

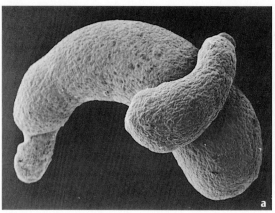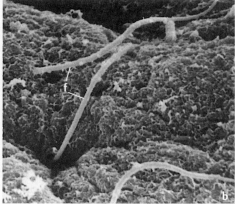

Fig. **39**   Two specimens of *Pelomyxa palustris* (**a**); flagella (f) of the uroid region (**b**). Magn.: a 100×, b 8000×.

several per cell and consists of a single flagellum and a single basal body.

The group contains *Pelomyxa palustris* (Fig. **39**), which formerly belonged to the specially erected order Pelobiontida (PAGE, 1976) within the "rhizopoda." Although known since about the middle of the last century, clear illustrative evidence has been provided in the last few years that the tail region (uroid) of the 1 to 5 mm sac-like cells bears several naked flagella. They have some typical structural characters of the Eukaryota flagellum, but they are obviously immotile. As with classic amoebae, the locomotion is substrate-bound and results from endoplasmic streaming with caudal ecto-endoplasm and frontal endo-ecto-plasm transformation.

The giant cells contain hundreds of nuclei and several simple karyomastigonts, but mitochondria and typical dictyosomes are missing. Hydrogenosomes are also not proven. In addition to large glycogen bodies, the cytoplasm bears numerous endobiotic bacteria belonging to three different (gram-positive, gram-negative or gram-variable) species, at least one of which is definitely a methanogen. They are often held in vacuoles surrounding the nuclei. The species has a very complex annual life cycle characterized by the appearance of small individuals with only two nuclei deriving from larger cells by plasmotomy and cysts with four nuclei. The diameters of the nuclei, the relative proportions between the bacterial populations, and the tolerance of dissolved oxygen all change simultaneously. Adult specimens can be found during summer and autumn in oxygen-deficient organically rich freshwater sediments.

Abnormal axonemal structures with irregular microtubular arrangements also occur in the genus *Mastigina* (Fig. **38**). The uninucleated species of the genera *Mastigella, Mastigamoeba,* and *Phreatamoeba,* however, possess apparently normal but functionally atypical flagellar apparatuses.

Examples: *Mastigella, Mastigina, Pelomyxa, Phreatamoeba*

Subkingdom **DIMASTIGOTA**
Hülsmann & Hausmann, 1994

With the doubling of flagellar structures, a process is started which can be followed up to the metazoans and terrestrial plants: If not secondarily reduced as in higher molds or flowering plants, the kinetosomes (or centrioles) appear only in pairs (Fig. **40**). Within the protozoans, there is only one known example for the secondary reduction to singular unpaired kinetosomes: in gametes of *Chlorarachnion,* a rhizopodially organized organism belonging to the chrysomonads. In the taxon, some trends are obvious regarding a more complex surface of the flagella (Fig. **41**), but also the multiplication of the flagella (e.g., four kinetosomes with four flagella). On the other side, there may also be a reduction (e.g., four kinetosomes with two flagella or two kinetosomes with one flagellum). In this way, several types of flagellation evolved (Fig. **42**). The principle of doubling favors the progress of a heterokonty, as documented by morphological differentiation into locomotive and nutritive flagella.

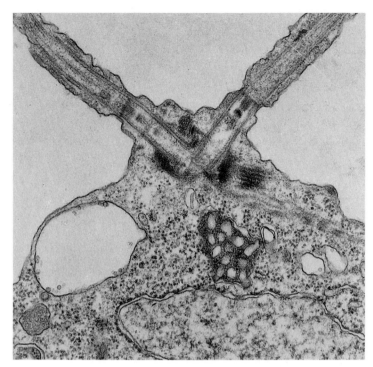

Fig. **40** Longitudinal section through the anterior of the heterotrophic flagellate *Aulacomonas submarina* (stramenopiles) with typical arrangement of the flagellar apparatus in dimastigote organisms (from Brugerolle and Patterson: Europ. J. Protistol. 25 [1990] 191). Magn.: 30 000×.

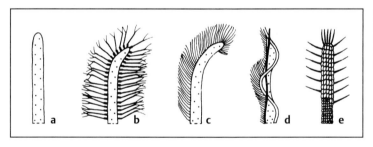

Fig. **41** Flagellar characters in chloromonads (**a**), chrysomonads (**b**), euglenids (**c**), dinoflagellates (**d**) and prasinomonads (**e**).

Superphylum **TETRAMASTIGOTA**
　　　　　Hülsmann & Hausmann, 1994

The flagellates belonging to this branch of the Dimastigota are characterized by a basic organization with four flagella arranged in two double-units. The singular recurrent flagellum and the associated microtubular root structures are seen as the most important homologous character. However, the taxa documented below also show divergent features. Therefore, some doubts about a monophyletic origin remain.

Phylum　　**Retortamonada**
　　　　　Grassé, 1952

The uni- or binucleated cells of the Retortamonada are characterized by one recurrent and three anterior flagella. The posteriorly directed flagellum runs through an invagination of the cell body which forms a cytostomal groove. The species have no mitochondria and occur in the intestines of other animals or, as free-living cells, in other anoxic microhabitats.

Class　　**Retortamonadea**
　　　　　Grassé, 1952

These small (mostly 5–20 μm) cells possess two pairs of kinetosomes positioned at the anterior in the vicinity of the nucleus and a large cytostome (Fig. **43**). From each pair originates at least one flagellum *(Retortamonas)*; in *Chilomastix* there are two, so that the members belonging to this genus are tetraflagellate. One flagellum always arises at the base of the cytostome, from

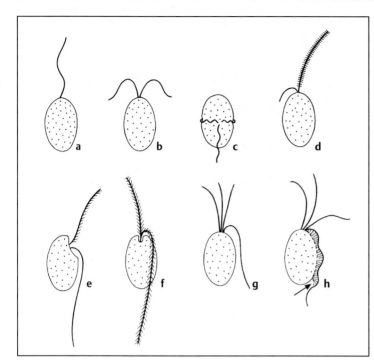

Fig. **42**   Flagellar numbers and insertion types. **a** uni-flagellate; **b** isokontbiflagellate; **c** biflagellate with transverse and longitudinal flagella; **d** heterokont-biflagellate, one flagellum bears mastigonemes; **e** biflagellate with one recurrent flagellum; **f** biflagellate with two fimbriated flagella originating from an apical invagination or gullet; **g** tetraflagellate with one recurrent flagellum; **h** tetraflagellate with undulating membrane (arrow).

Fig. **43**   Retortamonadea: *Retortamonas*, diagram (**a**) with indication of section levels figured in **b**. The recurrent flagellum with fin-like extensions passes through a groove covered by cytostomal lips. n = nucleus, fv = food vacuoles (adapted from Brugerolle). Magn.: a 7000×, b 15 000×.   ▽

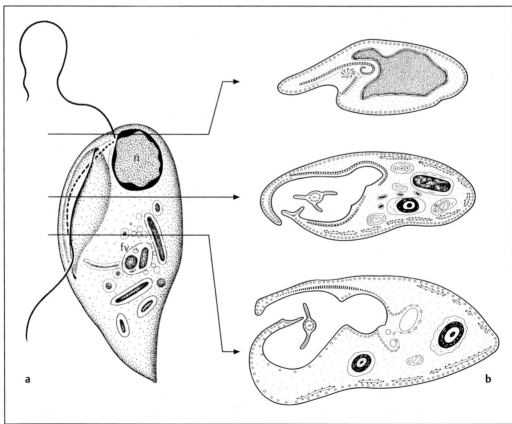

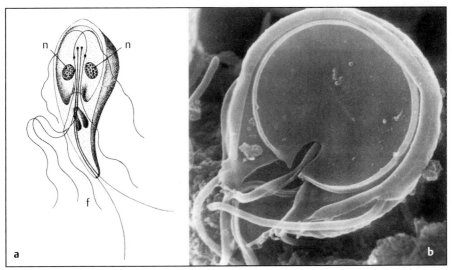

Fig. **44** Diplomonadea: **a** *Giardia* (= *Lamblia*) with two nuclei (n) and eight flagella (f). **b** adhesive disc in *Giardia* (a from Grell: Unterreich Protozoa, Einzeller oder Urtiere. In: Lehrbuch der Speziellen Zoologie, 4th edition, ed. H.-E. Gruner, Stuttgart 1980; b, courtesy of D. V. Holberton, Hull). Magn.: a 3000×, b 7000×.

which it projects and swings posteriorly. At the shaft it bears some seam-like projections optimizing the effectivity of stroke and serves for the whirling of food particles. The cytostome consists of a longitudinally running body furrow, roofed by the median overlapping of two velum-like cytostomal lips. It is stiffened, like the rest of the cell surface, by pellicular microtubules. The other flagella are posteriorly directed and function as a swimming device. Complicated rhizoplasts occur also in this group, whereas mitochondria and dictyosomes are absent.

The parasitic retortamonads have resting cysts used for transmission to new hosts (invertebrates and vertebrates). The excysted stages (trophozoites) which are active in feeding, live in the intestine and feed on bacteria. Besides harmless commensals, species have also been discovered that act as pathogenic agents for diarrhea (*Chilomastix mesnili* in humans, *Chilomastix gallinarum* in domestic fowl).

Examples: *Chilomastix, Retortamonas*

Class   **Diplomonadea**
Wenyon, 1926

These are aerotolerant anaerobes which lack mitochondria. Dictyosomes are also absent. The most derived species are most easily explained as double- organisms (diplozoic form).

The general characteristic of the monozoic organization is a group of up to four flagella: three of them are freely swinging, but the fourth is directed posteriorly along the ventral body side, usually within a furrow acting as a cytostome. This pattern resembles the organization of retortamonads. However, the kinetosomes of the monozoic diplomonads, as well as the entire microtubular and fibrillar systems form, together with the nucleus, a morphologically distinct complex (karyomastigont system). The characteristics of such systems are an indentation of the nucleus, including the kinetosomes, as well as extensive microtubular ribbons originating from the kinetosome of the recurrent flagellum.

In the diplozoic organization (Fig. **44**), two karymastigont systems are present. Therefore the cells have not only two cytostomes but also two nuclei and altogether eight flagella. The orientation of the system is radially symmetric; the morphology corresponds to a fusion of the dorsal sides of two monozoic units. This type is analogous to the development of Siamese twins. An additional character of diplozoic forms is the funnel-like construction of the cytostome, which can reach as far as the posterior part of the cell.

Order   Enteromonadida
Brugerolle, 1975

The enteromonads comprise a taxon presently containing not more than about 15 species. The morphological type is monozoic. They live as

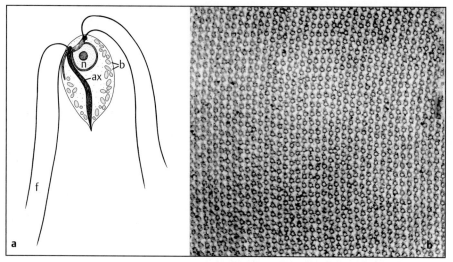

Fig. **45** Oxymonadea: **a** flagellate with four flagella (f), nucleus (n), intracellular bacteria (b), and motile axostyle (ax). **b** cross section of an axostyle with numerous microtubules (a from Radek: Arch. Protistenkd. 144 [1994] 373; b courtesy of R. A. Bloodgood, Charlottesville). Magn.: a 2500×, b 40000×.

nonpathogenic parasites in the intestines of humans (e.g., *Enteromonas hominis*) and of other vertebrates. The cysts are excreted with feces and transferred to other hosts. Within the group, a considerable reduction of free flagella can be observed.

Examples: *Caviomonas, Enteromonas, Trimitus*

Order   Diplomonadida
             Wenyon, 1926

There are about 100 species. Diplozoic forms are free-living in anoxic fresh water and endozoic as commensals or as obligate parasites in invertebrates and vertebrates. The pathogenic forms have specific mechanisms to anchor to the epithelium of distinct parts of the intestine. They hinder or disturb the uptake of nutrients by the host and cause severe bloody diarrhea when they occur as a mass development, as occasionally during immune deficiency (AIDS, etc.). In the most important genus *Giardia* (= *Lamblia*, see Fig. **44**), the axonemata run a considerable distance intracellularly before entering the flagellar shaft. About 50 species of this genus have a ventral adhesive disk. Most cannot be differentiated with morphological methods but must be distinguished by their host specificity or by molecular methods. The transfer to potential hosts is ensured by encysted stages.

Examples: free-living or endozoic: *Hexamita, Trepomonas, Trigonomonas* parasitic: *Giardia, Octomitus, Spironucleus*

Phylum   **Axostylata**
             Hülsmann & Hausmann, 1994

The members of this phylum possess an intracellular rod made of thousands of microtubules, called axostyle (Figs. **21 f**, **g**, **45 b**, **49**). Primarily, this organelle serves as a motile apparatus. However, during evolution, static and skeletal functions were also developed.

Class   **Oxymonadea**
             Grassé, 1952

In general, these flagellates (by some authors called Pyrsonymphidae) possess four flagella arising at the posterior end of the ovoid and partly distorted screw-like 50 μm long cells. Sometimes, the flagella are focally fixed with their shafts to the cell surface and in this way form an undulating membrane. A cytostome with an associated flagellum is missing. Characteristically, a motile rod-like organelle, the axostyle, is present. It traverses the cell body longitudinally and consists of several thousands of cross-linked microtubules originating from kinetosomes (Fig. **45**). Axostyle, kinetosomes, and nucleus form a functional morphological unit. This is documented during the formation of zygotes when not only the nuclei but also the axostyles fuse with each other. The motility of the axostyles manifests itself in the form of curving waves running from anterior to posterior and leading to a snake-like locomotion of the cells.

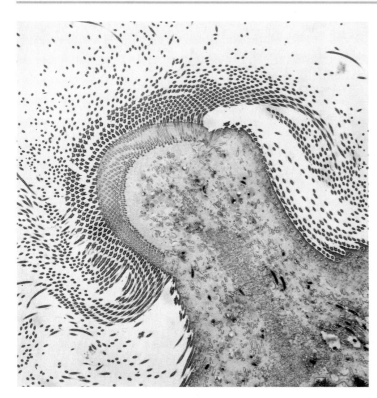

Fig. **46** Parabasalea: apical region of *Koruga bonita* (courtesy of R. Radek, Berlin). Magn.: 2500×.

Oxymonads live as obligate anaerobes exclusively in the intestinal diverticles of wood-eating insects (cockroaches, termites). They often occur in large quantities and represent an important part of the intestinal community to which also belong the Parabasalea (see below), fungi, and bacteria. Dictyosomes, mitochondria, and possibly also hydrogenosomes are missing. At the posterior end of the cells, phagocytosis of wood particles take place. It is not clear presently whether they have autonomous cellulolytic abilities, or whether intracellular bacteria and the spirochetes adhering to the cell surface are the responsible agents.

Examples: *Oxymonas, Pyrsonympha, Saccinobacculus*

Class **Parabasalea**
Honigberg, 1973

The mostly uninucleated Parabasalea appear as a heterogeneous taxon when only the numbers of flagella per cell (zero to several tens of thousands) are taken into account (Figs. **46** and **49**). A flagellation with four flagella is considered to be primitive; three of them are, as in the oxy-

monads, directed anteriorly, and one acts as a recurrent flagellum (*Monocercomonas* type). During evolution, the number of flagella has partly increased and partly decreased. A common apomorphic character of all representatives are so-called parabasal bodies or parabasal rods which represent aggregates of apparently very large dictyosomes with kinetosome-associated fibrillar systems. The morphological features of the dictyosomes, especially the extremely large number of up to 30 cisternae, are interpreted here as the result of an independent (convergent) process. As other typical characteristic axostyles appear which are nonmotile, in contrast to the situation in oxymonads. In more highly evolved groups, they are multiplied, reduced, or degenerated. Mitochondria are completely missing, but simply organized hydrogenosomes may be present in some groups. Distinct cytostomes do not appear, and phagocytosis takes place over most of the cell body. The Parabasalea are exclusively endozoic, feeding on bacteria and particulate material in the intestine. Some of them are important parasites; however, the mode of pathological interference with the host is at present not always completely understood.

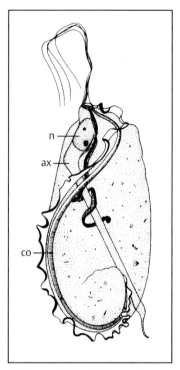

Fig. **47** Trichomonadida: *Trichomonas termopsides.* ax = axostyle, co = costa, n = nucleus (from Grell: Unterreich Protozoa, Einzeller oder Urtiere. In: Lehrbuch der Speziellen Zoologie, 4th edition, ed. by H.-E. Gruner, Fischer, Stuttgart 1980). Magn.: 3500×.

Order   Trichomonadida
               Kirby, 1947

Four to six flagella are typical for the relatively small (mostly 5–25 μm), trichomonads; however, some genera have only two flagella, or they are missing and the cell moves like an amoeba [*Histomonas* (causative agent of enterohepatitis in fowl), *Dientameoba* (in colon and appendix of humans)]. The axostyle is often elongate and terminates in a caudal process serving as an anchoring device. The multiflagellated forms have some other characters: a contractile rod (costa) enabling motility of the cells, and an undulating membrane formed by the recurrent flagellum and the cell surface (Fig. **47**). Within the more complicated types, the genus *Trichomonas* dominates, with several pathogenic species: *Trichomonas vaginalis* lives within the urogenital system of humans and is sometimes responsible for inflammations combined with slimy discharges. *T. hominis*, as the only intestinal species of

humans, is often the pathological agent of chronic diarrhea; *and T. foetus* causes cattle plague.

The termite-inhabiting species *Mixotricha paradoxa* shows some peculiarities because the cell surface also bears, besides some other prokaryotes, numerous and regularly arranged spirochetes. During locomotion, these are responsible for motility, whereas the four flagella of the cell serve only as steering organelles.

> Examples: *Dientamoeba* (no flagellum), *Monocercomonas* (3+1 flagella), *Pentatrichomonas* (5+1 flagella), *Tritrichomonas* (4+1 flagella)

Order   Polymonadida
               Dyer, 1990

The polymonads (= Calonymphidae) represent, as their name suggests, organisms with multiple cellular organelles. They are multinucleated and to each nucleus belong all the associated organelles typical for *Monocercomonas.* Therefore, the polymonads possess several groups of flagella, axostyles, and parabasal bodies. They live exclusively in the intestines of termites.

> Examples: *Calonympha, Coronympha*

Order   Hypermastigida
               Grassi & Foa, 1911

The representatives of the hypermastigids are characterized by the large number of flagella inserted at the anterior end or at the anterior periphery of the cell (Figs. **48** and **49**). Normally, they undergo wave-like movements. The flagella arise from concave deepened plates or they are ordered along longitudinal or spiral rows. In the same manner, the parabasal apparatuses also appear as multiple or bushy-branched units. Axostyles, however, are mostly present as singular organelles or are fused together. The dictyosomes are normally large and visible as long bodies, even in the light microscope. The mitotic spindle of dividing nuclei is located extranuclearly.

Polymastigids live exclusively in the intestines of wood-eating insects (cockroaches, lower termites), mostly in special fermentation chambers. They host numerous intracellular and extracellular bacteria and spirochetes responsible for the digestion of ingested cellulose and wood particles.

> Examples: *Barbulanympha, Joenia, Lophomonas*

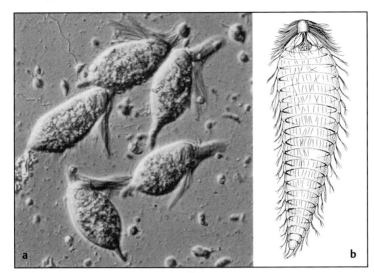

Fig. **48** Hypermastigida:
a polymastigid flagellates
from intestinal tract of the termite *Kalotermes flavicollis*.
b *Teratonympha mirabilis*
(from Grell: Unterreich Protozoa, Einzeller oder Urtiere. In:
Lehrbuch der Speziellen Zoologie, 4th edition, ed. by H.-E.
Gruner, Fischer, Stuttgart
1980).
Magn.: a 200×, b 280×.

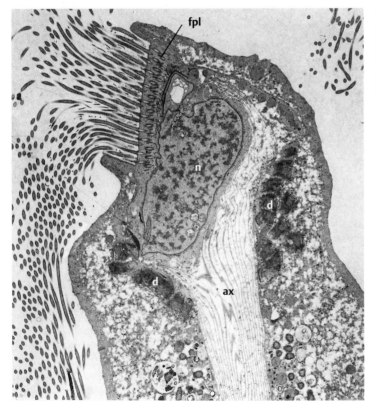

Fig. **49** Hypermastigida: axostyle (ax), flagellar plate (fpl),
nucleus (n) and dictyosomes
(= parabasal bodies) (d) in the
apical region of *Joenia annectens* (from Radek and
Hausmann: BIUZ 21 [1991]
160). Magn.: 3000×.

Superphylum **METAKARYOTA**
Cavalier-Smith, 1991

With the genesis of the Metakaryota, which
possess as autapomorphic characters not only
classical mitochondria (or secondarily: hydroge-
nosomes) but also typical dictyosomes, eukaryotic dimastigote organisms emerge, for the first
time exhibiting the typical image of a eucyte.
Whereas the mitochondria are considered the result of an endosymbiotic event, the evolution of
dictyosomes out of simpler elements of the en-

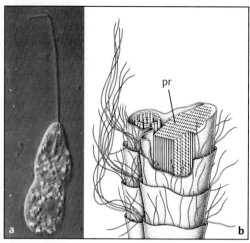

Fig. **50** Euglenoidea: **a** *Peranema* with clearly thickened flagellum. **b** flagellum with flaccid mastigonemes in *Anisonema*. pr = paraxial rod (b, adapted from Mignot). Magn.: a 500×, b 40 000×.

domembrane system is believed to be an anagenetic process which occurred in this stem line only once and which is not homologous to the corresponding structures in the Parabasalea. Primitively, the cells bear two heterokont flagella with mastigonemes (fimbriated appendices or flimmer hairs).

Within this taxon, a ramification occurs leading not only to the better known groups of the protozoans, but also to the plants, higher fungi, and animals. At present, however, the sister group relations are not completely understood: Molecular biological data, e.g., analysis of rDNA sequences, give the first indications of the temporal sequence of new subtaxa (see Figs. **29–31**).

Phylum   **Euglenozoa**
             Cavalier-Smith, 1981

The mitochondria of the Euglenozoa exhibit cristae which are of (or are derived from) the discoidal type. The organisms have cortical microtubular ribbons which stiffen the cell periphery and are responsible for a relatively constant cell shape. The axonemal microtubules of the flagella are normally accompanied by protein complexes (paraxial rod, see Figs. **24 b** and **50**); therefore, the diameter of the flagellar shaft is often unusually large. During mitosis, the nucleolus persists.

Subphylum   **Euglenida**
             Leedale, 1967

The two classes, Euglenoidea and Hemimastigophorea, are characterized by more or less longitudinally running microtubules stiffening the cortex or giving them the ability to perform a special movement, called euglenoid movement or metaboly. The number of flagella is two, but multiples can also be observed. The uninucleated cells are primarily heterotrophic.

Class   **Euglenoidea**
             Bütschli, 1884

The euglenids, represented by about 1000 species, have two heterokont flagella in their original organization. One of them can be extremely reduced in length, so that the proof of the real existence of two flagella can often only be obtained by electron microscopy. The flagella arise from an apical groove (the so-called flagellar sac, ampulla, or reservoir, Fig. **51**). The longer, swimming flagellum is normally thickened by a paraxial body with up to five times the normal diameter of about 0.2 µm. Due to this character, the euglenids can be easily recognized in the light microscope. This flagellum also bears one row of hair-like fimbria or mastigonemata (Fig. **50**) and exhibits, in phototrophic forms, a basal swelling (= paraflagellar body). This organelle is assumed to play, in cooperation with an extraplastidial stigma, an important functional role in the photosensory orientation of the cell. If projecting out of the lumen of the ampulla, the second flagellum is normally utilized as a trailing flagellum and serves for food catching or for gliding on, or fixing to the substrate.

The cell membrane is usually flanked by internal protein complexes ordered in tile-like overlapping and spirally arranged stripes (comp. Fig. **220**). Obviously, due to the involvement of microtubules, they can glide relative to each other and they are responsible for the so-called euglenoid or metabolic movement. This kind of crawling movement is found especially in cells which are able to resorb their flagella and invade the substrate, or which live on the substrate as phagotrophs or parasites. However, some genera are absolutely rigid *(Phacus)* or they assemble extracellular cell walls *(Trachelomonas)*, and others live as sessile and stalk-bearing organisms *(Colacium)*. In naked freshwater species, a contractile vacuole is present, discharging its contents into the lumen of the ampulla.

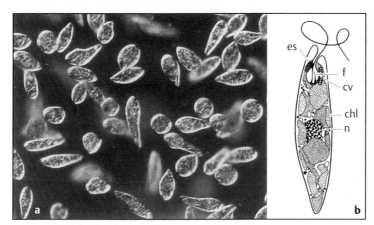

Fig. **51** Euglenoidea: **a** morphological variability (metaboly) in *Euglena gracilis*. **b** organization of *E. gracilis*. chl = chloroplast, cv = contractile vacuole, es = eyespot (stigma), f = flagella, n = nucleus (b adapted from Leedale).
Magn.: a 250×, b 1000×.

Even during interphase, the nuclei exhibit condensed chromosomes visible in the living cell. During mitosis, the nuclear envelope remains intact, and during metaphase, the chromatids are arranged parallel to the intranuclear spindle.

Only one third of euglenozoans is equipped with plastids. They are bordered by altogether three membranes, and do not show any spatial relations to the endoplasmic reticulum. They contain, besides other pigments, the chlorophylls a and b, thus they are comparable to the chloroplasts of the Chlorophyta. It is suggested that the plastids are relicts of a formerly free-living eukaryotic being, probably a green alga. As storage products there are, besides lipids, particularly glucanes (= paramylon), which are stored in slices or as granules, and deposited within the cytoplasm. From the phylogenetic point of view, it is important to argue that the postulated endosymbiosis between two eukaryotes was possible only after the evolution of green algae.

The other two-thirds of the Euglenozoa live exclusively as saprotrophs or heterotrophs on bacteria or eukaryotic unicells. The heterotrophic way of life (= primary phagotrophy) is considered primitive. However, there are indications that some of the species, which had adopted chloroplasts during their evolution and therefore had acquired the ability for photosynthesis, returned by the loss of plastids, convergently to the animal way of life (= secondary phagotrophy).

The occurence of the green Euglenozoa is limited to freshwater biotopes and brackish water. In Europe, an especially high number of species was identified in Lake Neusiedel (a soda lake in Austria). Mass developments of photo-

trophic forms occur frequently in organically polluted ponds. Phagotrophic euglenids are also important constituents of shallow marine benthic habitats.

Examples: phototrophic: *Euglena, Eutreptia, Phacus*
heterotrophic: *Anisonema, Astasia, Peranema*

Class   **Hemimastigophorea**
            Foissner et al., 1988

So far, only a few species of this class are known. The about 20 μm long colorless organisms have two rows of flagella (Fig. **52 a**). The arrangement of the cortical elements is rather similar to the situation found in the Euglenoidea and in cross section, a diagonal symmetry in the cellular architecture becomes obvious (Fig. **52 b**). Metabolic movement, typical for certain euglenids, has also been reported for one hemimastigophorean species. On the other hand, the root structures of the kinetosomes are reminiscent of opalinids and some other features indicate an early branching from the Euglenoidea during evolution.

Examples: *Hemimastix, Spironema*

Subphylum   **Kinetoplasta**
            Honigberg, 1963

The kinetoplastids (about 600 species) live as free-living bacterial feeders, as endobiotic commensals, or as harmful parasites. The name of the taxon refers to the typical character, the kinetoplast (Fig. **53**). This structural complex represents not a distinct organelle, but a segment of the unique mitochondrion with an unusually high content of so-called kDNA (Fig. **54**). In

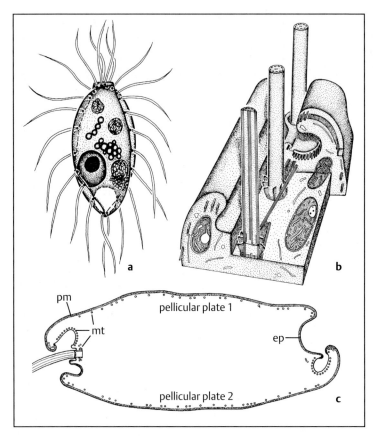

Fig. **52** Hemimastigophorea:
a *Hemimastix amphikineta.*
b 3-D reconstruction of pellic-ular and subpellicular or-ganelles in the middle region of *H. amphikineta.* **c** sche-matic transverse section of *H. amphikineta* resembling the gross cortical construction of euglenids. ep = epiplasm, mt = microtubules; pm = plasmamembrane (from Foissner et al.: Europ. J. Protis-tol. 23 [1988] 361). Magn.: a 2200×.

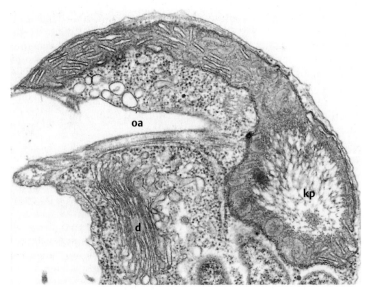

Fig. **53** Kinetoplasta: *Bodo* with intramitochondrial kine-toplast (kp). d = dictyosome, oa = oral apparatus. Magn.: 10 000×.

Crithidia fasciculata, for instance, the kDNA represents a network of about 5000 minicircles (~ 2,5 kb), and about 25 maxicircles (~ 37 kb) which are all topologically interlocked (Fig. **56**). This region is feulgen positive and therefore visible by light microscopy. The term kinetoplast derives from the preelectron microscopic era: The name refers only to the topological location nearby the two kinetosomes. The uninucleated cells possess two flagella, one of which is frequently reduced. They arise from an apical groove (flagellar sac). Normally, both flagella are stiffened by paraxial bundles and often constitute undulating membranes, which especially favor swimming in viscous media (blood, latex). Despite this congruence, both flagella are heterokont and heterodynamic; the anterior flagellum of the free-living Bodonida bears unilateral flimmer hairs (mastigonemata), the retrograde flagellum is naked and sometimes focally attached to the cell body (see Fig. **54**).

The pellicula of the mostly oblong cell body is stabilized by a microtubular cytoskeleton. In the same way, the rims of the oral apparatus contain microtubules. The amino acid sequence data of cytochromes b and c, as well as the nucleotide sequences of the 18s-RNA-subunits show that the kinetoplastids must have separated from their sister group, the euglenids, more than 1000 million years ago.

Class   **Bodonea**
Hollande, 1952

The bodonids are characterized by the primitve heterokont flagellation type. Both flagella arise in a flagellar groove (Fig. **55**). Bodonids are widely distributed, and they occur frequently as bacteria feeders in nutrient-rich habitats. The cytostome is well developed. Due to their small size, they are easily overlooked. However, some species typical of organically polluted waters can be easily distinguished because of their behavioral or morphological characters, e.g., *Bodo saltans* (with a dancing-springing motility), and *Rhynchomonas nasuta* (with a proboscis-like cell extension). Some bodonids are histophagic ecto- and endoparasites of fresh water fishes.

Examples: *Bodo* (free-living), *Cryptobia* (parasitic), *Ichthyobodo*

Class   **Trypanosomatidea**
Kent, 1880

The exclusively parasitic trypanosomatids possess one singular smooth flagellum which is

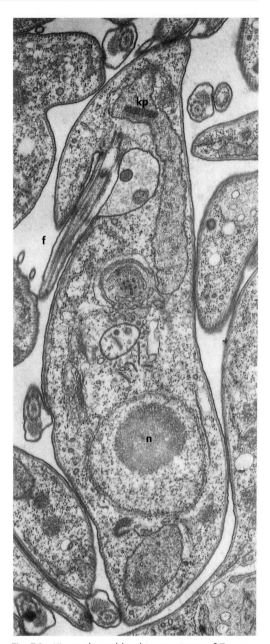

Fig. **54** Kinetoplasta: bloodstream stages of *Trypanosoma brucei*. f = flagellum, n = nucleus, kp = kinetoplast (courtesy of K. Vickerman, Glasgow). Magn.: 20 000×.

a homologue of the flimmer-bearing anterior flagellum of the bodonids. It swings freely or is connected by several adhesion spots to the cell surface, thereby forming an undulating membrane (Fig. **56**). Depending on the cell shape, rela-

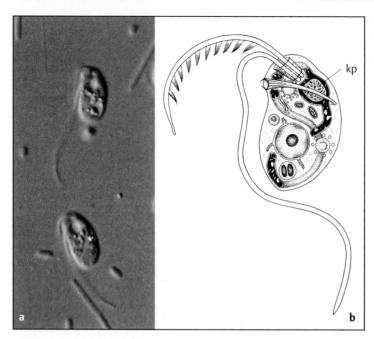

Fig. **55**   Bodonea: *Bodo* in life (**a**) and in schematic presentation (**b**). kp = kinetoplast (b, adapted from Vickerman). Magn.: a 1000×.

kp

a

b

tive positions, and morphological organization of the kinetoplast-kinetosomes flagellar groove-complex, several modifications can be distinguished (Fig. **57**). They are typical of distinct stages in the life cycle as well as of distinct genera. The following nomenclatural definitions are in use:

Kryptomastigote or amastigote (= *Leishmania* form): The flagellum inserts at the cell apex, does not project from the flagellar groove of the rounded cells, and remains therefore invisible at the light microscopic level.

Promastigote (= *Leptomonas* form): The flagellum arises from the anterior end of the slender cell.

Epimastigote (= *Crithida* form): The flagellum inserts in the middle of the slender cell.

Trypomastigote (*Trypanosoma* form): The flagellum inserts at the posterior pole of the slender cell.

Opisthomastigote: The flagellum arises at the posterior end, but emerges from the long flagellar groove at the anterior pole of the slender cell.

Choanomastigote: The rim of the flagellar groove is stretched out like a collar.

Sphaeromastigote: The flagellum arises from a cell with a compressed spherical outline.

The trypanosomatids are widely distributed in the subtropical and tropical regions of the Old and New World. They are successful parasites, attack warm- and cold-blooded vertebrates, invertebrates, protozoa and plants, and several are harmful pests (Table **2**). Besides monogenetic (monoxenic) forms, that attack only one host, there are also heterogenetic (heteroxenic) species that infect several hosts (Figs. **57** and **58**). Several blood-sucking insects are transmitters and intermediate hosts (vectors) of the *Trypanosoma*- and *Leishmania*-species of medical and veterinary importance (*Tabanus* [Tabanidae = horse flies], *Glossina* = tsetse fly [Muscidae = true flies], *Phlebotomus* [Psychodidae = butterfly-midges], *Triatoma* and *Rhodnius* [Reduviidae = bugs of prey], as well as blood-licking vampires (Desmodontidae). Within these vectors, several species (e.g., *Trypanosoma brucei*) are able to pass through special stages which are characterized, not only by fundamental morphological changes regarding the flagellar insertion type or the transformation of mitochondrial cristae from tubular to discoidal type, but also by radical alterations of the biochemical pathways of metabolism. The transmission of the parasites and the infection of the hosts is effected through contaminated saliva or vomited gut contents that fill during the sucking act, by oral uptake of partly encysted stages extruded with feces, by

Table **2**   Important *Trypanosoma* and *Leishmania* spp. parasitizing humans and domestic animals (after Mehlhorn)

Species	Hosts	Disease	Symptoms	Vectors	Geographic distribution
T. brucei brucei	equines, pigs, cattle, rodents	Nagana	stumbling gait, fever, lymphadenitis, meningoencephalitis	*Glossina* spp.	Tropical Africa
T. brucei gambiense	humans, monkeys, dogs, pigs, antelopes	sleeping sickness (mild variety)	fever, lymphoid hyperplasia, meningoencephalitis, coma	*Glossina* spp.	West and Central Africa
T. brucei rhodesiense	humans, wild game, pigs, rats (in experiments)	sleeping sickness (severe variety)	parasitemia, myocarditis, meningoencephalitis, coma	*Glossina* spp.	East and Central Africa
T. congolense	cattle, domestic animals	bovine trypanosomiasis	fever, anemia	*Glossina* spp.	Zaire, Kwazulu
T. cruzi	humans, domestic and wild animals	Chagas' disease	hypertrophy of heart, megaesophagus, megacolon, myocarditis, encephalitis	*Triatoma* and *Rhodnius* spp.	South and Central America
T. equinum	equines, cattle, water pigs	Mal de Caderas	fever, anemia	*Tabanus* spp.	South and Central America
T. equiperdum	equines	Dourine	genital swelling, anemia, nervous disorders	sexual transmission	Mediterranean countries, India, Java, Africa
T. evansi	ruminants, equines, dogs	Surra	fever, oedemas, anemia	*Tabanus* and *Stomoxys* spp.	India, Africa, Sibiria, Australia, Middle and Central America
L. braziliensis	humans	Espundia, mucocutaneous leishmaniasis	ulceration, lesions of skin, epithelia and cartilage	*Lutzomyia* spp.	Brazil, Mexico, Peru
L. donovani	humans, hamsters (in experiments)	Kala Azar, visceral leishmaniasis	hepatomegaly, splenomegaly, leukopenia	*Phlebotomus* spp.	Africa, Asia, Middle East, South America, India
L. tropica	humans	cutaneous leishmaniasis, oriental sore	ulceration, skin lesions	*Phlebotomus* spp.	Middle East

contact of skin lesions with feces, by venereal contact, or by medical blood transfusion.

The cell membrane of most trypanosomatids is covered with a thick (about 15 nm) glycoprotein surface coat. The chemical composition and the antigen behavior of this glycocalyx varies not only spatially, but also temporally (Fig. **58**). The variation is controlled by several hundreds (possibly more than 1000) of genes which comprise up to 40% of the total genome and which are recombined during sexual processes. Some of these genes are expressed at specific points during the life cycle, so that in the first step, after the infection and the first cell divisions of the parasites, a relatively homogeneous antigen homotype is present. This homotype is attacked by the immune system of the host with some success, but by genetic expression, more and more new antigen variants (antigen heterotypes) are produced, leading to new, infective

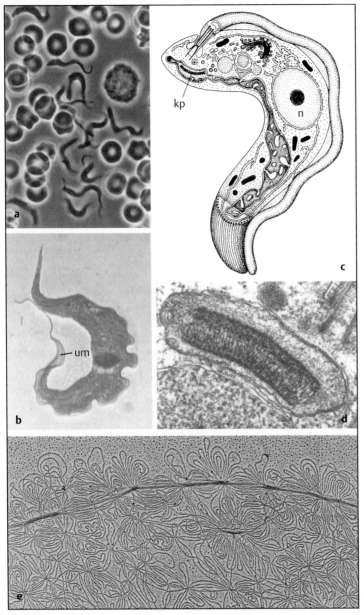

Fig. **56** Trypanosomatidea: **a** living bloodstream *Trypanosoma brucei* between erythrocytes. **b** view of entire *Trypanosoma fallisi*. **c** ultrastructural diagram of *T. congolense*. **d** kinetoplast (kp) of *Blastocrithidia triatomae*. **e** spread kinetoplast DNA from *Crithidia fasciculata*. n = nucleus, um = undulating membrane (b from Martin and Desser: J. Protozool. 37 [1990] 199, c adapted from Vickerman, d courtesy of H. Mehlhorn, Bochum, e from Pérez-Morga and Englund: J. Cell Biol. 123 [1993] 1969). Magn.: a 1000×, b 1100×, c 8000×, d 40000×, e 80000×.

populations of the parasite (chronic trypanosomiasis). The antibodies of the host are not able to control the population as a whole; their activity is limited to some selective effects on some of the parasite populations.

The pathogenic properties of the trypanosomatids are not based on the deprivation of nutrients, but on toxic metabolic byproducts (intravascular blood parasites) or on cell and tissue lesions (intracellular parasites) (Fig. **59**). In the case of Kala Azar, a tropical sickness frequently lethal for children, *Leishmania donovani* especially targets macrophages of the liver, brain, and bone marrow, resulting in fatal anemia. Some diseaes are chronic and lead to organic alterations (Chagas' disease, Fig. **60**). Several species of the genus *Phytomonas*, pathogenic to plants, and also attacking pathophagic Hemiptera, are economically important parasites of coconut palms and coffee plants.

Examples: *Leishmania, Phytomonas, Trypanosoma*

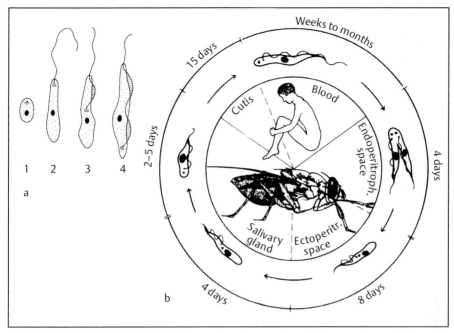

Fig. **57** Trypanosomatidea: **a** modification types of trypanosomatids. 1 amastigote (= krypto-mastigote, *Leishmania* type); 2 promastigote (*Leptomonas* type); 3 epimastigote (*Crithidia* type); 4 trypomastigote (*Trypanosoma* type). **b** Life cycle of the pathogen of African sleeping sickness, *Trypanosoma brucei gambiense* (b adapted from Dönges).

Subphylum **Pseudociliata**
Corliss & Lipscomp, 1982

The taxon comprises only four species belonging to one genus, *Stephanopogon* (Fig. **61**). Because of some superficial morphological similarities, they were long considered to be ciliates, and, because of the absence of nuclear dualism, as the most primitive ciliates (= Primociliata). However, due to new electron microscopical data, the necessity for a new systematic position became obvious. In *Stephanopogon apogon,* the most characteristic signs of ciliates, alveoli, and infraciliature remained undiscovered. Only the topographical position of the oral apparatus and the type of flagellation resembles that of the holotrich ciliates. The cortical system, however, resembles more the cortex of kinetoplastids and euglenids: The cell membrane is accompanied by numerous longitudinally arranged micro-tubules. The non-flimmered, smooth flagella arise from funnel-like grooves in the cell surface, stiffened by radially arranged microtubules (Fig. **61 b**). The identical (= homokaryotic) nuclei number from 2–16. Cell division takes place within cysts. Sexual events are not known.

Pseudociliates are small (20–50 μm) and they live in marine benthic habitats to a depth of about 100 m. They feed on diatoms, other flagellates, and bacteria.
Example: *Stephanopogon*

Phylum **Heterolobosa**
Page & Blanton, 1985

To this taxon belong some members of the former Sarcodina, which are characterized by the presence of nonmastigonemate flagella. In this group, several amoebae and acellular slime molds are subsumed.

Class **Schizopyrenidea**
Singh, 1952

There are about 100 species in this taxon. They are temporarily flagellated and otherwise amoeboid. Since important activities of the life cycle, such as division and food uptake, are present during the amoeboid, as well as during the flagellate stage, the systematic grouping was questionable in the past. At present, an evolutionary relationship with the Acrasea is favored: Autapo-

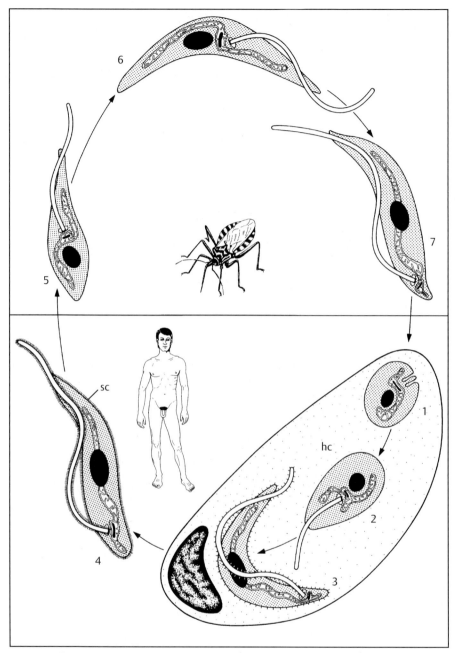

Fig. **58**   Life cycle of the pathogen of South American Chagas' disease, *Trypanosoma cruzi*.
1 amastigote or micromastigote inside host cell (hc); 2 and 3 transformation to epimastigote
(2) and trypomastigote (3); 4 free bloodstream stage with elaborated surface coat (sc); 5 transformation to epimastigote inside intestinal tract of reduviids; 6 epimastigote division form inside rectum; 7 infective trypomastigote stage within rectum or feces of bugs (adapted from
Mehlhorn).

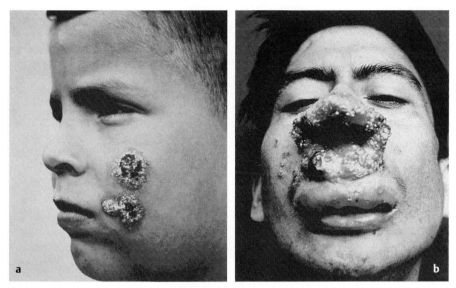

Fig. **59**   Subcutaneous lesions caused by *Leishmania tropica* (**a**) and *L. brasiliensis* (**b**) (from Frank & Lieder: Taschenatlas der Parasitologie, Franckh-Kosmos, Stuttgart, 1986).

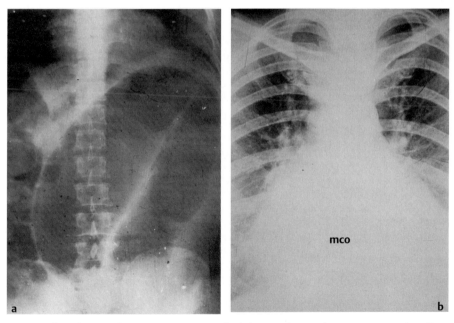

Fig. **60**   Clinical cases *of Trypanosoma cruzi*-evoked chronic Chagas' disease. **a** myocarditis (fibrosis of myocardial fibers), **b** anomalies of intestinal tract (megacolon, mco) (courtesy of WHO-TDR Image Library, Geneva, 1990).

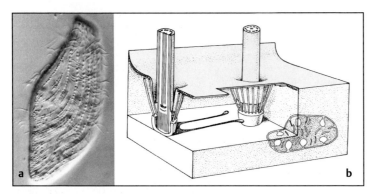

Fig. **61**   Pseudociliata: *Stephanopogon* (**a**) and schematic presentation of the cortical system (**b**). (a courtesy of D. J. Patterson, Sydney, b adapted from Lipscomb and Corliss). Magn.: a 800×.

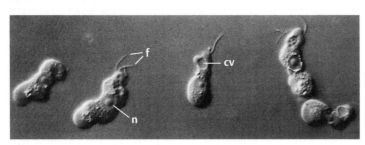

Fig. **62**   Schizopyrenidea: flagellate stage of *Naegleria*. cv = contractile vacuole, f = flagella, n = nucleus. Magn.: 1000×.

morphic features, common to both, are the eruptive manner of pseudopod expansion and the ability to produce flagella and cysts. Special mouth openings typical for some genera such as *Tetramitus* during the flagellate stage, provide justification for an independent group with predominantly flagellate characters.

The schizopyrenids are normally soil-borne organisms, but they also occur in marine and freshwater sediments. They are monopodial and have only one nucleus (Fig. **62**). The formation of the two or four smooth flagella takes place within a few minutes after sudden changes in the surrounding medium, such as a lowering of temperature or loss of electrolytes. In the case of desiccation and food deprivation, protective cysts are formed.

Besides bacteria feeders or harmless endobionts such as *Vahlkampfia* or *Tetramitus*, there also exist facultatively pathogenic forms, especially within the genus *Naegleria*. The potentially harmful species (*Naegleria fowleri* and, possibly, *Naegleria australiensis*) are thermophilic, and prefer naturally or artificially warmed fresh water (e. g. bathing establishments). Their optimum temperature is about 40° C. Infection, e. g., during swimming in contaminated water, is via the human nasal epithelium and olfactory system, then to the brain. A mass development starts, that can be fatal within a few days (Primary Amoebic Meningo-Encephalitis = PAME).

Molecular biological data indicate that the species belonging to the genus *Naegleria* exhibit greater interspecific genetic differences than representatives of the metazoa. Frequently, no morphological differences between species can be ascertained; the species are differentiated particularly by immune cytochemical parameters.

Examples: *Naegleria* (with two flagella), *Tetramitus* (with four flagella), *Vahlkampfia* (up to date no flagella observed), *Willaertia*.

Class   **Acrasea**
Blanton, 1900

The representatives of this class produce fruiting bodies (sorocarps) which develop from pseudoplasmodia (Fig. **63**). The myxamoebae of the acrasea exhibit characters homologous with those of the schizopyrenids: the eruptive lobopodia, and stages of the life cycle (in some genera e. g., formation of flagellated stages).

Examples: *Acrasis, Pocheina*

Phylum   **Dictyostela**
Olive, 1970

The members of this taxon, which formerly belonged to the so-called higher slime molds

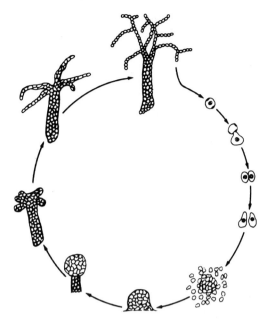

Fig. **63**   Acrasea: life cycle of *Acrasis rosea* (from Olive et al.: J. Protozool. 8 [1961] 467.

(Eumycetozoa), represent a group in which the typical flagellate characters are completely reduced. They undergo a complex life cycle (Fig. **64**). At the starting point of their development, they are naked filopodial cells, often designated as myxamoebae. They live on bacteria in humus soils or leaf litter and have a short generation time. In the following mass development, cell aggregates appear with a noticeable degree of differentiation. Under the effect of so-called acrasines (organic compounds such as e. g., cAMP), which are produced by the cells themselves, a pseudoplasmodium develops by aggregation and accumulation of single cells (Fig. **65**). This is visible to the naked eye and consists of several thousands of myxamoebae which form, at the end of this stage, a multicellular, more or less branched, and cellulose-containing fruiting body (sorocarp, sporangium). After agglomeration to the fruiting body, the cells which had formed the base and the stalk become necrotic and die, whereas the other surviving cells encyst, and later develop into spores. Under special conditions, nonflagellate gametes appear, fusing to produce zygotes. The diploid stages grow up, sometimes by cannibalism (phagocytosis of destroyed cells), to giant cells and encyst. In such macrocysts, meiosis takes place, producing haploid cells.

Examples: *Acytostelium, Dictyostelium, Polysphondylium*

Phylum   **Protostela**
              Olive, 1970

As in dictyostelids, the trophic phase of the life cycle of protostelids is characterized by the appearance of filopodial myxamoebae. However, in contrast to these, such amoebae are able to fuse to true (acellular, multinucleated) plasmodia. The fruiting bodies are relatively small; they mature from singular amoebae or plasmodia fragments, and they contain only one or a few spores. Habitats are soil, plant surfaces, and fresh water mosses. It is possible that some members of the marine Acarpomyxea also belong to this taxon.

Examples: *Cavostelium, Ceratiomyxa, Protostelium*

Phylum   **Myxogastra**
              Olive, 1970

The true (acellular) plasmodia of the species subsumed in this taxon live in humus soil or on leaf litter or dead wood; they are macroscopically vis-

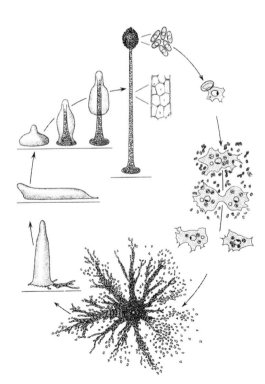

Fig. **64**   Dictyostela: life cycle of *Dictyostelium discoideum* (adapted from Gerisch).

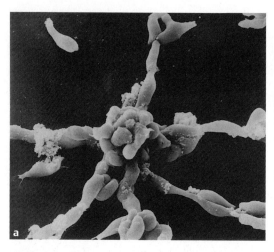

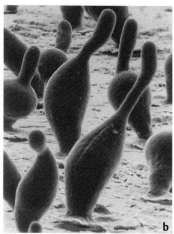

△
Fig. **65**   Dictyostela: **a** aggregation of single amoebae; **b** developing sorocarps (a courtesy of H. Claviez, Martinsried; b from Wu et al: Dev. Biol. 167 [1995] 1). Magn.: a 700×, b 100×.

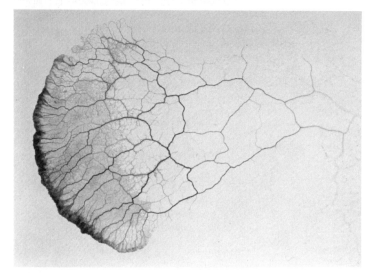

Fig. **66**   Myxogastra: plasmodial stage of *Physarum confertum* (from Stiemerling: Cytobiologie 1 [1970] 399). Magn.: 2×.

ible (see Fig. **28**) and frequently yellow or reddish colored. Under most favorable, anthropogenically influenced circumstances, they can reach a size of several decimeters or even meters. The plasmodia consist of networks of protoplasmic veins (Fig. **66**) 1–2 mm maximum thickness, exhibiting a special kind of protoplasmic streaming, the so-called shuttle streaming. They serve as models for studies on cell motility and amoeboid movement.

The diploid plasmodia develop after fusion of two haploid cells which are biflagellated or amoeboid. Following subsequent synchronous nuclear divisions, they grow into net-like plasmodia. Under unfavorable conditions (desiccation, food exhaustion), stalked sporangia are produced in which meiosis and the formation of

spores takes place. Because haploid single amoebae and diploid plasmodia are able to divide into daughter organisms (by mitosis or by plasmotomy), a heterophasic alternation of generations occurs.

Examples: *Echinostelium, Physarum, Trichia*

Phylum   **Chromista**
Cavalier-Smith, 1987

This phylum includes highly variable organisms, all combining the primitive status of a heterokont flagellation, with the synapomorphic character of (tripartite) mastigonemes, or other Golgi or endoplasmic reticulum derivatives. Originally, their representatives were heterotrophic predators, but most of the recent forms also live, after uptake of rhodophyceae or, other putative symbionts, as photosynthetically active organisms, exhibiting a colorful pigmentation. Due to the presence of highly evolved eukaryotic endosymbionts, the pigmented species are more phylogenetically recent. The relationships between the three subphyla, however, remain questionable at present.

Subphylum   **Prymnesiomonada**
            Hibbered, 1976

The mainly marine representatives of this monadial taxon, formerly called Haptophyta, are very important in systematic, paleontological, and ecological respects. Presently, about 500 species are known.

At the apical pole, the uninucleated cells bear two (or in rarer cases four) partly heterokont, partly more isokont flagella, which are mostly covered, as is the rest of the cell surface, with cellulose-containing and sometimes calcified scales (coccoliths) (Fig. **67**). The species-specific scales are formed within the endoplasmic reticulum. Flimmer hairs are not present. Between the two (more or less) isokont flagella, a third filiform flagellum-like structure, the haptonema, is inserted (Fig. **68**). It represents a device for adhesion, locomotion, or food uptake and exhibits a fine structural composition completely different from that of flagella. In cross section, it appears as a bundle of 6–8 singular microtubules enveloped by cisternae of the endoplasmic reticulum. The haptonema performs slow, swinging movements or fast, screw-like upcoiling. The functional significance, however, is not always completely clear: some species make use of it for fixing the cell body to the substratum or for gliding on surfaces, others for food uptake (comp. Fig. **234**). In some species, the haptonema exceeds by about 100 μm the length of the flagella, but in other species it is stump-like, shortened, or more or less completely reduced.

The plastids of the prymnesiomonads are, as in other flagellates such as the chrysomonads, enveloped by an endoplasmic reticulum cisterna. A girdle lamella is missing. The plastids contain the chlorophylls a and c, and additionally some accessory pigments such as fucoxanthin or diatoxanthin, both responsible for the yellow-brown coloration. The polysaccharides paramylon and chrysolaminarin serve as reserve material; they are collected in vacuoles outside of the chloroplasts. A few representatives are colorless and phagotrophic.

Fig. **67** Prymnesiomonada: siliceous scales on the cell surface of *Emiliana huxleyi* (courtesy of A. Kleijne, Amsterdam). Magn.: 15 000×.

Some species are capable of mass development. In spring and summer, *Phaeocystis* is frequently observed at the channel coasts of England and France or in the North Sea, producing slimy-spumy waste products toxic or harmful to fish (Fig. **295**). Also, *Chrysochromulina* is said to be responsible for fatal diseases of fish, clams and seals, as during the early summer of 1988 at the Danish–German North Sea coast. The pathological effects are evoked by exotoxins and slimy substances. However, due to their abundance in the nanoplankton of the oceans, the coccolithophorids play an especially important role in the fixation of carbon and, at least since the Jurassic Era, in sediment formation.

The main taxonomic groups are distinguished from each other by the presence and length of the haptonema, by isokonty or heterokonty of flagella, and by the composition of scales. Possibly, they do not represent natural assemblages.

Examples: *Emiliana, Pleurochrysis* (isokont, haptonema reduced, or missing), *Coccolithus, Umbellosphaera* (with coccoliths, haptonema still partly present), *Chrysochromulina, Prymnesium* (isokont and with distinct haptonema), *Diacronema, Pavlova* (heterokont and with reduced haptonema)

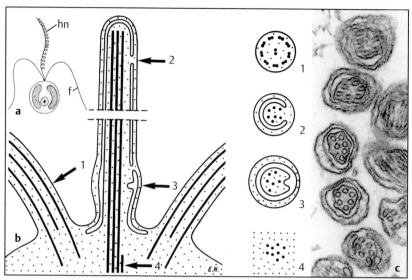

Fig. **68**   Prymnesiomonada: **a** with haptonema between the two flagella. **b** scheme of the apical region: 1 cross-sectioned flagellum with 9×2+2 arrangement; 2–4 cross section of haptonema with irregularly arranged microtubules and er-cisternae. **c** cross and oblique sections of a coiled haptonema (a and b adapted from several authors, c courtesy of ø. Moestrup, Copenhagen). Magn.: c 70 000×.

Subphylum   **Cryptomonada**
            Senn, 1900

Cryptomonads have two flagella that differ from each other in length and structural composition. The longer flagellum bears two rows of about 1.5 μm long mastigonemes, each with one terminal filament. The shorter flagellum has only one row of hairs, but each possesses two terminal filaments (Fig. **69**). Both flagella arise at the flange of an apically positioned deep invagination of the cell body (= vestibulum). The uninucleated cells exhibit a relatively sharply defined form: They are more or less oval or bean shaped with an oblique flattening at the opening of the vestibulum. The stiffness of the cell body derives from a nearly gapless layer of organic protein-containing plates (= periplast) accompanying the cell membrane either intra – or extracellulary, and giving the cell surface a hexagonal or rectangular appearance. As extrusomes, two similar types of ejectisomes are apparent: smaller ones at the corners of the periplast plates, bigger ones at the brim of the vestibulum. Freshwater species have a contractile vacuole, opening into the lumen of the vestibulum (Fig. **200**).

The chromatophores of the cryptomonads contain the chlorophylls a and c, as well as some accessory pigments (carotenoids, alloxanthin, phycocyanin, and phycoerythrin). The relative quantities of these is species specific; this leads to a cell coloration ranging from olive green to brown-red or even blue. Colorless forms with degenerate plastids and a heterotrophic way of life also occur. In contrast to chloroplasts of the Chlorophyta or Biliphyta, the plastids do not lie freely in the cytoplasm, but together with other organelles such as the pyrenoid or vacuoles and nucleomorph, surrounded by additional double membranous envelopes, the so-called plastidal endoplasmic reticulum, and therefore covered by four membranes altogether (Fig. **69**). The morphological complex composed of chloroplast with two membranes, extraplastidal pyrenoid, vacuoles, nucleomorph, and the outer third membrane is understood to be the relic of a reduced eukaryotic endosymbiont. The nucleomorph (see Fig. **19**) is interpretable as a formerly autarchic nucleus, the third membrane as the cell membrane of a formerly autarchic red alga. The fourth and outermost membrane therefore represents an endoplasmic reticulum membrane (vacuole membrane) of the host cell.

Examples: *Chilomonas, Cryptomonas, Pyrenomonas*

Subphylum   **Heterokonta**
Luther, 1899

The main groups subsumed under this taxon reflect extremely different developmental trends.
On the one hand, there are typical plant-like representatives such as brown algae or diatoms; on
the other hand there are lower fungi and rhizopodial forms exhibiting classical animal behavior and nutrition. The similarities between
the seemingly very different forms become clear
when considering the fine structural uniformity
of the monadial (= flagellate) organized unicells,
or the zoospores and gametes of the more highly
evolved forms.

Important and non-plastid-derived characteristics of all members are
- bilaterally symmetrical organization of the
  motile unicells or developmental stages
- heterokont flagellation (with a spiral body
  and mastigonemes in the longer flagellum,
  and a corresponding basal swelling in the
  shorter flagellum)
- formation of mastigonemes in Golgi vesicles
- endoplasmic reticulum bound localization of
  the plastids
- similar spatial relations between nucleus,
  Golgi system, and flagellar rootlets

In the following descriptions of the individual
classes, only the unicellular and/or flagellated
representatives are taken into account. Regarding the brown algae, diatoms, and oomycetes,
the reader may be referred to corresponding
botanical text books.

Within the unicellular organized Heterokonta,
organisms emerge with extracellular, tripartite
mastigonemes (stramenopiles), which communicate, via the cell membrane, with axonemal microtubules (comp. Fig. **212 d**). At the anterior
flagellum, they are connected with two opposed
microtubule doublets of the axoneme, and hence
they assemble into two rows of flimmer hairs.
The mastigonemes are formed in cisternae of the
endoplasmic reticulum or Golgi system, and are
attached, after exocytosis, outside the cell membrane to the flagellum. These mastigonemes are
regarded as an autapomorphy for the Heterokonta, and the name stramenopiles was proposed by Patterson in 1991. Another common
characteristic is the presence of so-called transitional helices (spiral bodies) which mark, as protein complexes, the transitional region between
kinetosome and flagellar shaft.

The starting point of the evolutionary radiation is obviously represented by a group of flagel-

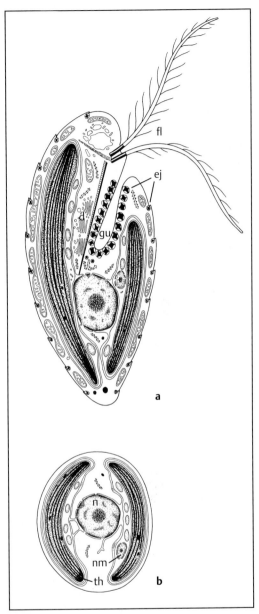

Fig. **69** Cryptomonada: longitudinal (**a**) and cross
section (**b**) of *Cryptomonas ovata*. showing spatial relationships between nucleomorph (nm) and thylakoid
(th). d = dictyosome, ej = ejectisomes, fl = fimbriated
flagella (flimmer), n = nucleus, gu = vestibulum or gullet (adapted from several authors). Magn.: 5000×.

lates with stramenopiles not positioned at the
flagella, but on the posterior part of the cell
body. Such appendages are designated somatonemes. From this group, the proteromonads,

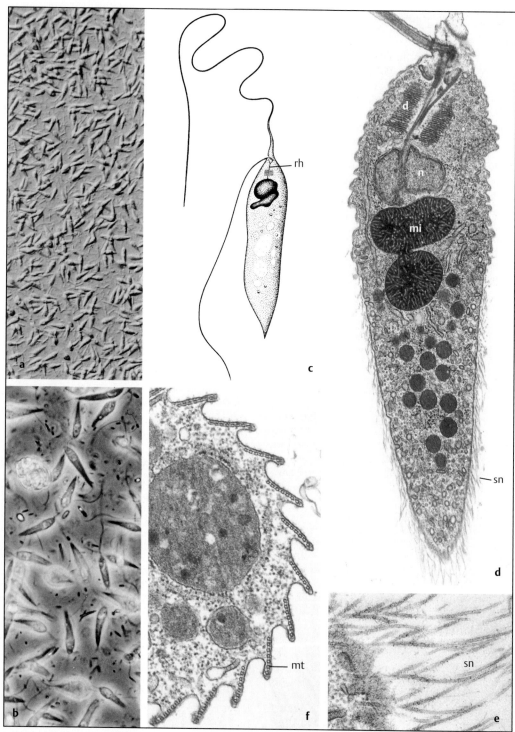

Fig. **70** Proteromonadea: *Proteromonas lacertae-viri-dis* in life (**a** and **b**), in schematic representation (**c**) and in ultrathin sections (**d** and **e**). **f** cross section of *Keratomorpha bufonis* with subpellicular microtubular ribbons (mt). d = dictyosome, n = nucleus, mi = mito-chondrion, rh = rhizoplast, sn = somatonemes (c–f courtesy of G. Brugerolle and J. P. Mignot, Clermont-Ferrand). Magn.: a 350×, b 850×, c 3500×, d 10000×, e 50000×, f 40000×.

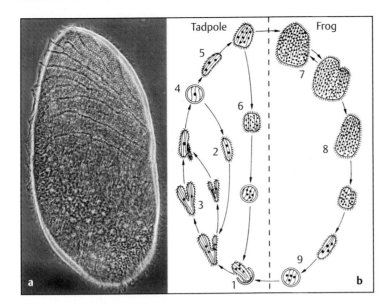

Fig. **71** Opalinea: **a** *Opalina ranarum* from rectum of frog. **b** life cycle of *Opalina*: (1) multinucleate cell escaping from cyst and transforming, after division, into gamont (2) or into uninucleate gametes (3) which form an encysted zygote (4); after transfer of such cysts to another tadpole, *Opalina* cells excyst, and, after several divisions, become small trophic forms (6) undergo encystment, or grow to normal large trophic stages (7); after multiplication (8) such trophic stages encyst (9) and are extruded from the frog together with the feces (b adapted from Wessenberg). Magn.: 300×.

the other heterotrophic and autotrophic taxa, could be derived.

Class   **Proteromonadea**
Grassé, 1952

The taxon contains only two genera. These are characterized by one or two pairs of heterokont and heterodynamic flagella. They arise at the anterior end of the 10–30 μm long uninucleated cells. The thickened base of the long anterior flagellum of *Proteromonas* contains, besides the axoneme, a bundle of loosely connected microfibrils and is connected with the corresponding part of the trailing flagellum by a special gap-junction complex. Mastigonemes are missing, but the kinetosomes show, as in chrysomonads, a transitional helix at the transitional region to the flagellar shaft. They are connected via a band-like rhizoplast made of microtubular and filamentous elements, with the surface of the single mitochondrion. In its course through the anterior part of the cell, the rhizoplast crosses through the circular dictyosome and also through the nucleus. Other constituents of the cytoskeleton are microtubules which, singularly or in rows, support the cell surface and cause a typical longitudinal or oblique striping of the pellicle. The posterior part of the cell body of *Proteromonas* (Fig. **70**) bears hair-like appendices, the so-called somatonemes which, as structures homologous to the mastigonemes of the chryso-

monad-flagellum, are also constituted in Golgi vesicles and also connected with subpellicular microtubules. The distinct expression of a pellicular cytoskeleton made of rows of microtubules is interpreted as a primitive character and not common in the more highly evolved taxa such as chrysomonads.

The proteromonads are cosmopolitan and live as endobionts in the intestinal tracts of amphibia, reptiles, and mammals. The transfer to new hosts is effected by cysts egested with feces. Pathogenic effects are not known.

Examples: *Kerotomorpha, Proteromonas*

Class   **Opalinea**
Wenyon, 1926

In the preelectron microscopic era, this taxon, with altogether about 400 species belonging to only four genera, was ranked with the ciliates. However, due to the absence of nuclear dualism, alveolae and typical infraciliature they must be separated from this group. Some similarity in the construction of the cortex indicates that they may be closely related to the proteromonads.

The multinucleated, often flattened, and large (occasionally up to 3 mm) cells are covered with thousands of relatively short cilia-like flagella. They are ordered in dense rows (kineties), and beat in metachronal waves, a type of flagellation and movement loosely resembling that in ciliates (Fig. **71**). However, the kineties run in a helical manner and originate, at least in their majority, from a group of flagellar rows (falx) at the anterior pole. Between the individual flagellar rows, some folds rise from the cell surface; these are supported by numerous microtubules. A special cytopharynx or any other devices for the uptake of particular food are missing; the nutrients are absorbed by pinocytosis over the entire cell surface. The division plane runs between the flagellar rows (interkinetal), and divides the falx into two equal parts. Although some geometrical complications exist, the cytokinesis is a symmetrically longitudinal division. This is quite different to the corresponding situation in the Ciliophora, in which division is perpendicular to the kineties.

Another typical characteristic of the Opalinea is their exclusive occurrence in cold-blooded vertebrates, mostly in frogs, occasionally also in urodeles or fish, seldom in reptiles. They live as nonparasitic endocommensals in the terminal parts of the intestine. Encysted stages enable transmission between hosts.

The life cycle of opalines is relatively complicated (Fig. **71 b**). The trophic stages can be found almost continually in adult hosts. During the mating period of frogs, the protozoans multiply by a rapid sequence of longitudinal or oblique cell divisions without cell growth. The resulting smaller individuals have only a few nuclei and remnants of kineties. This rapid multiplication is said to be induced and controlled by hormones of the host.

The microforms of the opalines, with only three to six nuclei, encyst and are egested with feces. The cysts are able to survive for several weeks in freshwater; they release swarmers when ingested by tadpoles. In the intestine, slender microgametes and bigger macrogametes differentiate during several divisions combined with meiosis. After sexual fusion, the diploid zygotes encyst and are egested. In the subsequent intestinal passage through tadpoles or adult hosts, the development of a new generation of gametes or a morphogenetic maturation to large trophic forms takes place. Thereafter, a generation of cysts arises by an asexual path. The several developmental alternatives at this stage of the life cycle ensure an effective spread in tadpoles.

Examples: *Cepedea, Opalina, Protoopalina, Zelleriella*

Class   **Chrysomonadea**
         Engler, 1898

There are roughly 1000 species of autotrophic or heterotrophic chrysomonads. They are relatively small, only seldom being larger than 5 to 20 μm (Fig. **72**). They occur preferentially in freshwater. In their typical organization, the chrysomonads have two anisokont (= heterokont) flagella arising from the anterior pole. The longer flagellum is directed forward, and functions as a propelling device; it is equipped with two rows of stiff tripartite mastigonemes (flimmers), and is often therefore designated a pleuronematic flagellum. The second flagellum is, if present, significantly shorter and smooth; it runs backwards along the cell body and bears a basal swelling facing a slightly concave invagination at the cell flange. In this region, inside the chloroplast, the red lipid granules of the eyespot are found.

The single nucleus is connected by a cross-striated rootlet (rhizoplast) to the basal body of the pleuronematic flagellum. One or two pulsating vacuoles are located beside the nucleus, as well as one large or several smaller dictyosomes in which mastigonemes are synthesized. Normally, the cell body is naked; however, in some genera *(Synura, Mallomonas, Paraphysomonas)* the cell surface is found to be covered with delicately sculptured silicious scales (Fig. **73**). They are formed within vacuoles beside the chloroplasts. Some genera such as *Dinobryon* live in loricae. Among the types of extrusomes are, especially, discobolocysts. For some genera (e. g., *Dinobryon*) sexual processes (isogamy) are typical. Two unicells fuse with each other to form a zygote, without formation of gametes. By endogenous development of silicious capsules, such diploid protoplasts encyst and survive as duration stages. Endogenous duration cysts can also be formed asexually.

The one or two chloroplasts of the photosynthetically active representatives have a characteristic golden-yellow to golden-brown coloration (designation "gold algae"). The accessory pigment responsible, fucoxanthin, superimposes the chlorophylls a and c. Inside the chloroplasts lay stacks, each with three thylakoids. At their periphery, a girdle lamella occurs, made also of three thylakoids. The polysaccharide chrysolaminarin and several lipids serve as reserve products accumulated in vacuoles. The chloroplast surrounds the nucleus, and is morphologically connected to it by a common envelope of the endoplasmic reticulum. The chloroplasts are therefore sheathed by an additional mem-

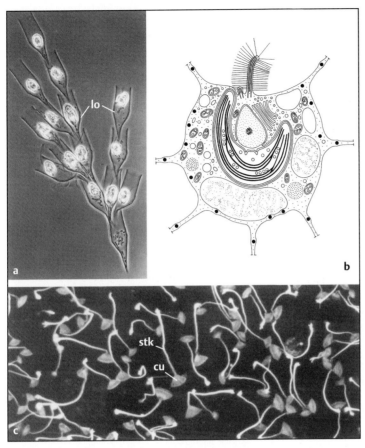

Fig. **72** Chrysomonadea: **a** colony of *Dinobryon,* each cell living in its own lorica (lo). **b** *Chrysamoeba radians* with typical chrysomonadal organization and filopodial extensions. **c** loricae of *Poterioochromonas malhamensis* with stalk (stk) and cup (cu) (a and c courtesy of W. Herth, Heidelberg, b adapted from Hibberd). Magn.: a 450×, c 750×.

Fig. **73** Chrysomonadea: siliceous plates of *Synura petersenii* (courtesy of B. S. C. Leadbeater, Birmingham). Magn.: 500×.

branous system, in the region of contact with the nucleus by the inner nuclear envelope, or by the outer nuclear envelope. This morphological situation is interpreted as the ancient artifact of a formerly independent eukaryotic cell.

Besides the monadoid type with flagellated unicells or colonial aggregates, there exist also several organization types representing transition stages to a higher evolved status. The trend towards a rhizopodial organization, with a secondary phagotrophic way of life is very important and leads to the reduction of plastids and flagella, and the development of different pseudopodial types. Representatives include, besides unicellular forms, such as *Chrysamoeba*, large plasmodial assemblages, such as *Chrysarachnion*.

Ecologically, the chrysomonads represent an important part of the nanoplankton. The relevance of the photoautotrophic representatives is derived from their capacity in primary production. Because they often occur in mass development, especially during the colder months, they can cause problems in commercial fish cultiva-

Fig. **74** Chrysomonadea: colony of *Anthophysa* (from Zölffel and Hausmann: Mikrokosmos 76 [1987] 353). Magn.: 230×.

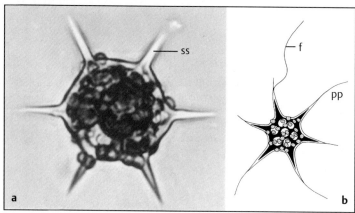

Fig. **75** Silicoflagellida: *Distephanus speculum* with siliceous skeleton (ss) (**a**), flagellum (f) and pseudopodial extensions (pp) (**b**) (a from Drebes: Marines Phytoplankton. Thieme, Stuttgart 1974). Magn.: 1200×.

tion by secreting ketones and aldehydes. The collapse of mass developments frequently pollutes freshwater reservoirs.

Systematically, three groups are established:

Order   Chrysomonadida
              Pascher, 1912

In this order, especially monadial genera are subsumed, and occur as freeswimming or sessile forms (Fig. **74**). Some have a rhizopodial organization.
   Examples: monadial: *Dinobryon, Ochromonas, Synura,*
   rhizopodial: *Chrysamoeba, Chrysarachnion, Chrysostephanosphaera*

Order   Pedinellida
              Pascher, 1910

The sessile, chloroplast-containing representatives of this order possess at the apical region,

close to a rudimentary trailing flagellum, a long flagellum surrounded by a collar-like arrangement of stiff cellular projections. This organization type resembles the situation in choanoflagellates; however, there is no phylogenetic relation. The similarities are the result of convergent development.
   Example: *Oikomonas, Pedinella, Pteridomonas*

Order   Silicoflagellida
              Borgert, 1891

The representatives of this order, nowadays with only few species, are of great geological importance due to the deposits of silicious skeletons they produced in the past. The earliest fossil records are known from the Cretaceous Period and their maximum abundance occured in the Tertiary Period. The few, rather unfamiliar recent forms bear only a single flagellum, but many slender pseudopodia (Fig. **75**). The intracellular and extracellular skeletal elements exhibit a starlike shaping. Apart from their ability to metabolize silicious derivatives, the ranking with the chrysomonads is based on similarities in chloroplast morphology and is, due to their endobiotic nature, rather questionable.

Examples: *Dictyocha, Distephanus*

Class   **Heteromonadea**
Leedale 1983

The bilaterally symmetrical representatives of the heteromonads resemble the chrysomonads insofar as they also bear heterokont flagella, and show mostly identical ultrastructural characters. However, the stump-like shortened trailing flagellum, arising near the long propelling flagellum is completely naked and ends with a terminal filament. It is therefore designed as an akronematic flagellum and also exhibits, close to a plastidal eyespot, a basal swelling. An important divergence from the chrysomonads is the complete absence of fucoxanthin, which, together with the presence of β-carotene and xanthophylls, produces a more yellow-greenish coloration of the cells and the morphological construction of the endogen cysts.

Besides the better known trichal algae *(Tribonema, Vaucheria)*, a few flagellate species exist. At present, the best investigated forms are the rhizopodial representatives. Given their generally low abundance and distribution, the ecological importance of the unicells appears to be limited.

Examples: monadial: *Chloromeson*
rhizopodial: *Rhizochloris, Reticulosphaera*

Class   **Labyrinthulea**
Cienkowski, 1867

This taxon contains only the single order Labyrinthulida with the two families Labyrinthulidae and Thraustochytriidae. Only few species are known. The organisms occur mostly in marine coastal and estuarine environments, more rarely in freshwater or in saline soils. They are mostly found associated with algae or vascular plant debris or in detritus sediments.

During their trophic phase, the organisms move as wandering plasmodia (Fig. **76**). Spindle-shaped "cell-bodies," stiff or variable in contour, exhibit a gliding cell motility within a variable network of pseudopodia-like tubes. The ectoplasmic network is bordered by a common cell membrane; each cell body is covered by a double membrane envelope containing scales and communicates with the plasmic content of the tubes by special openings (Fig. **77**). These

openings have special designations such as bothrosome, sagenogenetosome, or sagenogen. They represent, in combination with the double membranes, an apomorphic character. The cell bodies do not delineate a true cellular constellation, but a distinct kind of demarcation of individual nucleated cytoplasmic fragments from the common tubular matrix, which is interpretable as a pseudopodial (filopodial) formation.

The mechanism of motility is unknown, but thought to be associated with actin-microfilaments located within the tubes. Multiplication occurs by binary fission within the network, or by plasmotomy of the plasmodium. During the incompletely understood, but obviously sexual life cycle of some better known species, multicellular cysts, sporangia (sori) as well as biflagellate swarmers with tripartite mastigonemes (zoospores), occur (Fig. **75**). An important character for classification of some zoospores is, apart from the heterokont flagellation type, the presence of a pigmented eyespot beneath the posteriorly directed flagellum.

Examples: *Labyrinthula, Labyrinthuloides, Thraustochytrium*

Class   **Raphidomonadea**
Heywood & Leedale, 1983

About ten exclusively monadial organized, generally dorso-ventrally flattened species of raphidomonads (= Chloromonadea) are known. Rather atypical for "phytoflagellates," the scale- and wall-less cells reach a size of up to 90 μm. Due to a rigid pellicle, the cell shape is relatively constant (Fig. **78**). Mucocysts and variants of spindle trichocysts of unknown function are located in subpellicular regions. The heterokont flagella arise at a ventral groove at the anterior cell pole; their kinetosomes are connected with the nuclear surface via kinetids. The long, posteriorly directed, and mostly inactive flagellum is smooth and bears no basal swelling. The shorter and very rapidly propelling anterior flagellum has stiff tubular mastigonemes produced in cisterns of the endoplasmic reticulum or dictyosome. Stigmata are not present. A layer of Golgi elements covers the apical part of the large nucleus in a characteristic manner. At least in the freshwater genus *Vacuolaria*, some of the large Golgi vesicles continuously convert to a contractile vacuole.

The raphidomonads are photoautotrophic. The numerous plastids are bright green and contain the chlorophylls a and c, and several acces-

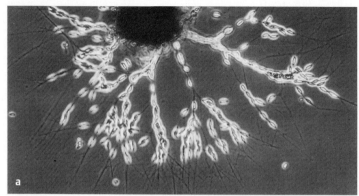

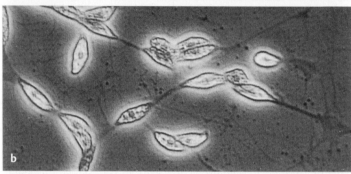

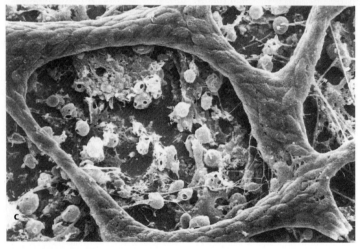

Fig. **76**   Labyrinthulea: **a** plasmodial network of *Labyrinthula coenocystis.* **b** spindle-like cellular units migrating inside tubular filopodial extensions. **c** tightly packed cellular units within the tubular filopods. (a and b from Stey: Z. Zellforsch. Mikrosk. Anat. 102 [1969] 387). Magn.: a 230×, b 1000×, c 400x.

sory pigments such as carotenoids, xanthophylls or fucoxanthins. They contain stacks of three thylakoids and also exhibit a girdle lamella. Lipid droplets serve as nutritional reserves.

Examples: *Chattonella, Gonyostomum, Vacuolaria*

Class   **Plasmodiophorea**
Cooke, 1928

The members of this taxon exist as obligate intracellular parasites in root organs of vascular plants, forming there multinucleated plasmodia and also sori. The fruiting bodies contain spores which are discharged to the soil after decomposition of their hosts. The diploid plasmodia of the

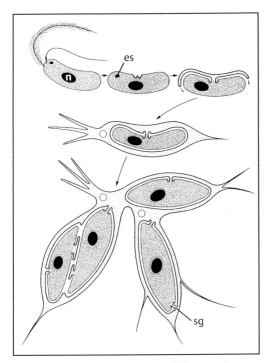

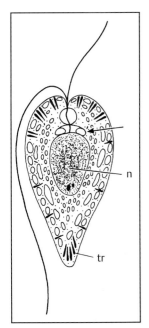

Fig. **77** Model of development of the membranous systems in labyrinthulids. es = eyespot (stigma), n = nucleus, sg = sagenogenetosome (or bothrosome).

Fig. **78** Raphidomonadea: *Gonostomum semen* with typical ring-like dictyosome (arrow) covering the nucleus (n). tr = trichocysts. Magn.: 2500×.

next generation propagate after sexual fusion of free-living zoospores. These are heterokont and biflagellate, but not mastigonemate. During the invasive attack of host cell walls, particular organelles emerge, the stachel and rohr, interpretable as morphological differentiations of the endoplasmic reticulum (Fig. **79**). During karyokinesis, the unicells and the plasmodia exhibit a characteristic intranuclear cross, formed by the chromatids arranged in the metaphase plate and the poleward-stretched nucleolus (cross of the Plasmodiophorea). A phylogenetic relationship to hyphae-forming taxa within the so-called lower fungi is suspected.

Examples: *Plasmodiophora, Sorosphaera, Spongospora, Tetramyxa*

Class   **Bicosoecidea**
             Grassé & Deflandre, 1952

The representatives of the 40 or so species of this taxon live as single cells or in colonies which are characterized by the presence of vase-like shells or loricae made of chitin, in contrast to some choanoflagellates or chrysomonads in which the loricae are siliceous. The uninucleated cells are heterokont biflagellate. The smooth flagellum runs posteriorly in a furrow of the cell body and serves to attach the cell within the lorica and for retracting it. The second, longer flagellum is directed anteriorly. It bears one or two rows of stiff mastigonemes and functions as the collector of food, such as bacteria or other microbes (Fig. **80**).

Cells measure about 5 μm. They move freely or they become fixed by their shells to various substrata. Bicosoecids belong to the heterotrophic nanoplankton community of marine and freshwater habitats, and probably play an important role in microbial foodwebs.

Examples: *Bicosoeca, Cafeteria, Pseudobodo*

Phylum   **Alveolata**
             Cavalier-Smith, 1992

Molecular biological data, especially the distance matrix trees inferred from rRNA sequences, demonstrate that three large groups of the classical system, the Ciliophora, Apicomplexa, and Dinoflagellata, form a common monophyletic taxon. Thus, some structural complexes formerly thought to have evolved independently, now appear as homologous formations derived from a common antecedent. So, the amphiesmal vesicles of the biflagellate dinoflagellates, the inner membrane complexes of the

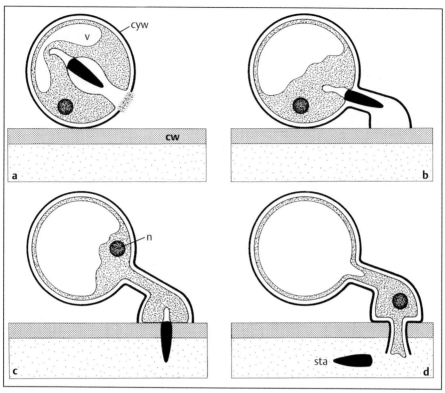

Fig. **79**   Plasmodiophorea: penetration apparatus of *Plasmodiophora* during attack on a plant cell. cw = cell wall of the host cell, cyw = cystic wall, n = nucleus, sta = stachel, v = cacuole (adapted from Keskin and Fuchs).

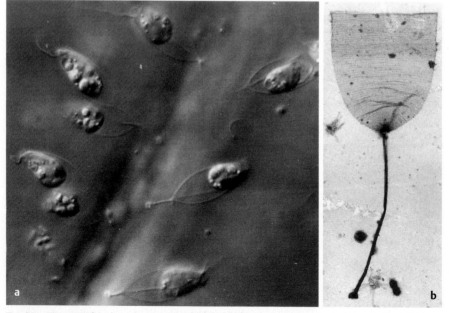

Fig. **80**   Bicosoecidea: living specimens (**a**) and single lorica (**b**) of *Bicoeca* (a courtesy of H. Schneider, Landau). Magn.: a 1000×, b 3000×.

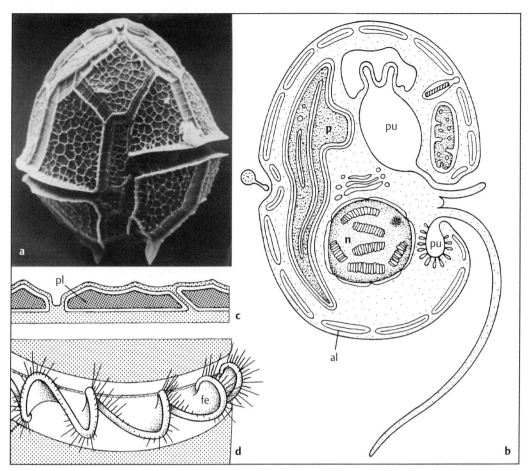

Fig. **81** Dinoflagellata: **a** thecal plates of *Peridinium bipes*. **b** scheme of internal organization. **c** amphiesmal vesicles (alveoli, al) with cellulose plates (pl). **d** cingulum flagellum with mastigonemes and flagellar extension (= paraxial hem) (fe). n = nucleus with condensed chromosomes, p = plastid, pu = pusules (a courtesy of R. M. Crawford, Bristol, b adapted from Taylor, c adapted from Gaines and Taylor). Magn.: a 1200×.

aflagellate apicomplexans, and the alveoli of the ciliated Ciliophora represent a system inherited from on single ancestor which was living about 900 million years ago as a biflagellate protozoan and had acquired an unusual pellicular vacuolar system. Also, the parasomal sacs of ciliates, the pusules of the dinoflagellates, and the micropores of the Apicomplexa are interpretable as homologous structures with similar or comparable functions.

The flagella of the hypothetical founder species were heterokont and, at least partly, mastigonemate. This situation, however, has survived among all recent descendants only in the dinoflagellates.

Subphylum   **Dinoflagellata**
Bütschli, 1885

Dinoflagellates normally have two heteromorphic and heterodynamic flagella. In primitive forms, they insert apically; in more highly evolved species they arise ventrally, and operate in a transversally running constriction (= cingulum, girdle) and in a longitudinal groove of the posterior cortex (= sulcus) (Fig. **81**). This so-called dinokont flagellation type can also be found in multinucleated and multiflagellated organisms. The transverse undulating flagellum winds to the left (= counterclockwise) around the cell, and bears a striated strand (= paraxial rod) and one row of flimmer hairs. The trailing flagellum normally projects beyond the posterior end, and is naked or equipped with two rows of stiff flimmers. The insertion type and functional-morphological characters of the flagella produce a screw-like swimming movement.

About 4000 species of dinoflagellates are known, and they are morphologically very diverse. Shape ranges from rounded to rodlike or starlike unicells, to cylindrical multinucleated cells, and from naked to armored cell bodies. The protective covering (Fig. **81 c**) contains one or several cellulose plates lying immediately beneath the plasmalemma in flattened vacuoles (= alveoli = amphiesmal vesicles). The intracellular closely fitting wall plates often form two scaly envelopes (= thecae), joining each other in the region of the cingulum: The posterior hemisphere is known as the hypotheca, the anterior, the epitheca. Amphiesmal vesicles are also found in unarmored naked cells.

The haploid nucleus, also called dinokaryon, exhibits some peculiarities which were previously interpreted as primitive characters, but presently as autoapomorphies: The chromosomes are condensed, even during interphase, and appear as distinct structures in the light microscope. At the ultrastructural level, they show a fibrillar composition. In contrast to the remainder of all Eukaryota, the chromosomes contain few, if any, complexing basic nucleoproteins (histones). During mitosis, they have no direct contact with spindle microtubules; they are bound to the persisting nuclear envelope.

The chloroplasts are sheathed by triple-membraned envelopes, which are not components of the endoplasmic reticulum. They have threefold stapled thylakoids. As in the other cases, they are interpreted as extremely reduced relics of a eukaryotic endosymbiont. Possible ancient or recent cytobionts include: cyanobacteria (phaosomes), red algae, prymnesiomonads (*Gyrodinium* spec.), cryptomonads (*Gymnodinium cyaneum, G. acidotum*), and putative chrysomonads (e. g., *Peridinium balticum*). Besides chlorophylls a and c, they contain pigments such as β-carotene, dinoxanthin, and the xanthophyll peridinin, causing the typical yellow-brown or brown-red color of dinoflagellates. Storage products are starch and oil, produced exteriorly to the plastids. Some photosynthetic forms are known to live as intracellular symbiotic "zooxanthellae" (e. g., *Zooxanthella, Symbiodinium*) in a great range of hosts. In colonial radiolarians, foraminifers, molluscs, and especially in coelenterates such as hard corals, they are responsible for the brownish coloration. In addition, there also exist several heterotrophic genera living as phagotrophs, such as *Noctiluca*, or as parasites in protozoa, algae, copepods, or fish *(Blastodinium, Stylodinium, Oodinium)*.

Contractile vacuoles are not present, but there are invaginations (= pusules, comp. Fig. **202**) of reasonable size lying beside the flagellar basis and communicating with the outside. They are suggested to fulfill osmoregulatory functions. As extrusomes, spindle trichocysts are normally present as well as, in some members, so-called nematocysts, structures which are not related to the cnidocysts of coelenterates (comp. Fig. **196**).

The ecological importance of the dinoflagellates is very great. Particularly in coastal and oceanic waters, they often represent the principal component of the phytoplankton. However, they are also responsible for natural phenomena such as the kilometer-long tracks of so-called red tides occurring regularly, especially along the African coasts of the Atlantic, the Pacific shores of America and Japan, and sometimes, during the summer, also along European shores. Some species (e. g., *Protogonyaulax tamarensis, Protogonyaulax catenella, Gymnodinium veneficum*) produce alkaloids (e. g., saxitoxin or brevetoxin-complexes) which accumulate in fish, shellfish, and crayfish and which are sometimes harmful to them. Consumption of these by humans can lead to gastro-intestinal diseases and also to irreversible paralysis: the so-called shellfish poisoning. However, not all red tides are toxic; some are responsible for the natural spectacle of the bright phosphorescence of the sea at night. This kind of emission is caused by a luciferin-luciferase system localized in intracellular spherical bodies. Typical bioluminescent representatives include, especially in the North Sea, the heterotrophic *Noctiluca miliaris (scintillans)* and *Gonyaulax*.

Numerous dinoflagellates survive unfavorable conditions by forming duration cysts. Such cysts are known as microfossils, possibly from the Precambrian Era, but more likely from the Silurian Era, 400 million years ago.

The subphylum can be divided into two classes: Diniferea and Syndinea. Whereas the Diniferea possess the above mentioned derived characters, the Syndinea represent intracellular symbionts or parasites exclusively without armor plating;during mitosis they exhibit an extranuclear mitotic spindle with centrioles.

Examples: *Ceratium,   Gymnodinium, Merodinium, Noctiluca, Peridinium, Syndinium*

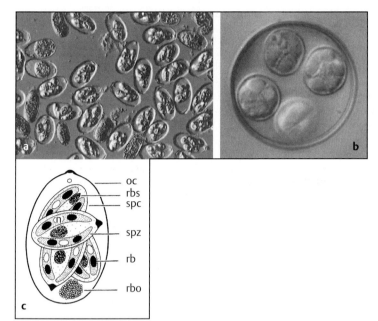

Fig. **82** Apicomplexa:
a oocysts of *Eimeria stiedai*.
b oocyst of an *Eimeria* species
isolated from rectum of a
gekko. **c** oocyst in schematic
presentation. n = nucleus, oc
= envelope of oocyst, rb = refractile body, rbo = residual
body of oocyst, rbs = residual
body of sporocyst, spc =
sporocyst, spz = sporozoite
(b courtesy of B. Bannert, Berlin). Magn.: a 300×, b 1600×.

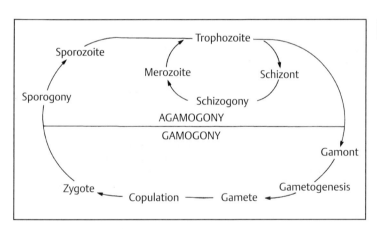

Fig. **83**
Life cycle of Apicomplexa.

Subphylum **Apicomplexa**
Levine, 1970

Like all representatives of the former taxon
Sporozoa, the Apicomplexa are also obligate parasites which have a spore stage in the life cycle
(Fig. **82**). Presently, they comprise at least 2500
species. Most are pathogenic, and many are of
great importance to human and animal health.
Typically, they have a double or triple-phase alternation of generations, each of which has typical infection, growth, multiplication, and sexual
stages (Fig. **83**). The infection is evoked by sporozoites transferred to new hosts within
sporocysts or oocysts. The designation of the
taxon goes back to the so-called apical complex —
an agglomeration of organelles at the posterior
region of sporozoites or merozoites developing
from them.

The apical complex of the 2–20 µm long,
spindle-shaped sporozoites and merozoites normally consists of three distinct morphological
components (Figs. **84** and **85**): (1) the cone-
shaped conoid made up of helically running microtubules with two preplaced conoidal rings,
(2) the pole ring complex, interpreted as a

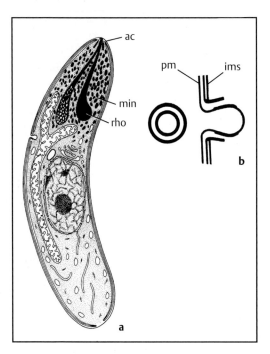

Fig. **84**   Apicomplexa: **a** sporozoite. ac = apical complex, min = micronemes, rho = rhoptry. **b** micropore in cross section (left) and longitudinal section (right). pm = plasma membrane, ims = inner membraneous system (alveoli) (adapted from Scholtyseck and Mehlhorn).

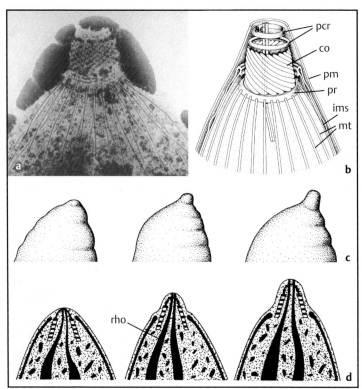

Fig. **85**   Apicomplexa: conoid regions of sporozoites. **a** total spread preparation of conoid complex. **b** schematic reconstruction; co = conoid with microtubules and mt-associated structures, pm = plasma membrane, ims = inner membrane system (alveoli), pr = polar ring with emerging microtubules (mt), pcr = preconoidal rings. **c, d** penetration activity of the apical complex in top view and cutaway drawing. rho = rhoptry (a from Nichols et al.: J. Protozool. 34 [1987] 217; b–d adapted from Scholtyseck and Mehlhorn).
Magn.: a 35 000×.

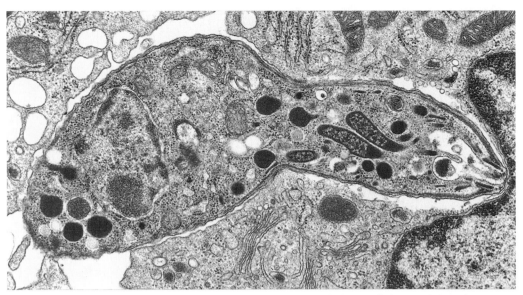

Fig. **86** Apicomplexa: *Toxoplasma gondii* sporozoite during invasion of a host cell (from Nichols and O'Connor: Lab. Invest. 44 [1981] 324). Magn.: 20 000×.

MTOC, with emerging subpellicular and posteriorly running microtubules and, normally (3) two bottle-like secretion organelles, so-called rhoptries, which run through the pole ring and the conoid, and discharge their contents at the anterior end. Besides these elements, there also exist filiform micronemes considered as derivates of the Golgi system, filled with enzymes. The whole elaborate system is interpreted as a penetration apparatus enabling the parasites to invade their host cells (Figs. **85** and **86**). However, it should be noted that the aforementioned components do not necessarily all occur in all extant apicomplexans (e. g., after secondary loss of the conoids) and that some primitive Apicomplexa living as extracellular parasites make use of the apical complex only for attachment purposes.

An identical organization of the cortical structures is common to all Apicomplexa. Apart from the longitudinal microtubular system arising at the pole ring complex, three membranes belong to it: the peripheral cell membrane and two subplasmalemmal membranes which are designated the inner membrane complex, and represent a flat vesicular or alveolar system. This system is interpreted as a structural homologue for the amphiesmal vesicles of the Dinoflagellata and the alveoli of the Ciliophora. It surrounds the cytoplasm completely, with some exceptions at the anterior and posterior ends of the cells for exocytotic, or lateral for endocytotic processes.

The one or more lateral invaginations of the cell surface are called micropores. They serve for food uptake.

The life cycles of the Apicomplexa are relatively intricate. A generalized and simplified mode of metamorphosis can be given as follows (Fig. **87**):

The uninucleated sporozoites which initiate the infection process, derive from a multiplication step following meiotic reduction fissions (sporogony). Inside the intestine of the new host, the sporocysts emerge, invade epithelial cells, and grow. In the most simple case, they develop into gamonts. These divide repeatedly before transforming to gametes and fusing to zygotes (gamogony). After such a biphasic metagenesis, a new generation of haploid sporozoites derives again from zygotes.

In more highly evolved genera, the lag between sporogony and gamogony is filled with additional asexual multiplication stages (schizogonies, merogonies) leading to the formation of numerous merozoites (Fig. **88**). This phase is induced by sporozoites differentiating to so-called trophonts or feeding cells. They grow up to large multinucleated cells inside the host cell, and are normally characterized by the reduction of their apical complex; due to the formation of additional micropores, or even oral apparatuses, they are predestined for food uptake. At the end of the growing phase, they transform to schizonts

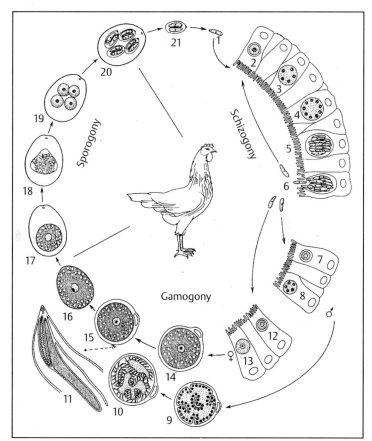

Fig. 87 Apicomplexa: life cycle of *Eimeria maxima*. 1 Infection of the host by sporozoites; 2–6 schizogony in intestinal epithelium of chicken; 7–16 gamogony; 12–14 development of macrogametes; 8–11 development of microgametes; 16 zygote (oocyst); 17–20 sporogenesis inside the oocystic envelope; 21 sporocyst with two sporozoites (adapted from Grell).

(meronts), and release merozoites during multiple plasmotomy (schizogony). The meronts again exhibit an apical complex, attack host cells, develop to trophonts, and lead, at the end, to a severe parasitemia of the host. The merozoites are also influential at the start of the gamogony phase, under circumstances coupled with an interchange of host. The gamogony takes place mostly as an oogamy, which is illustrated by the appearance of macrogametes transforming directly, without intermediary mitosis, from macrogamonts, and of microgametes developing from microgamonts after some additional fissions. The individual stages of the life cycle are typified by distinct cytological characters (comp. Fig. **258**).

Apical complexes are formed only during sporogony or schizogony (merogony), and therefore they appear in their typical configuration only in sporozoites or merozoites. The sporogony serves for reimbursement of a haploid status and, simultaneously, by the additional mitotic cell divisions for a more effective

establishment within potential new hosts. The fissions during schizogony and gamogony, however, influence the number of gametes and zygotes, and therefore of sporocysts or oocysts, and increase therefore the probability of new infections.

The systematics of the Apicomplexa is presently rather questionable. A division into three classes is commonly favored.

Class   **Gregarinea**
Dufour, 1828

The primary and primitive character of this taxon, containing about 500 species, is that male as well female gamonts undergo multiple fissions and therefore produce nearly the same number of gametes. The gametes can be of different (anisogamy) or of similar morphology (isogamy). Originally, the gametes develop separately and independently from each other; the gamonts of the higher evolved taxa, however, mate to a syzygy before finishing their in-

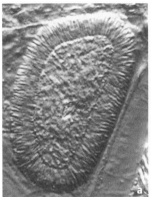

Fig. **88** Apicomplexa: early (**a**) and later stage (**b**) of schizogony in *Eimeria canadensis* (from Müller et al.: J. Protozool. 20 [1973] 293). Magn.: a 800×, b 1000×.

dividual growing phases (Fig. **89**), and encyst later by secretion of a common envelope (gamontocyst). After gamontogamy, the formation of gametes and zygotes takes place. From zygotes originate sporocysts, frequently containing 4–16 haploid sporozoites.

The trophic stages arise from sporozoites and grow up to large (occasionally up to 10 mm long) gamonts. They normally perform a typical gliding movement, facilitated by numerous longitudinal surface folds stiffened by microfilaments (comp. Fig. **227**). The gamonts commonly consist of two parts: an anterior protomerite and a posterior larger and nucleated deutomerite. Both parts are segregated from each other by a ring-like furrow. The anterior ends of protomerites normally bear a special appendage, the epimerite, which is formed by the conoid. Epimerites are species-specific in their construction; they serve for anchoring to the epithelial cells of their hosts. They are abandoned after formation of syzygies when the protomerite of the posterior cell (= satellite) is affixed to the deutomerite of the anterior cell (= primite).

Apart from the early stages of development, gregarines live mostly as extracellular parasites. They inhabit the intestine or the body cavities especially of articulates, molluscs, echinoderms, and tunicates. The most important hosts are arthropods. However, pathogenic effects are known only in a few insects. An alternation of hosts does not occur.

The systematic classification comprises four orders (Blasto-, Archi-, Eu-, and Neogregarinida) and is based on the presence or absence of gamontocysts, and on cytological details regarding septation and epimerite morphology.

Examples: *Gregarina, Mattesia, Monocystis, Selinidioides, Theydleckia*

Class   **Coccidea**
          Leuckart, 1879

The life cycle of the coccidians differs from that of gregarines in several aspects. The macrogamonts have, in contrast to the microgamonts, no division stages. They develop directly to oocytes. Triflagellate microgametes typically arise from the microgamonts (Fig. **87**). The zygote forms a zygotocyst (oocyst), and normally divides during sporogony into 4–32 (or more) sporoblasts. The sporoblasts normally form an individual envelope (sporocyst) and produce, by mitotic cell division in each case, 2–8 (or more) sporozoites.

With the exception of some cases in the order Coelotrophida, the coccidians develop as intracellular parasites within their hosts. The separation into subordinated taxa is based on the differences in the life cycles. The most important systematic character is, especially in the higher evolved groups, the number of sporozoites per sporocyst.

Order   Coelotrophiida
          Vivier, 1982

The representatives of this taxon do not pass through schizogonies. Trophonts, as well as gamonts, mature as extracellular parasites. They occur in the intestine or in body cavities of marine annelids.

Examples: *Eleutheroschizon, Grellia*

Order   Adeleida
          Léger, 1911

As in gregarines, the metamorphosis of the gamonts of adeleids involves close mutual associations (formation of syzygies). However, the ga-

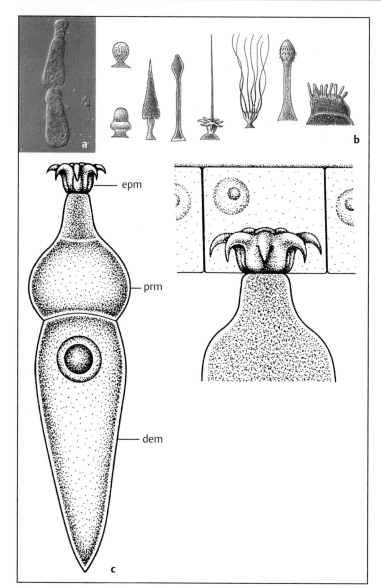

Fig. **89** Gregarinea: **a** pair of gamonts forming syzygy. **b** different types of epimerites. **c** typical appearance of epimerite (epm), protomerite (prm) and deutomerite (dem) and scheme of intracellular attachment by epimerites (b and c adapted from Léger). Magn.: a 100×.

epm

prm

dem

monts are of different sizes and therefore clearly discernible as macro- and microgamonts (sexual dimorphism). The microgamonts, which laterally mate with macrogamonts, produce mostly two to four microgametes. Before differentiating, they pass through one or several schizogonies.

The Adeleida especially parasitize the intestinal epithelium, and glandular and fat cells of invertebrates (nematodes, annelids, arthropods, molluscs, sipunculids). Some species of the genus *Haemogregarina* attack the blood cells of fish and amphibians, and cause commercially important injuries. During alternation of host they are transmitted by blood-sucking vectors such as leeches or mites. Species of the genus *Klossia* can be also detected in the kidneys of mammals. *Cryptosporidium parvum* is recognized as a significant cause of diarrheal disease in animals and humans.

Examples: *Adelea, Cryptosporidium, Haemogregarina, Klossia*

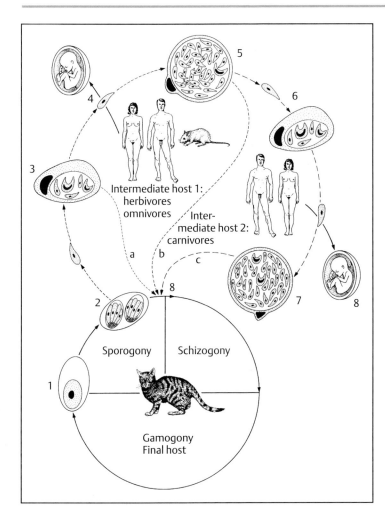

Fig. **90** Coccidia: Eimeriida, life cycle of *Toxoplasma gondii*. 1 = nonsporulated oocyst from Felidae feces; 2 = development of two sporocysts, each with four sporozoites which are set free in the intestine of the inter- mediate host; 3 = genesis of intracellular pseudocysts filled with merozoites; 4 = mero- zoites (tachyzoites) from bursting pseudocysts, causing fetal infection or forming re- sistant tissue cysts (5) with cystic merozoites (cysto- zoites, bradycytes) inside muscular or brain tissues; 6 = recapitulation of stages 3–5, but inside carnivores or- ganisms and combined with development of tissue cysts (7) or causing congenital toxo- plasmosis in fetus (8); the du- ration of oocyst transforma- tion (gamogony) in cats de- pends on the route of infec- tion: after ingestion of pseu- docysts (a) 9–11 days, after in- gestion of tissue cysts (type I) (b) 21–24 days and of tissue cysts (type II) (c) 3–5 days (adapted from Mehlhorn).

Order   Eimeriida
            Léger, 1911

The path of development is characterized by the separate differentiation of the macro- and micro- gamonts. The microgamonts always produce a large number of microgametes. The sporogony runs in two phases: The formation of oocysts with the production of sporoblasts is followed by the formation of sporocysts with the produc- tion of sporozoites. The genera can be distin- guished by the presence or absence of an alterna- tion of hosts.

The more than 1000 species of the genera *Eimeria* (four spores each with two sporozoites, see Fig. **87**) and *Isospora* (two spores each with four sporozoites) develop within a single host (monoxenic cycle) and are normally very host- specific. They are transmitted via feces and are

ingested orally by the next host. The double en- velope of the oocyst and sporocyst wall allows not only sporogony outside the host, but repre- sents also a sufficient protection against desicca- tion. This permits the oocysts to remain infec- tious for several months. The same is valid for heteroxenic species which are characterized by the facultative or obligatory alternation of hosts *(Toxoplasma, Sarcocystis, Frenkelia)*.

In the case of *Toxoplasma gondii* (two sporocysts each with four sporozoites), cats are the final hosts. In their intestinal epithelia, all three generation phases can be achieved (Fig. **90**). However, by oral uptake of oocysts, not only other cats, but also omnivores or her- bivorous mammals can be infected. The prey of felids, but also humans, are possible intermedi- ary hosts. Inside the intermediary hosts, the sporozoites perforate the intestinal epithelium

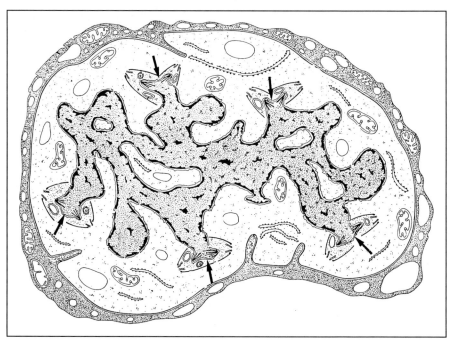

Fig. **91**   Coccidia: Schizogony of *Sarcocystis suihominis*. Arrows indicate duplications of the apical complex (from Heydorn and Mehlhorn: Zentralbl. bakt. Mikrobiol. Hyg. 1. Abtl., Orig. A, 240 [1978] 123).

and in particular invade the cells of the lymphatic system. Through subsequent binary fissions (here designated as endodyogeny), numerous parasite cells arise and can be found within a parasitophorous vacuole (pseudocyst). In later stages of the infective progress, so-called tissue cysts with strengthened walls between the host-parasite interface are formed in brain or muscular tissue of the host. They contain numerous banana-shaped cystomerozoites, also arisen from endodyogeny, which represent resting stages. When muscle tissue that is infected in such a way with tissue cysts is consumed by an intermediary host, the endodyogeny cycle starts again. But not before pseudocysts or tissue cysts arrive in the intestine of the final host (cat), is the life cycle completed by continuing through schizogony (Fig. **91**) and gamogony.

Humans who are infected, usually by cat feces or from raw meat, rarely demonstrate pathological symptoms apart from diseases of the lymph nodes (toxoplasmosis of adults). However, via a possible third path of infection, the placental transmission to the fetus, severe or fatal infections can arise (congenital or infant toxoplasmosis), which may coincide with hydrocephaly, brain calcification, or choriorenitis. This kind of toxoplasmosis appears only when women are in-fected for the first time during the last third of their pregnancy.

Obligatory alternations of hosts occur also, for example, in *Sarcocystis, Aggregata*, and *Frenkelia* species. Prey animals or herbivorous animal for slaughter function here as intermediary hosts. The phases of gamogony and sporogony, however, take place only in carnivores (humans, beasts of prey, birds of prey, snakes, cephalopods) and are normally combined with less pathological accompaniments. The ecological relationships between prey and predator, or between intermediary host and final host serve, commonly, as epithets for the scientific naming of species (e. g., *Sarcocystis suihominis* [pig-men], *S. equicanis* [horse-dog] or *S. ovifelis* [cow-cat]). During schizogony, normal merozoites arise, by multiplicative fission of trophonts with a single and possibly polyploid giant nucleus, which then start an additional phase of schizogony. The merozoites of the second generation are called metrocytes; they invade the muscular tissue. In the host cells, they form multicellular, large, and macroscopically visible tissue cysts by endodyogeny.

Examples: *Eimeria, Frenkelia, Isospora, Sarcocystis, Toxoplasma*

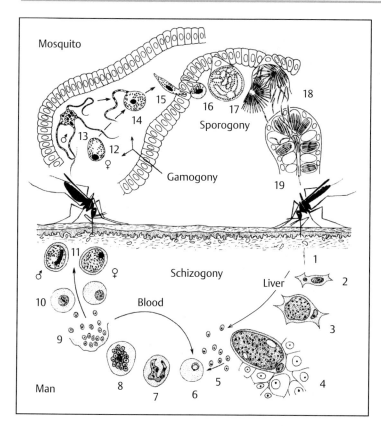

Fig. **92** Haematozoea: Haemosporida, life cycle of *Plasmodium vivax*. Stages in man: 1–5 schizogonies in liver parenchyma; 6–9 schizogonies in erythrocytes; 10–11 formation of gametes; stages in *Anopheles*: 12 macrogamete transformation; 13 genesis of microgametes; 14 fertilization; 15 ookinete; 16–18 sporogenesis; 19 immigration of sporozoites into salivary glands (adapted from Grell).

Class **Haematozoea**
Vivier, 1982

The apical complex of sporozoites and merozoites of the haematozoans is reduced; the conoid and occasionally also the conoidal ring system are missing. No sporocysts emerge from the motile zygotes (= wandering zygotes, kinetes or ookinetes) except sporozoites directly. The encysted stages are therefore suppressed, and the infective stages are consequently transferred to intermediary hosts only in liquid milieu, such as spittle. The Haematozoa are blood parasites with an obligatory alternation of hosts between vertebrates (intermediary host) and arthropods (final host). The classification into the Haemosporida and Piroplasmida is preferentially based on the different systematic positions of final and intermediary hosts.

Order **Haemosporida**
Danilewsky, 1885

The sporozoites of the haemosporidians are transferred by blood-sucking Diptera and in-

jected into their intermediary host along with saliva. In the case of the most medically important malaria parasite *Plasmodium*, reptiles, birds, and mammals (here mostly rodents and primates) serve as intermediary hosts. Midge genera such as *Anopheles, Aedes,* and *Culex* function as final hosts, in which gamogony and sporogony are performed. The sporozoites of *Plasmodium* attack, in a first step, parenchymatic cells of the liver (Fig. **92**) and develop here into large (up to 1 mm) schizonts. They can generate several thousands of merozoites (primary or preerythrocytic schizogony). In the second step of infection, the merozoites invade erythrocytes (Fig. **93, 94**), and multiply here by schizogonies, differing from the former by the smaller release of daughter merozoites (secondary or erythrocytic schizogony). After several days, the divisions inside the red blood cells are synchronized, so that the intermediary hosts suffer from periodic parasitic attacks. The merozoites, localized within a parasitophoric vacuole, satisfy their requirements for protein by consuming hemoglobins, incorporated by phagocytosis via the micropores. Following the lysis of attacked erythro-

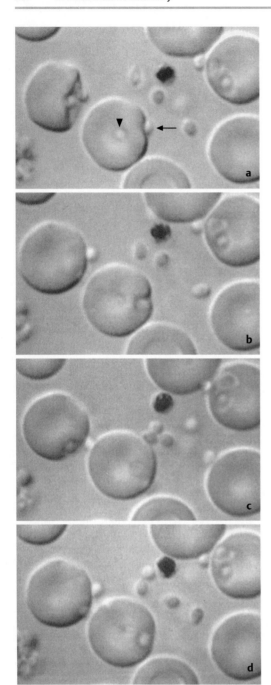

Fig. **93** Haematozoea: Haemosporida **a–d**, stages of invasion of *Plasmodium* sporozoites (arrow) into a preinfected erythrocyte (arrowhead) (courtesy of B. Enders, Marburg, and G. Kerlen, IWF Göttingen). Magn.: 4000×.

cytes, cellular fragments and undigested residual bodies (so-called pigments or hemozoines) appear. They are responsible for the characteristic intermittent fever attacks (malaria, paludism). After about ten days in the intermediary host, the first male and female gamonts develop from merozoites, but they do not pass through gamogony, and produce gametes before being transferred to the intestine of mosquitos. The macrogametes are fertilized by flagellated microgametes and, as ookinetes, they invade intestinal tissue, where they are encapsulated within a layer produced by the intestine cells. Such pseudo-oocysts are the place of development of sporozoites, which become infective elements after bursting out of the envelope and after migrating via hemolymph to the saliva glands of the host. Because such developments can be completed only at temperatures above 16 °C, malaria diseases occur mostly in warmer regions of the earth. Of the roughly 160 presently known *Plasmodium* species, only 11 are important for human or epizootic health. Some characteristics of the four species dangerous for humans are listed in Table **3**.

Malaria (from Old Italian mala aria = bad air) is not only a tropical, subtropical, or Mediterranean disease, but also an epidemic disease. Designations such as "cold fever," "intermittent fever," "Sumpffieber" (German), "marshy land fever," or "Butjadinger Seuche" (German), are testament to this. In mid-Europe, people were affected in the Dutch and German coastal and marshy regions. The diseases were possibly caused by *P. vivax* and were of epidemic character, as was the case between 1858 and 1869 when, during the construction of the German marine port Wilhelmshaven, about 18 000 persons were affected. Not until the present century could malaria be controlled in moderate climates, and in the Mediterranean regions, mostly by the drainage of marshy and swampy regions, and the use of pesticides against mosquitos. However, endemic malaria centers can arise, caused by infected persons returning from typical malaria zones. In tropical coastal regions, malaria is still prevalent. It is mostly maintained by the evolution of therapy-resistant strains of *P. falciparum*, as well as the by insufficient resistance in the inhabitants. It is likely that more than 200 million people are currently suffering from symptoms of pernicious malaria. According to the World Health Organization (WHO), about one million malaria-infected children die each year in Africa.

Examples: *Haemoproteus, Leucocytozoon, Plasmodium*

Table **3**   Characteristics of *Plasmodium* spp. pathogenic for humans

Species	Type of malaria	Time of incubation	Periodicity of fever	Mortality
P. vivax	malaria tertiana	8–16 days	48 h	–
P. ovale	malaria tertiana	about 15 days	48 h	+/–
P. malariae	malaria quartana	20–35 days	72 h	+/–
P. falciparum	malaria tropica	7–12 days	irregular	+

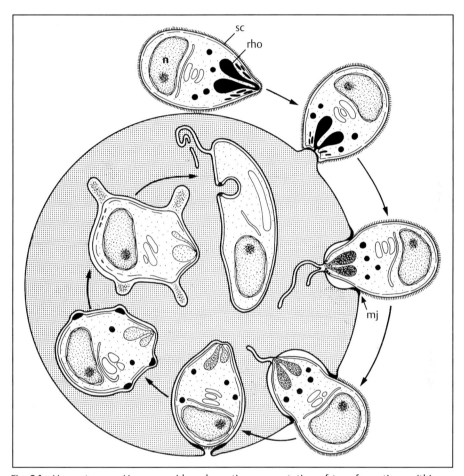

Fig. **94**   Haematozoea: Haemosporida, schematic representation of transformations within an invading merozoite of *Plasmodium knowlesi*. mj = moving junction passing backward over the invading parasite, n = nucleus, rho = rhoptry in different stages of depletion, sc = surface coat (adapted from Bannister).

Order   **Piroplasmida**
             Wenyon, 1926

In the sporozoites and merozoites of the globally distributed Piroplasmida, the conoid, and partly also the pole rings, the subpellicular microtubules, and the micronemes or rhoptries are present only as rudimentary forms, or they are completely lost. They live as parasites in lymphocytes, erythrocytes, and other blood and blood-forming cells of cold- and warm-blooded vertebrates. However, they release few, if any, pigments. Besides multiplicatory fissions (schizogonies), binary divisions occur also. During gamogony, aflagellate microgametes appear; the function of flagella is accomplished, as in trophozoites of the species *Entopolypoides macaci*, by axopodia-like protuberances of the cell surface.

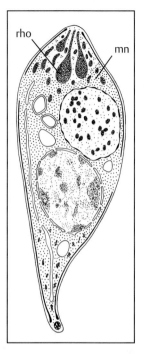

Fig. **95** Haematozoea: Piroplasmida, *Babesia bigemina,* merozoite from erythrocyte without conoid, mn = micronemes, rho = rhoptry (adapted from Scholtyseck).

The life cycle is similar to that of the Haemosporida: As far as is known, the transmission to the intermediary hosts is realized via the saliva of ticks (Ixodidae). In their intestinal epithelia, and also in the alveoli of their salivary glands, the stages of gamogony and sporogony are completed. The zygotes of some *Babesia* species are enabled to invade not only the salivary glands, but also other organs of the ticks. In the case of invasion of ovaries and eggs, a new transovarial transmission pathway to the offspring is opened. Therefore, even young nymphs, which never had the chance to suck the blood of vertebrates, can function as vectors. The resulting danger potential of ticks is reflected in many pests of veterinary and medical importance. For example, Texas fever, evoked by *Babesia bigemina* (Fig. **95**) in cattle, leads to a mortality rate of up to 50%. Three *Babesia* species (*B. bovis, B. divergens, and B. microti*) also infect humans, causing severe or lethal diseases, and not only in those with acquired immune deficiency syndrome (AIDS).

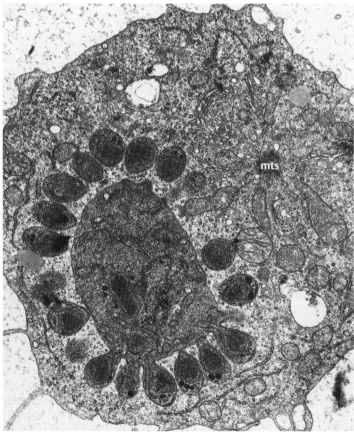

Fig. **96** Haematozoea: Piroplasmida, *Theileria annulata* schizont of the merozoite stage inside a bovine lymphocyte. The division of the host cell was stimulated by the presence of the parasite. mts = mitotic spindle (courtesy of H. Mehlhorn, Bochum). Magn.: 8000×.

It is generally assumed that at least some Piroplasmida fill the ecological niche of attacking ruminants, which are immune to the malaria agent *Plasmodium*. This is probably so, especially for representatives of the genus *Theileria* (Fig. **96**), which in cattle, sheep, and goats causes the feared and often lethal diseases known as theilleriosis (= African east coast fever = Mediterranean coast fever).

Examples: *Babesia, Dactylosoma, Piroplasma, Theileria*

Subphylum   **Ciliophora**
            Doflein, 1901

The taxon was formerly called Ciliata (Perty, 1852) or Infusoria (Bütschli, 1887). It presently comprises more than 8000 species, whose common principal characters lie in their morphological organization and life cycles. Apart from the typically very numerous cilia (Fig. **97**), the general characteristics can be reduced to three features: (1) the construction of the cortex, (2) the presence of nucelar dualism, and (3) the phenomenon of a conjugation stage during the sexual phase of the life cycle. It is very important to emphasize that only with this combination of features, the ciliates can be defined as a monophyletic taxon.

The cortex (Figs. **98** and **99**), also called cortical plasm or cortical layer, is responsible for the relative constancy of shape of individual species. It comprises, with a total thickness of about 1–4 μm, two components: (1) the pellicle, and (2) the root-structures of cilia (= kinetids) which form, in their total arrangement, the so-called infraciliature. Additional structural elements can occur.

The cell membrane (plasmalemma) and, in some cases, also an additional perilemma as an extracellular component belong to the pellicle. Close to the cilia, regularly arranged invaginations occur (= parasomal sacs), representing the sites of pinocytotic activity. A system of alveoli is located immediately beneath the plasmalemma. The flattened vacuoles are arranged in a mosaic-like pattern, and they nearly always exhibit a species-specific arrangement. In several cases, proteinaceous plates (*Euplotes*, Fig. **100**) or calcified polysaccharide plates (*Coleps*, comp. Fig. **121**) are internal constituents of the alveoli, which leads to greater stability of the cortical plasm. Besides this function, there are indications that they also play an important role in ionic regulation. The pellicle, in the widest sense, includes a

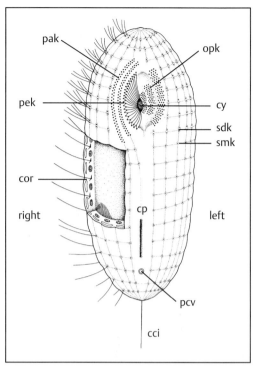

Fig. **97** Ciliophora: characters of the ventral side of an idealized ciliate. cci = caudal cilium, cor = cortical region with alveoli, mitochondria and filamentous network, cp = cytoproct or cytopyge, cy = cytostome, opk = oral polykinetids, pak = paroral kinety, pcv = pore of contractile vacuole, pek = perioral kineties, sdk = somatic dikinetid, smk = somatic monokinetid (adapted from Lynn).

proteinaceous layer, called the epiplasm, underlying the alveolar vacuoles (Fig. **101**), and longitudinal ribbons of microtubules running parallel to the epiplasm (supraepiplasmatic and subepiplasmatic microtubules). Both components probably provide a stabilization function as well as having a role in morphogenesis of the cortex.

The kinetids are the fundamental constituents of the infraciliature. There are basal bodies (= kinetosomes), combined as singular, double, or mutliple components with several associated rootlet structures connected to a superior structural complex (hence somatic monokinetids, somatic dikinetids, or oral polykinetids). It is not necessary for each kinetosome to bear its own cilium. The associated root structures comprise the following components: a mostly anteriorly directed kinetodesmal fibril consisting of filamentous subunits, as well as two microtubular

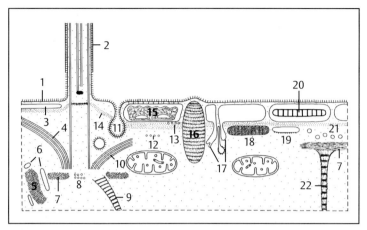

Fig. **98** Generalized representation of structures in the cortex of ciliates: 1 plasma membrane, 2 perilemma, 3 alveolus, 4 postciliary microtubules, 5 myoneme, 6 cisterna of endoplasmic reticulum, 7 filament layer, 8 subkinetal microtubules, 9 kinetodesmal fiber, 10 transverse microtubules, 11 parasomal sac, 12 and 13 longitudinal microtubules, 14 epiplasm, 15 calcified alveolar plates, 16 extrusome, 17 alveolocysts, 18 polysaccharide lates, 19 cisterna of rough endoplasmic reticulum, 20 alveolar glycoprotein plates, 21 vesicles, 22 cross-striated filament bundle (adapted from Bardele).

Fig. **99** Ciliophora: cortex of *Paramecium* (**a, b**) and *Pseudomicrothorax* (**c, d**). al = alveolus, ep = epiplasm, ialm = inner alveolar membrane, ks = kinetosome, oalm = outer alveolar membrane, pm = plasma membrane, ps = parasomal sac. Magn.: a 50 000×, b 150 000×, c 160 000×, d 25 000×.

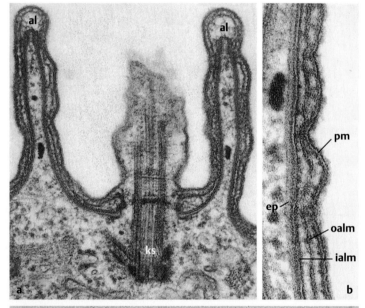

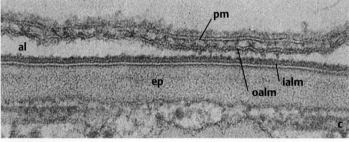

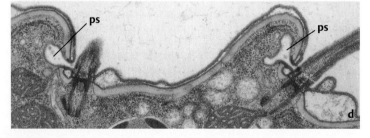

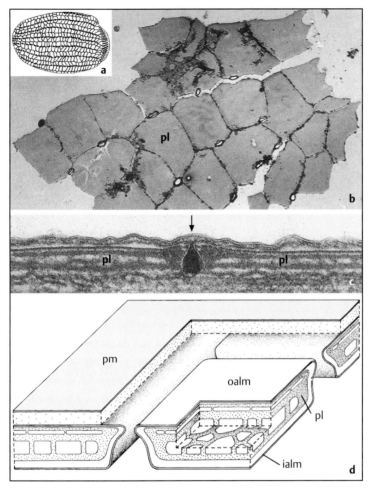

Fig. **100** Ciliophora: alveolar plates in euplotid ciliates. **a** silver line system. **b** isolated plates in original connex. **c** section of plates (pl) located within two alveoli bordering at the arrow. **d** schematic representation of the pellicular region. ialm = inner alveolar membrane, oalm = outer alveolar membrane, pl = plates, pm = plasma membrane (from Hausmann and Kaiser: J. Ultrastruct. Res. 67 [1979] 15). Magn.: a 350×, b 3000×, c 120 000×.

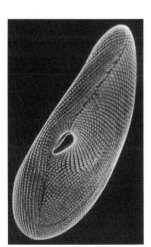

Fig. **101** Ciliophora: epiplasmic protein plates of *Paramecium* visualized by immuno-fluorescence confocal microscopy (from Jeanmaire et al.: Europ. J. Protistol. 29 [1993] 311). Magn.: 300×.

ribbons running as transverse microtubules to the lateral side, or as postciliary microtubules running obliquely to the posterior end of the cell (Fig. **102**). The relative size, the formation, and the orientation of the individual structures of the somatic kinetids are of systematic importance; the corresponding classes have their own typical configurations (comp. Fig. **111**).

The kinetosomes are not necessarily combined with cilia. They are generally connected to each other by basal (= subkinetal) microtubular ribbons, such as the postciliary and kinetodesmal fibrils, for an exact longitudinal orientation of themselves, or of the cilia when present. Therefore, the somatic cilia, which serve for locomotion, are arranged in longitudinal rows (= kineties). This regular pattern is only interrupted in the oral region by so-called perioral kineties (Fig. **103**). In some special cases, groups of cilia are positioned very closely to each other but

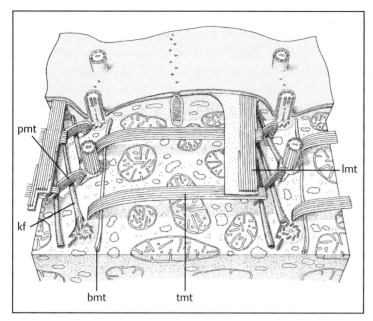

Fig. **102**   Ciliophora: root structures of the kinetosomes of *Tetrahymena*: bmt = basal microtubules, kf = kinetodesmal fiber, lmt = longitudinal microtubules, pmt = postciliary microtubules, tmt = transverse microtubules (adapted from Allen).

Fig. **103**   Ciliophora: arrangement of cilia in *Paramecium* (**a**), *Frontonia* (**b**), and *Climacostomum* (**c**). Magn.: a 275×, b 1400×, c 200×.

without any obvious sign of fusion (Fig. **104**), except for the only known case sofar, *Glaucoma ferox*, in which the tips of the membranellar cilia are morphologically connected by specially constructed junctions. Cilia might form clusters (= cirri) or flat-spread units (= membranelles). Cirri serve as extremity-like organelles for walking movements, membranelles mostly for swirling of food particles.

The cortex carries other organelles, particularly extrusomes and contractile fibrils (= myonemes), but also mitochondria, cisternae of the endoplasmic reticulum, and several types of vesicles. Between the cortex and the endoplasm, a border might be present made of a flat-spread or net-like layer with cross-striated filament bundles.

The highly organized and fine-meshed structural organization of the cortical plasm does not influence its contractility, but generally impairs the exchange of materials with the environment. At the ventral side of many ciliates, a rather long gap in the cortical pattern (= suture) exists, which is reserved for import and export purposes. In this region, the typical ciliate frequently bears a deeply invaginated oral funnel, whose bottom is involved in phagocytosis and food vacuole formation. In a similar way, dis-

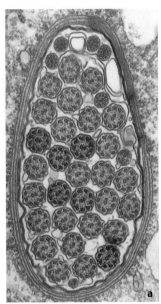

Fig. **104**   Ciliophora: **a** basal region of a cirrus of *Euplotes vannus*. **b** membranelles of *Euplotes vannus* (a from Hausmann and Kaiser: J. Ultrastruct. Res. 67 [1979] 15). Magn.: a 28 000×, b 2000×.

tinct structural complexes also exist here for exocytotic events, involving the fusion of the vacuolar membranes with the plasmalemma: one cytoproct (= cytopyge) for excretion of particular food residuals, and one or more pores for the elimination of the fluid contents of the contractile vacuoles. The special architecture of the cortex also influences the mode of mitotic division insofar as the plane of fission normally runs perpendicularly to the direction of kineties. During this process, the posterior cell half (opisthe) develops a new oral region, whereas the anterior cell half (proter) has to form a new cytoproct region.

An impression of the complexity of the cortical organization can be gained even at the light microscopical level with aid of certain silver impregnations (Fig. **105**). Although it is not always clear which elements of the pellicle or kinetidal structures are really visualized by these techniques, the results are reproducible and species specific. For systematic or morphogenetic investigations, the analysis of silver line systems is still a dependable method (Fig. **106**). However, the full complexity of the cortex can be elaborated only by electron microscopical investigations of serial ultrathin sections, and very exact three-dimensional reconstructions of the cortical organization exist for several species (Fig. **107**).

Apart from the ciliature and infraciliature, the next conspicuous character of the ciliates is the presence of a heterokaryotic nuclear condition. There exist one or more somatic nuclei (= macro-

Fig. **105**   Ciliophora: silver line systems of *Colpidium* (**a**) and *Euplotes* (**b**) (a courtesy of W. Foissner, Salzburg). Magn.: a 450×, b 450×.

nuclei), but also one to several generative nuclei (= micronuclei) (Fig. **108**). The micronuclei are diploid and remain, with a diameter of about 2–5 μm, relatively small, whereas the macronuclei, in accordance with their highly amplified genes, are much larger and polymorphic (Fig. **109**). The macronuclei are the site of RNA synthesis and they fulfill the tasks of normal cell metabolism. The micronuclei, however, are the site of genetic recombination. The shape of the macronuclei is quite variable; there exist roundish, branched, pearl string–like, and even fragmented types. Only the micronucleus performs a regular mitotic or meiotic fission (Fig. **110a**), during which an intranuclear spindle is formed. For

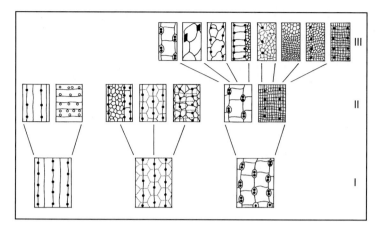

Fig. **106** Ciliophora: the hierarchy of silver line systems (I–III) can be reduced to basal types, which may give clues to phylogenetic relationships (courtesy of W. Foissner, Salzburg).

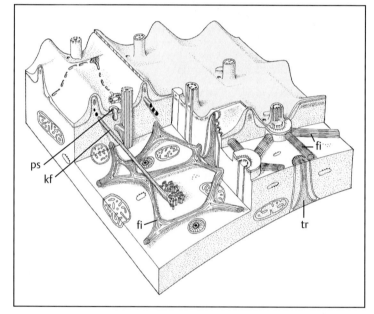

Fig. **107** Ciliophora: cortical structures of *Paramecium caudatum.* fi = filaments, kf = kinotodesmal fiber, ps = parasomal sac, tr = trichocysts (adapted from Allen).

karyokinesis of the macronuclei, however, the separation mechanism does not involve the participation of spindles, and it is not completely understood. During the phase of sexual reproduction, and in the case of the Karyorelictida, also during normal binary fission, macronuclei normally undergo degeneration, to be later reestablished from modified micronuclei.

In Ciliophora, the sexual phenomena are encompassed by the phenomenon of conjugation, when two conspecific cells fuse with each other (Fig. **110 b**). During this partial fusion, which may last several hours, haploid nuclei derived from micronuclei are mutually exchanged. After this nuclear migration, the partners separate. The so-called exconjugants be-

come, at least potentially, the starting points for new vegetatively reproducing clones.

The systematic subdivision of the Ciliophora has been altered often since the era of Bütschli and Kahl. Originally, the type of ciliation was the chief criterion. The larger groups such as the Holotricha, Chonotricha, Peritricha, Spirotricha, and Suctoria could be identified rather easily. Since about 1970, morphogenetic parameters have been included, especially those concerned with the formation and morphogenesis of the oral apparatus (Table **4**). The Levine et al. system, published in 1980, distinguished three classes: Kinetofragminophorea (oral ciliation derived from body ciliation), Oligohymenophorea (oral apparatus with only few membranelles), and

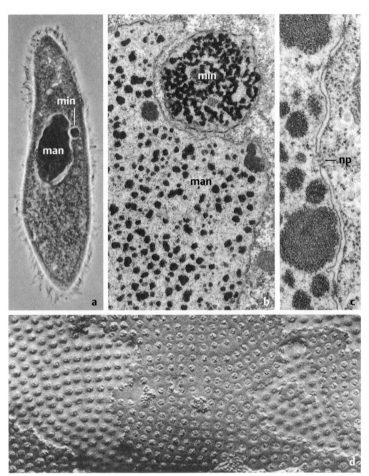

Fig. **108** Ciliophora: nuclear dualism in *Paramecium* (**a**) and *Colpidium* (**b**, **c**); **d** arrangement and frequency of macronuclear pores in *Paramecium*. man = macronucleus, min = micronucleus, np = nuclear pore (d courtesy of R. D. Allen, Hawaii). Magn.: a 400×, b 15 000×, c 60 000×, d 40 000×.

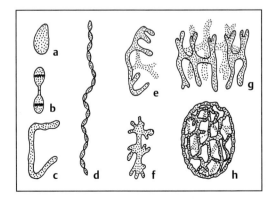

Fig. **109** Ciliophora: morphological variability of macronuclei: **a** *Paramecium*; **b** *Stylonychia*, **c** *Vorticella*; **d** *Spirostomum*; **e** *Ophryodendron*; **f** *Conchophthirius*; **g** *Ephelota*; **h** *Metaphrya* (adapted from Grell).

Polyhymenophorea (oral apparatus with numerous membranelles). Presently, a subdivision conforming with the characters of the somatic and oral infraciliature is favored (Fig. **111**). This system is based on the fine structural characters of the cortex, but it also includes stomatogenesis and other morphogenetic details, the life cycles, and, so far as it is known, the molecular biological data. Because ultrastructural and molecular genetic investigations are time-consuming, new changes to the system should be expected. According to present knowledge, however, it seems certain that organisms with somatic dikinetids and a less pronounced nuclear dimorphism represent the starting point of radiation. Eight classes have evolved from such organisms, which are normally combined in three superclasses (Table **5**).

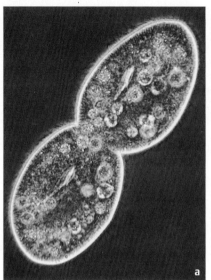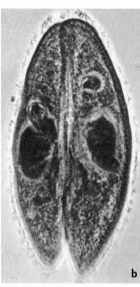

Fig. **110**   Ciliophora: *Paramecium* during mitosis (**a**) and during conjugation (**b**). Magn.: a 350×, b 400×.

Bütschli, 1887–1889	Kahl, 1930–1935	Honigberg et al, 1964	Levine et al, 1980
Classis Infusoria	Superclassis Ciliophora		
Subclassis Ciliata	Classis Euciliata	Classis Ciliatea	Phylum Ciliophora
Ordo Holotricha ————	Ordo Holotricha ————	Subclassis Holotrichia	Classis Kinetofragminophorea
Subordo Gymnostomata ——	Subordo Subordo Gymnostomata ——	Ordo Gymnostomatida	Subclassis Gymnostimatia
		Ordo Chonotrichida	Subclassis Hypostomatia
	Subordo Apostomea	Ordo Apostomatida	
Subordo Astomata	Subordo Astomata	Ordo Astomatida	
Subordo Trichostomata ———	Subordo Trichostomata ———	Ordo Trichostomatida	SubclassisVestibuliferia
	Subordo Thigmotricha	Ordo Thigmotrichida	
	Subordo Hymenostomata	Ordo Hymenostomatida	
			Subclassis Suctoria
	Ordo Chonotricha		Classis Oligohymenophorea
		Subclassis Peritrichia	Subclassis Hymenostomatia
	Ordo Peritricha ———	Ordo Peritrichida	Subclassis Peritrichida
			Classis Polymenophorea
Ordo Spirotricha———	Ordo Spirotricha ———	Subclassis Spirotrichia	Subclassis Spirotrichia
	Subordo Entodiniomorpha	Ordo Entodiniomorphida	
Subordo Heterotricha——	Subordo Heterotricha ——	Ordo Heterotrichida	Ordo Heterotrichida
Subordo Oligotricha——	Subordo Oligotricha ——	Ordo Oligotrichida	Ordo Oligotrichida
		Ordo Tintinnida	
Subordo Hypotricha——	Subordo Hypotricha ——	Ordo Hypotrichida	Ordo Hypotrichida
Subordo Peritricha —	Subordo Clenostomata ———	Ordo Odontostomatida	Ordo Odontoslomatida
Subclassis Suctoria———	Superclassis Suctoria ——	Subclassis Suctoria—	

Fig. **111**   Ciliophora: The kinetosome-associated structural elements of the somatic kinetids of the eight classes of ciliates. Associated with the kinetosomes are the anterior directed kinetodesmal fibers (kf), the posterior directed postciliary (pmt) and the transverse microtubular ribbons (tmt) as well as the parasomal sacs (ps) (courtesy of D. Lynn, Guelph): Karyorelictea: *Geleia* (1) and *Protocruzia* (2); Spirotrichea: *Stylonychia* (3) and *Climacostomum* (4); Colpodea: *Sorogena* (5) and *Pseudoplatyophrya* (6);

Phyllopharyngea: *Hypocoma* (7) and *Trichophrya* (8); Nassophorea: monokinetid (9) and dikinetid (10) of *Nassula*; Oligohymenophorea: *Colpidium* (11) and *Ichthyophthirius* (12); Prostomatea: monokinetid (13) and dikinetid (14) of *Coleps*; Litostomatea: *Lepidotrachelophyllum* (15) and *Isotricha* (16)

▷

Fig. **111**

Table **5**   The system of Ciliophora after Lynn and Corliss (1991), based on three superclasses and eight classes, demonstrating the fate and current position of some of the higher well-known taxa of older systems

Designation of superclasses (I–III) and classes (1–8)	Taxa adapted from the Levine et al.-system and characteristic genera
I. Postciliodesmatophora	
1. Karyorelictea Corliss, 1974	Gymnostomatia (Karyorelictida): *Loxodes*
2. Spirotrichea Bütschli, 1889	Spirotrichia: *Stentor, Euplotes*
II. Rhabdophora	
3. Litostomatea Small & Lynn, 1981	Gymnostomatia (Prostomatida, Haptorina): *Didinium* Gymnostomatia (Prostomatida, Pleurostomatida): *Litonotus* Vestibuliferia (Trichostomatida): *Balantidium* Vestibuliferia (Entodiniomorphida): *Entodinium*
4. Prostomatea Schewiakoff, 1896	Gymnostomatia (Prostomatida, Prostomatina): *Holophrya* Gymnostomatia (Prostomatida, Prorodontina): *Coleps*
III. Cyrtophora	
5. Phyllopharyngea de Puytorac et al., 1974	Hypostomatia (Cyrtophorida): *Chilodonella* Hypostomatia (Chonotrichida): *Spirochona* Hypostomatia (Rhynchodida): *Gargarius* Suctoria: *Discophrya*
6. Nassophorea de Puytorac et al., 1974	Hypostomatia (Nassulida): *Nassula, Pseudomicrothorax* Hymenostomatia (Peniculina): *Paramecium, Frontonia*
7. Oligohymenophorea de Puytorac et al., 1974	Hymenostomatia (Tetrahymenina): *Tetrahymena* Hymenostomatia (Astomatida): *Anoplophrya* Hymenostomatia (Scuticociliatida): *Pleuronema* Peritrichia: *Vorticella*
8. Colpodea de Puytorac et al., 1974	Vestibuliferia (Colpodida): *Colpoda*

Superclass   **Postciliodesmatophora**
Gerassimova & Seravin, 1976

Both classes of this taxon have the same type of somatic dikinetids. The kinetodesmal fibrils, each arising only from the posterior kinetosome, are mostly distinct, and run as singular strands to the anterior, but partly also to the posterior end of the cells, as in the Oligotrichia and Stichotrichia. The postciliary microtubules are directed posteriorly, and superimpose with their equivalents arising from more anteriorly positioned kinetosomes of the same kinety. In this way, combined microtubular bundles (= postciliodesmata), which are visible as fibers at the light microscopical level, are the most important morphological character. Beside kineties with dikinetids, fields with monokinetal ciliary rows also occur. Parasomal sacs and also alveoli are sometimes less developed. As extrusome types, mucocysts, and rhabdocysts occur.

Class   **Karyorelictea**
Corliss, 1974

The extant members of this class (Fig. **112**) have a primitive nuclear type. Micronuclei and macronuclei have nearly the same diploid DNA content. Moreover, the macronuclei, of which normally two per cell occur, are not able to undergo karyokinesis. They undergo degeneration and, after each mitosis, they are restored by an additional fission of the daughter micronucleus. Only in *Protocruzia* do the macronuclei arise by fission from themselves. This case also offers a remarkable exception insofar as the macronuclear divisions are realized by a true mitotic karyokinesis with elaboration of a spindle apparatus.

The Karyorelictea, which were formerly counted within the Holotricha, are mostly regularly ciliated and have an elongate cell body. Most of them live in the marine interstitial. The only freshwater relatives belong to the

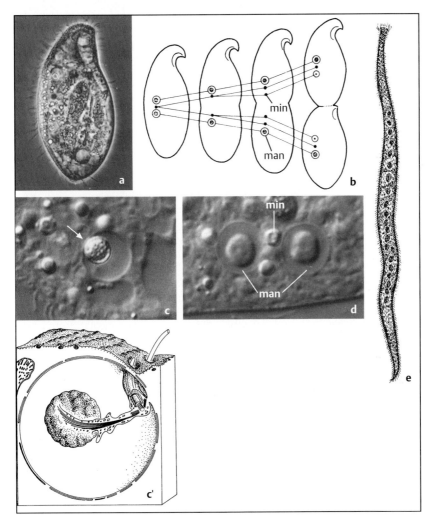

Fig. **112** Karyorelictida: **a** *Loxodes*. **b** the nondividing macronucleus (man) is derived from the micronucleus (min). **c** Müller body in *Loxodes* (arrow). **c'** schematic presentation of Müller's body. **d** macronucleus (man) and micronucleus (min) of *Loxodes*. **e** *Trachelocerca* (b adapted from Raikov, c' adapted from Fenchel and Finlay, e from Dragesco: Trav. Biol. Roscoff 12 [1960] 1). Magn.: a 200×, c 2000×, d 2000×, e 250×.

Loxodidae: they have special organelles, so-called Müller's bodies (Fig. **112**) which are probably gravity receptors, comparable to the statocysts of metazoa.

Examples: *Geleia, Loxodes, Protocruzia, Remanella, Trachelocerca*

Class   **Spirotrichea**
Bütschli, 1889

The somatic ciliation of the spirotrich ciliates (= Polyhymenophorea) appears in two variations: kineties as well as cirri. The kineties are represented by linear di- or polykinetids, with kinetosomes which are either both ciliated or from which only the anterior one bears an axoneme. Cirri are located mostly on the ventral side. The postciliodesmata, which are also known from the older literature (as so-called Km-fibers) are mostly well developed. In their oral region, the cells have serial-arranged polykinetids running clockwise to the cytostome (adoral zone of membranelles = AZM). Even at the light microscopic level, this character leads to an easy identification as spirotrichs. Four subclasses are generally accepted.

Subclass   Heterotrichia
Stein, 1859

The Heterotrichia are characterized by somatic dikinetids or polykinetids and by adoral right-winding membranelles (Fig. **113**). This morphological principle reflects the prototype of a heterotrich (= different) ciliation. The oral polykinetids are formed especially at the anterior left

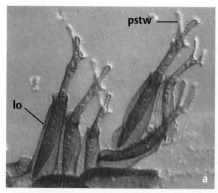

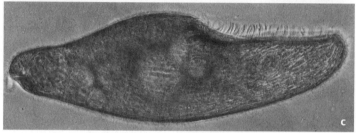

Fig. **113** Spirotrichea, Hetero-trichia, Heterotrichida: **a** *Eufol-liculina* living in lorica (lo) and whirling food particles with membranelles located on per-istomal wings (pstw). **b** so-matic cilia (sci) of *Eufolliculina* are shorter than membranel-lar cilia (mci). c *Blepharisma*. Magn.: a 75×, b 2500×, c 300×.

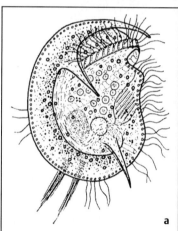
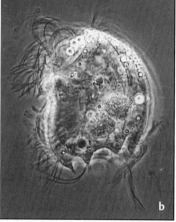

Fig. **114** Spirotrichea, Hetero-trichia, Odontostomatida: *Dis-comorpha pectinata* (**a**) and *Sa-prodinium putrinum* (**b**) with conspicuous somatic processes (a adapted from Streble and Krauter, b cour-tesy of G. Schrenk, Tübin-gen). Magn.: a 500×, b 450×.

side of the cell body. The taxon is divided into seven orders from which here only the following are considered.

Ordo   Heterotrichida
              Stein, 1859

Among the heterotrich ciliates are rather large representatives, reaching a length of more than 2 mm. In such cases, the cytoplasm is highly vacuolized. Some genera, such as *Stentor* and *Spirostomum*, are very contractile. This contractil-

ity is based, as it is in the stalks of the sessile Peritrichia on the activity of a system which differs from actomyosin. The cells are often brightly colored by pigment granules. Certain *Blepharisma*-species are rose-colored or red; *Stentor coeruleus* has bluish, *Stentor igneus* red-violet pigments. The Heterotrichida are widely distributed in marine and freshwater habitats.

Some Heterotrichida are sessile and live in self-constructed loricae such as the Folliculin-idae. In the members of this family, which settle frequently in colonies, the peristomal field is expanded into two wing-like structures bordered by membranelles (Fig. **113**).

Examples: *Blepharisma, Folliculina, Spirostomum, Stentor*

Order   Odontostomatida
Sawaya, 1940

The wedge-like cell bodies of the Odontostomatida are laterally compressed and often ornamented with spiniform or corset-like processes (Fig. **114**). The body ciliation is secondarily limited to a few regions, and even the adoral membranelles have only 8–10 polykinetids. Odontostomatida are all anaerobic and commonly encountered in freshwater sediments.

Examples: *Discomorphella, Epalxella, Myelostoma, Saprodinium*

Order   Clevelandellida
de Puytorac & Grain, 1976

The members of this order live exclusively as endosymbionts in the intestines of arthropods and some vertebrates. The kinetids exhibit additional nonmicrotubular fibrils. Usually, two sutures are formed.

Examples: *Clevelandella, Nyctotherus*

Order   Phacodiniida
Small & Lynn, 1985

The laterally compressed cell bodies of the phacodinids have a somatic ciliation, characterized by rows of densely packed polykinetids (Fig. **115**).

Example: *Phacodinium*

Subclass   Oligotrichia
Bütschli, 1887

In oligotrichs, the well developed (and up to 40 μm long) membranelles surrounding the apical oral aperture are used for collecting food particles and for swimming. The somatic cilia are almost totally reduced. The cytoproct is also missing, possibly because many of these ciliates are semi-autotrophic or mixotrophic by cultivating chloroplasts sequestered from previously ingested algae.

Order   Choreotrichida
Small & Lynn, 1985

The polykinetids of the choreotrich ciliates surround the oral region in a coronal form. During their swinging activity, they seem to "dance"

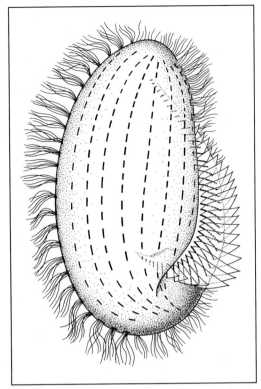

Fig. **115** Spirotrichea, Heterotrichia, Phacodiniida: *Phacodinium metchnikoffi* (adapted from Dragesco). Magn.: 800×.

around the oral aperture. The representatives of this taxon occur, despite a few limnic species, chiefly in the open water of oceans. The species which belong to the Tintinnina have loricae made of an organic matrix with incorporated foreign particles (Fig. **116**). Fossilized shells are known since the Ordovician. Besides the Tintinnina, shell-less forms also exist (*Strombidinopsis, Strobilidium*).

Examples: *Codonella, Favella, Tintinnidium*

Order   Oligotrichida
Bütschli, 1887

The regularly shell-less oligotrichs have bipartite membranellar ribbons which fulfill both locomotory and nutritional functions (Fig. **117**). As remnants of the body ciliature, cirrus-like compositions of cilia can be found. The predominantly small cells are easy to identify by their peculiar, often erratic locomotory behavior.

Examples: *Halteria, Strombidium, Tontonia*

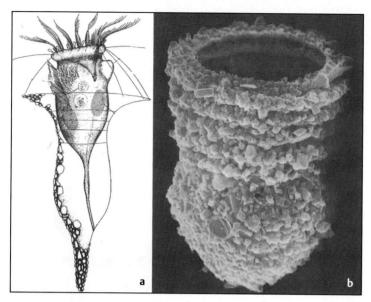

Fig. **116**   Spirotrichea, Oligotrichia, Choreotrichida: **a** *Tintinnopsis* with transparently-drawn lorica. **b** lorica of *Codonella* (a adapted from Mackinnon and Hawes; b from Barnatzky et al.: Zool. Scripta 10 [1981] 81). Magn.: a 900×, b 1000×.

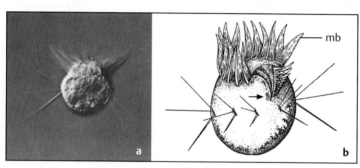

Fig. **117**   Spirotrichea, Oligotrichia, Oligotrichida: *Halteria* alive (**a**) and schematic (**b**). The large membranelles (mb) serve for swimming, the shorter ones for whirling food particles into the mouth (arrow) (b adapted from Mackinnon and Hawes). Magn.: a 500×.

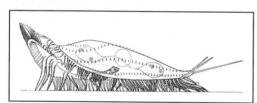

Fig. **118**   Spirotrichea, Stichotrichia: *Stylonychia* in lateral view (courtesy of H. Machemer, Bochum). Magn.: 200×.

Subclass   Stichotrichia
            Small & Lynn, 1985

Morphological characteristics of the mostly dorsoventrally flattened stichotrich ciliates are numerous cirri (Fig. **118**), arranged in the ventral somatic region, in longitudinal straight or zigzagging rows. Dikinetids with postciliodesmata are located on the dorsal side. Their kinetodesmal fibrils run, as also sometimes in the Oligotrichia,

but in contrast to the other Spirotrichia, not anteriorly, but in a curve to the posterior end. The oral apparatus is found on the left side of the anterior cell pole.

Stichotrich ciliates occur with numerous genera and species in marine, freshwater, and terrestrial habitats.

Examples: *Keronopsis, Stichotrichia, Stylonychia, Urostyla*

Subclass   Hypotrichia
            Stein, 1859

As the only taxon within the Spirotrichea, the hypotrich ciliates have postciliary microtubular ribbons with obviously rudimentary desmos formations. Their kinetids are very similar to those of the Nassophorea (see below); this was the reason for a transitional systematic combination of both groups. The other morphological characters can be compared with those of the Stichotrichia: There are ventrally located cirri, combined in

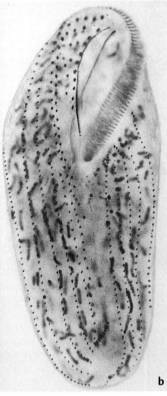

Fig. **119** Spirotrichea, Hypotrichia: *Urostyla grandis* (**a**) after protargol impregnation (**b**) (from Foissner et al.: Taxonomische und ökologische Revision der Ciliaten des Saprobiensystems. Band I: Cyrtophorida, Oligotrichida, Hypotrichia, Colpodea. Informationsberichte des Bayr. Landesamtes für Wasserwirtschaft, München 1991). Magn.: a and b 260×.

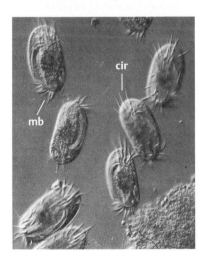

Fig. **120** Spirotrichea, Hypotrichia: *Euplotes* with cirri (cir) and membranelles (mb). Magn.: 200×.

groups, and dorsal dikinetids (Figs. **119** and **120**). Also, the oral apparatus is similar in formation, and comprises pronounced adoral membranelles. It is suggested that a sister group relation exists between the hypotrich ciliates and the Stichotrichia and Oligotrichia. In older systems, Stichotrichia and Hypotrichia were combined within the latter taxon.

The hypotrich ciliates are predominantly bacterivorous, and some occur frequently in high abundance in sewage sludge.

Examples: *Aspidisca, Euplotes*

Superclass **Rhabdophora**
Small, 1976

The taxon is divided into two classes (Litostomatea and Prostomatea) and comprises both free-living and endobiotic ciliates with a somatic ciliature arising from monokinetids. In the region of the oral aperture alone, dikinetids act as the origin for very conspicuous transversal microtubular ribbons. These microtubular bundles are known as nematodesmata, or as nematodesmal fibrils. Nematodesmata strengthen the cytostome and they sometimes form a basket-like structure (= rhabdos). During mitosis, the oral structures are transmitted, more or less unchanged, to the anterior cell (proter). As extrusive organelles, toxicysts especially occur. After discharge, these toxic extrusomes cause immobilization or even death of the prey.

Fig. **121** Prostomatea: *Coleps hirtus* (courtesy of I. Huttenlauch, Göttingen). Magn.: 1000×.

Class **Prostomatea**
Schewiakoff, 1896

The oral aperture of the prostome ciliates is located, suggested by their name, at or near the anterior end of the cell (Fig. **121**). The monokinetids of the somatic ciliature have transverse microtubular ribbons with basal parts projecting radially from the kinetosomes.

Examples: *Coleps, Holophrya, Prorodon*

Class **Litostomatea**
Small & Lynn, 1981

The litostome ciliates are characterized both by transverse microtubular ribbons, which project tangentially from the kinetosome, and by laterally directed kinetodesmal fibrils and convergent postciliary microtubules. Two subclasses exist: the exclusively free-living and carnivorous Haptoria (with toxicysts) and the principally endobiotic Trichostomatia (without toxicysts).

Subclass Haptoria
Corliss, 1974

The position of the cytostome can be lateral, ventral, or posterior. The oral region has no conspicuous additional structures and lies immediately beneath the cell surface. Mainly predatory ciliates belong to this taxon (Fig. **122**). *Mesodinium*

rubrum, sometimes the causative agent of red tides, however, has changed to a semi-phototrophic way of life with the aid of endobiotic dinoflagellates.

Examples: pleurostome: *Amphileptus, Litonotus, Loxophyllum*
hypostome: *Dileptus, Trachelius*
prostome: *Didinium, Homalozoon, Lacrymaria, Mesodinium*

Subclass Trichostomatia
Bütschli, 1889

For this taxon, a densely ciliated oral region is typical. The body ciliature is limited mostly to individual zones and often arranged in ribbons or bunches (Fig. **123**). A conspicuous layer of microfilaments stabilizes the frequently bizarre body shapes.

The representatives of the Trichostomatia live as endobionts in the intestines of numerous animals; they can be found regularly in the rumen of ruminants. Here they feed on bacteria, cellulose particles, and on other ciliates. The degradation of the cellulose takes place, as in termite flagellates, via endobiotic bacteria. The ciliate fauna of the rumen is transmitted to the next ruminant generation by regurgitation and feeding of the intestinal content. None of the Trichostomatia have mitochondria. Instead, they are equipped with hydrogenosomes, which may be derived from mitochondria.

Examples: *Balantidium, Entodinium, Ophryoscolex, Troglodytella*

Superclass **Cyrtophora**
Small, 1976

The four classes of the Cyrtophora are combined by their unique kinetid structure (see fig. **111**). Other common characters can be observed during cell division, when the oral apparatus of the proter dedifferentiates and two oral apparatuses are newly formed. Their dikinetids possess postciliary microtubules running in the direction of the cytostome.

Class **Phyllopharyngea**
de Puytorac et al., 1974

The ciliates belonging to this taxon have normal somatic monokinetids. Regarding the architecture of the infraciliature, the transverse microtubular ribbons are only rudimentary. The distinctly arranged kinetodesmal fibrils project laterally. Responsible for the formation of the

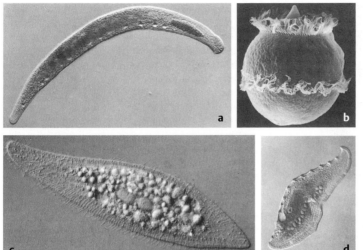

Fig. **122**   Litostomatea, Haptoria: **a** *Homalozoon*; **b** *Didinium*; **c** *Litonotus*; **d** *Loxophyllum* (b courtesy of E. Small, Maryland). Magn.: a 130×, b 200×, c 400×, d 100×.

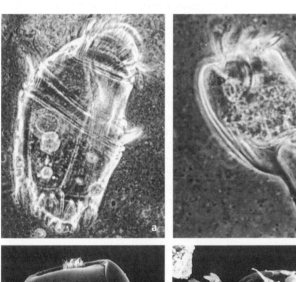

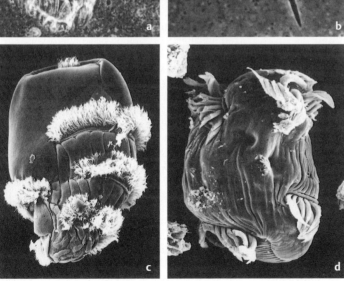

Fig. **123**   Litostomatea, Trichostomatia: **a** *Ophryoscolex*; **b** *Entodinium*; **c** *Troglodytella gorillae*; **d** *Tetratoxum parvum* (c and d courtesy of S. Imai, Tokyo). Magn.: a 250×, b 500×, c 300×, d 600×.

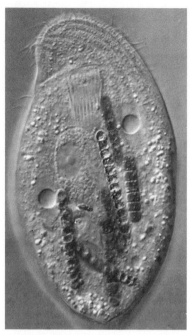

Fig. **124** Phyllopharyngea, Phyllopharyngia: *Chilo-donella* with ingested cyanobacteria (courtesy of D. J. Patterson, Sydney). Magn.: 750×.

longitudinally running kineties, are subkinetal microtubules running to the anterior or to the posterior end, and connecting neighboring kinetosomes with each other. In the region of the cytopharynx, leaflet-like orchestrated microtubular ribbons (phyllae) arise, according to which the taxon is named. These ribbons can be surrounded by nematodesmata emerging from kinetosomes and representing the framework of the basket-like ingestion apparatus (cyrtos). This morphological complex seems to be homologous with the corresponding structure in the Nassophorea. Three subclasses are recognized.

Subclass   Phyllopharyngia
              de Puytorac et al., 1974

Both free-swimming and sessile (or endobiotic) organisms belong to this taxon. The ciliature is present mainly on the ventral side, and the cells have a true oral apparatus. They also frequently have a dorsoventral flattening. *Chilodonella* (Fig. **124**) occurs mostly in detrital sediments.

Examples: *Chilodonella, Chitonella, Crebricoma*

Subclass   Chonotrichia
              Wallengren, 1895

In the adult and sessile representatives of this taxon, the body ciliature is missing. Only the juvenile swarming cells (swarmer), derived by budding from the mother cell, temporarily bear a holotrich ciliation used for migration to new habitats. The adults live, almost without exception, on crustaceans, such as on the gill plates of Gammaridae. They have a helically formed and funnel-like collar, covered inside with several ciliary rows. The cilia produce a water stream by which food particles are transported to the oral aperture (Fig. **125**).

Examples: *Heliochona, Spirochona, Stylochona*

Subclass   Suctoria
              Claparède & Lachmann, 1858

The suctorians represent an aberrant strongly modified taxon, but they are still affiliated with the ciliates by their life cycle characteristics (Fig. **126**). Like the Chonotrichia, the suctorians are ciliated only after separation, by endogenous or exogenous budding, from the sessile mother cell. During this short period, they swarm around and settle on new, mostly species-specific substrates. However, the internal characters, mostly discernible only in the electron microscope, such as infraciliature (kinetosomes and associated rootlets), alveoli, or nuclear dimorphisms, remain unchanged. Therefore, no doubts remain concerning the systematic affiliation of the suctoria.

The adult semaphoronts are predators specialized on ciliates or even on other suctorians. They are characterized by catching and feeding tentacles (Fig. **127**). The tentacles used for food ingestion bear a knob-like thickening at the top of each one. Here, haptocysts are localized serving for contacting, holding, and immobilizing the prey. The tentacles contain tube-like configurations of microtubules, which are homologous with the nematodesmal fibers of the cyrtos (Fig. **128**). Food transport is realized by microtubular activity, a phenomenon which superficially resembles a sucking act, as implied by the trivial name of suctorians, sucking infusoria. However, pressure gradients between prey and predator, which could explain the mechanical basis of the transport phenomena, can be refuted because the tube diameters are too small. The motive force is thought to be generated at the interface between microtubules and the newly formed food vacuole.

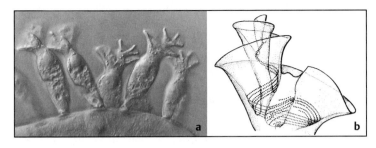

Fig. **125** Phyllopharyngea, Chonotrichia: **a** *Spirochona gemmipara* on gill plate of *Gammarus.* **b** complex oral region (a courtesy of J. Fahrni, Geneva; b from Grell: Unterreich Protozoa, Einzeller oder Urtiere. In: Lehrbuch der Speziellen Zoologie, 4th edition, ed. by H.-E. Gruner, Fischer, Stuttgart 1980). Magn.: a 200×.

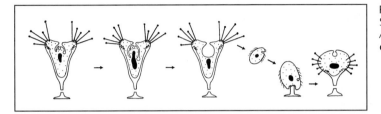

Fig. **126** Phyllopharyngea, Suctoria: metamorphosis in *Acineta* (adapted from Bardele).

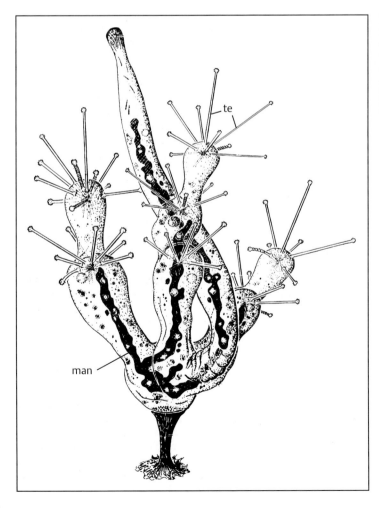

Fig. **127** Phyllopharyngea, Suctoria: branching trophont of *Dendrosomides grassei* which lives on the extremities of crabs. man = macronucleus, te = tentacles) (from Batisse: Protistologica 22 [1986] 11). Magn.: 750×.

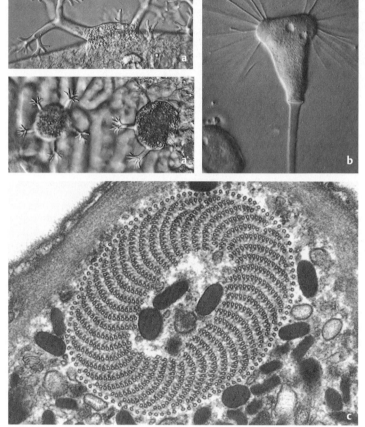

Fig. **128** Phyllopharyngea, Suctoria: *Dendrocometes* with branching tentacles (te) in lateral view (**a**) and dorsal projection (**a'**). **b** *Tokophrya* with tentacle bundle (teb). **c** inner structure of catching tentacle in *Acinetopsis* (b courtesy of J. Fahrni, Geneva; c from Grell and Meister: Protistologica 18 [1982] 67). Magn.: a 200×, (a') 150×, b 400×, c 36 000×.

The majority of the species occur in freshwater. Two species colonize the intestines of warm-blooded vertebrates.

Examples: *Dendrocometes, Discophrya, Ephelota, Tokophrya*

Class   **Nassophorea**
           Small & Lynn, 1981

The representatives of the nassophorid ciliates have a body ciliature with mono- or dikinetids. A typical characteristic is the presence of transverse microtubular ribbons oriented tangentially to the kinetosomes; in dikinetids, however, they occur only in connection with the anterior kinetosome. The kinetodesmal fibrils are normally well developed, overlapping themselves parallel to the course of the kineties. In the region of the oral apparatus, several nematodesmal fibrils are found, arising from the kinetosomes of the oral field and surrounding the cytopharynx. Sometimes the Nematodesmata form a basket-like cyrtos.

After exclusion of the Hypotrichia, which now are subordinated in the Spirotrichea, the taxon Nassophorea embraces only five orders, which are covered in the only subclass Nassophoria. Here only the most important orders are presented.

Order   Nassulida
           Jankowski, 1967

Most representatives are relatively large (>100 μm) and regular-densely ciliated. The cyrtos made of palisade-like arranged nematodesmata is clearly formed and easy to recognize with the light microscope (Fig. **129**).

Examples: *Furgasonia, Nassula*

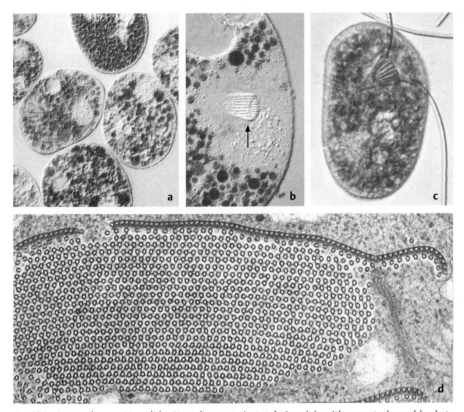

Fig. **129** Nassophorea, Nassulida: *Nassula ornata* in total view (**a**), with nematodesmal basket-like nasse or cyrtos (**b**, arrow) and during food uptake (**c**). **d** microtubular aggregates of nematodesmata. Magn.: a 100×, b 300×, c 200×, d 40 000×.

Order   **Microthoracida**
Jankowski, 1967

In contrast to the Nassulida, the normally slightly smaller Microthoracida possess relatively few ciliary rows, separated from each other by broad interkinetal gaps. The posteriorly positioned cyrtos is also well developed.
  Example: *Microthorax*

Order   **Propeniculida**
Small & Lynn, 1985

The two genera resemble the Microthoracida and therefore also have only few kineties. The tube-like cyrtos is the dominant intracellular organelle; it opens at the anterior region of the cell and extends for about one-third of the body (Figs. **236–238**). As in the Nassulida, the cyrtos serves for the ingestion of cyanobacteria.
  Examples: *Leptopharynx, Pseudomicrothorax*

Order   **Peniculida**
Fauré-Fremiet, 1956

The oral apparatus of the peniculid ciliates appears as a small, but very elastic slit, which is surrounded, on the left side, by three oral polykinetids (= peniculus) and, on the right side, by a sometimes reduced paroral dikinetid. This oral complex resembles the situation in some Oligohymenophorea; in the past, this analogous character suggested a direct evolutionary relationship between *Paramecium* (see Figs. **103 a** and **130**) and e. g., *Tetrahymena* (Fig. **131 a**). A distinct cyrtos is missing; however, some nematodesmal fibrils support the oral region. To the order belong large predatory species, but also several bacterivores.
  The best known member of the order is the slipper animalcule *Paramecium*. This is due to its wide distribution and easy cultivation.

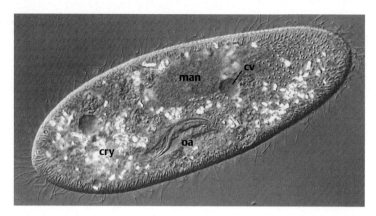

Fig. **130** Nassophorea, Peniculida: *Paramecium caudatum*. cry = crystals, cv = contractile vacuole, man = macronucleus, oa = oral apparatus. Magn.: 450×.

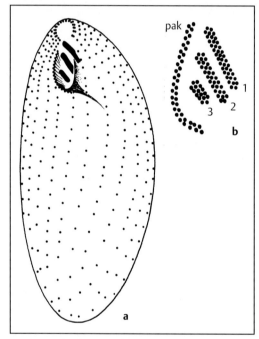

Fig. **131** Oligohymenophorea, Hymenostomatida: *Tetrahymena*. **a** somatic ciliature. **b** oral ciliature comprising three polykinities (1–3) and one undulating membrane (= paroral kinety) (pak). Magn.: a 900×.

Class   **Oligohymenophorea**
          de Puytorac et al., 1974

This taxon represents the most widely distributed class and contains the best known species. The scientific name is based on the presence of few (maximum three) oral polykineties, which are located at the left side of the oral apparatus where they, as a hymenoid velum, face an undulating membrane (Fig. **131 b**). Normally, the cytostome is situated at the bottom of a

more or less deeply invaginated ventral indentation of the cell surface. The somatic cilia are arranged mostly in longitudinal rows; they are of monokinetidal or dikinetidal origin. In contrast to the Nassophorea, the transverse microtubular ribbons run not tangentially, but radially from kinetosomes. The kinetodesmal fibrils are anteriorly directed, whereas the divergent postciliary ribbons run to the posterior end.

The life cycle is often characterized by polymorphic stages. For example, the following semaphoronts can appear sequentially: trophonts (feeding cells), tomonts (cyst-forming division stages), tomites (nonfeeding cells after division), and theronts (excysted migration stages, later transforming to trophonts). The class is divided into four subclasses.

Subclass   Hymenostomatia
             Delage & Hérouard, 1896

The ciliature of the oral apparatus of the hymenostome ciliates has a typical arrangement: The (normally three) oblique membranelles, formed as polykinetids, are associated on one side by a uni- to tripartite oral dikinetid (= undulating membrane = endoral membrane = paroral kinety). The membranelles draw food particles, which are lead by the undulating membrane to the cytostomal region (Fig. **131**). The body ciliation is holotrich. The subclass contains two orders.

Order   Hymenostomatida
          Delage & Hérouard, 1896

The characters of the Hymenostomatida include both single-segmented oral dikinetids and a mostly monokinetidal body ciliature. The life cycle normally includes polymorphic stages, e. g., in *Ophryoglena*, trophonts, pretomonts, to-

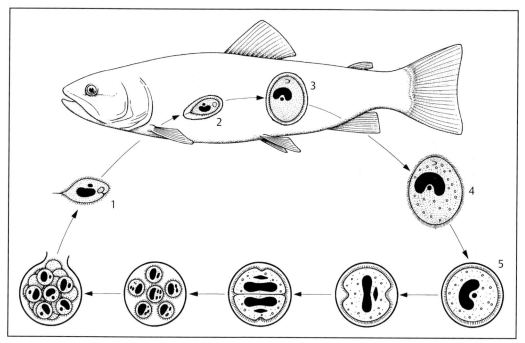

Fig. **132** Oligohymenophorea, Hymenostomatida: life cycle of *Ichthyophthirius multifiliis*: 1 swarmer; 2 encystment inside host tissue; 3 trophozoite cyst; 4 release; 5 encystment in the sediment and consecutive fissions (adapted from Mehlhorn).

monts, tomites, and theronts can be distinguished from each other.

The hymenostomatid ciliates have broad ecological distributions. Besides free-living species, they include also those which are ecto- or endobiotic, and besides ecologically important bactivores, there occur also parasitic histophages *(Ichthyophthirius)*, and predatory histophages *(Ophryoglena)*. Important representatives of the free-living forms are *Tetrahymena pyriformis* (Fig. **131**) and *T. thermophila*. Because of their rapid reproduction and easy cultivation, these species are especially favored objects for physiological, biochemical, and molecular biological investigations.

The causative agent of the so-called semolina grain disease of freshwater fishes, *Ichthyophthirius*, invades the skin epithelia of the host and forms here multiplication cysts (Fig. **132**). In advanced cases, severe illness and fish deaths occur. Sometimes aquarium fish also suffer from this disease.

Examples: *Colpidium, Ichthyophthirius, Ophryoglena, Tetrahymena*

Order   Scuticociliatida
               Small, 1967

In contrast to the Hymenostomatida, the somatic ciliature of the scuticociliates is composed of dikinetids, and the undulating membrane of three segments. In the light microscope, however, only few morphological differences are obvious. During stomatogenesis, a transient scutica, i. e., a hook-like or whip-lash configuration, appears, located near the posterior termination of the presumptive infraciliary base of the paroral membrane in both the proter and the opisthe.

Examples: *Cyclidium, Pleuronema, Uronema*

Subclass   Peritrichia
               Stein, 1859

The peritrich ciliates are presented by about 1000 species. They are characterized by an adoral ciliary ring made of a polykinety. It runs in a counter-clockwise helical turn from the brim of the oral field (= peristome) to the cytostome and splits, inside the oral cavity, into

Fig. **133**   Oligohymeno-phorea, Peritrichia, Sessilida: Colony of *Epistylis* exposing the oral ciliature (courtesy of H. Schneider, Landau-Godram-stein). Magn.: 250×.

three membranelles (Fig. **133**). The body cilia are more or less reduced and occur merely in the special configuration of an aboral girdle (telotroch). The content of the contractile vacuoles is discharged into the oral funnel and from there, released by the ciliary activity.

Order   Sessilida
Kahl, 1933

The stages involved in food uptake (trophonts) are normally sessile, and live on various sub-strates, sometimes also as epizoic symphoriants, on metazoans such as copepods, snails, or aquatic beetles. In many cases, e. g., in *Vorticella* (Fig. **134 a**), and *Carchesium*, contractile stalks are produced (Fig. **134 b**), and are used for sud-

den retraction in unfavorable situations. Besides these, there exist also numerous genera with rigid stalks in which only the cell bodies remain contractile (e. g., *Epistylis*). The stalks derive from a special organelle, the scopula.

Some sessile peritrichs live in loricae *(Vagini-cola, Cothurnia)* or within a gelatinous matrix *(Ophrydium*, Fig. **135**). Colonial societies are widely distributed (Fig. **136**). Both under un-favorable conditions, and during mitosis or con-jugation, a basal ciliary corona (telotroch) develops, enabling the cells, after separation from the stalk, to migrate and to settle new habi-tats. Multiplication takes place as a modified longitudinal fission or by budding.

The capacity to occur as mass developments and the possession of an effective filtration ap-

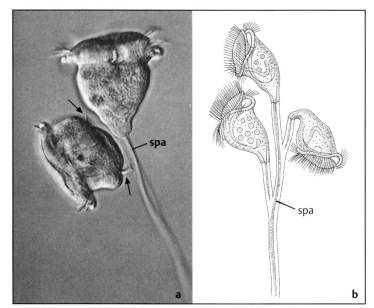

Fig. **134** Oligohymeno-
phorea, Peritrichia, Sessilida:
a *Vorticella* after fission,
daughter cell (left) still connected to the mother cell by
a little stem, but already with
a small ciliary ring or
telotroch (arrows) for independent swimming. **b** colony of
Carchesium. spa = spas-
moneme (b adapted from
Grell). Magn.: a 200×.

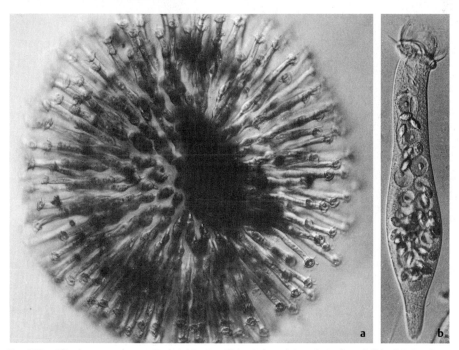

Fig. **135** Oligohymenophorea, Peritrichia, Sessilida: Colony (**a**) and single cell (**b**) of *Ophrydium*
(from Hausmann and Ostwald: Mikrokosmos 76 [1987] 129). Magn.: a 45×, b 250×.

paratus enable the Peritrichia to fulfill important
tasks in the biological purification of sewage.
    Examples: *Carchesium, Cothurnia, Epistylis,
Ophrydium, Vaginicola, Vorticella*

Order   Mobilida
         Kahl, 1933

The motile, but seldom free-swimming members of the Peritrichia are characterized by a per-

Fig. **136**   Oligohymenophorea, Peritrichia, Sessilida:
colony of *Zoothamnium arbuscula* (from M. Müller in
Foissner et al.: Taxonomische
und ökologische Revision der
Ciliaten des Saprobiensy-
stems, Band II: Peritrichia,
Heterotrichida, Odontostomatida. Bayer. Landesamt für
Wasserwirtschaft, München
1992). Magn.: 100×.

manent basal ciliary corona which is connected
to an intracellular adhesive apparatus (Fig. **137**).
This organelle affords the cell temporary fixation to the epithelial surface during its wandering over a host organism. The cells of *Trichodina
pediculus* live on freshwater metazoans such as
cnidarians, bryozoans, amphibian larvae, and
fish.

They are normally bacterivorous, but they
also sometimes damage epithelial cells due to
the action of their adhesive disc. *T. pediculus*, and
others, such as *T. cyprinis*, sometimes produce
severe ulcerations of the skin of the host, especially in the gill region. Later, this parasitism
can cause secondary infections by fungi or bacteria which lead finally to the so-called tricho-
diniasis, a disease typical of cultivated fish.

Examples: *Trichodina, Urceolaria*

Subclass   **Astomatia**
              Schewiakoff, 1896

The astomate ciliates are mouthless endobionts
(Fig. **138**). They live as harmless commensals in
the intestines of aquatic and terrestrial animals.

Particular hosts are annelids, but sometimes also
turbellarians, gastropods, and amphibians. The
systematic grouping within the Oligohymenophorea is due to homologies in the arrangement
of the somatic cilia.

Dissolved nutrients are absorbed over the
whole surface. Some genera have evolved complex adhesive devices. Cell fission takes place
mostly by budding, which leads occasionally to a
chain of several adhering cells.

Examples: *Anoplophrya, Haptophrya, Radiophrya*

Subclass   **Apostomatia**
              Chatton & Lwoff, 1928

The apostomate ciliates occur as epi- and endobionts of crustaceans, annelids, and coelenterates. The cytostome-cytopharynx apparatus is
strongly reduced. Only in the tomites three
polykinetids can be found. Another specific
character is the so-called rosette, an ultrastructurally very complex organelle of unknown function. The body cilia are arranged in helical rows
(Fig. **139**).

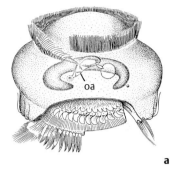

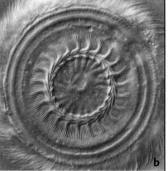

a

b

Fig. **137** Oligohymeno-
phorea, Peritrichia, Mobilida:
Trichodina. **a** general organization; **b** adhesive apparatus
(suction disc); **c** cross section
with contractile vacuole (cv),
food vacuoles (fv), macronucleus (man), micronucleus
(min) and oral apparatus (oa)
(a adapted from Mackinnon
and Hawes; c from Hausmann
and Hausmann: J. Ultrastruct.
Res. 74 [1981] 131). Magn.:
a 800×, b 1000×, c 1200×.

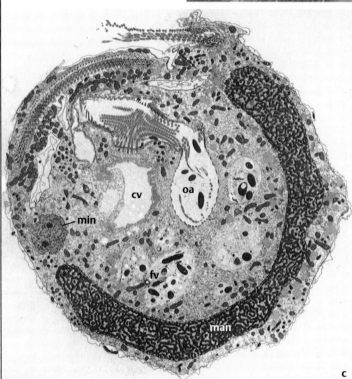

c

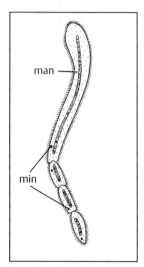

Fig. **138** Oligohy-
menophorea, Asto-
matia: chain forma-
tion in *Radiophrya*.
man = macronu-
cleus, min = micronu-
cleus (adapted from
Kudo). Magn.: 75×.

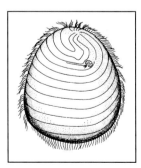

Fig. **139** Oligohymeno-
phorea, Apostomatia:
Foettingeria (from Corliss:
Trans. Amer. Microsc. Soc.
97 [1978] 419). Magn.:
300×.

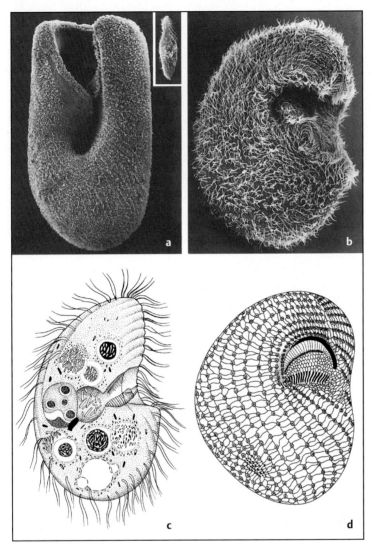

Fig. **140**   Colpodea: **a** *Bursaria truncatella* (with *Paramecium* for comparison); **b** *Bresslauides australis*; **c** *Colpoda* (only the peripheral cilia are drawn); **d** *Hausmanniella* (protargol impregnation) (a from Foissner: Class Colpodea (Ciliophora). In: Protozoenfauna, Vol. 4/1, ed. by D. Matthes, Fischer, Stuttgart 1993; b from Platt and Hausmann: Arch. f. Protistenk. 143 [1993] 297; c and d adapted from Foissner). Magn.: a 90×, b 480×, c 550×, d 500×.

During the complicated life cycle, endobiotic stages (trophonts, tomonts, and tomites) alternate with an ectobiotic stage (phoronts). An alternation of hosts is possible, but not obligatory.

Examples: *Foettingeria, Hyalophysa, Spirophrya*

Class   **Colpodea**
         Small & Lynn, 1981

The Colpodea are named according to their best known representative, the type-genus *Colpoda* O. F. Müller, 1773. Altogether, about 150 species can be combined in the taxon using only their ultrastructural characters (see Fig. **111**). Typical

characters of the somatic ciliature are the spiralling kineties, composed of dikinetids, and transverse microtubules running from the posterior kinetosome of the dikinetids to the posterior of the ciliate, and overlapping each other in a so-called Lkm fibril (or transversodesmal fibril). During stomatogenesis, the cilia of the oral apparatus derive from somatic cilia of the mother cell. Mitosis occurs predominantly inside cysts.

According to their shape, with a mostly kidney-like outline, the Colpodea represent a relatively homogenous taxon (Fig. **140**). However, the oral apparatus is extremely variable, and a number of genera resemble other types of ciliates and can be easily confused with them. The

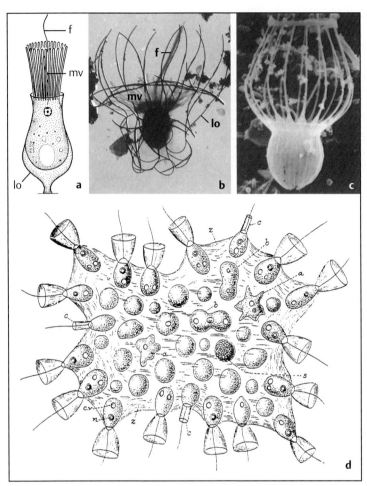

Fig. **141** Choanoflagellata: **a** *Salpingoeca amphoroideum* within lorica (lo); **b** *Acanthoecopsis apoda* with lorica composed of siliceous needles; f = flagellum, mv = collar-forming microvilli. **c** *Diplotheca costata*; **d** social colony of about 40 cells of *Proterospongia haeckeli* (a adapted from Grell; b and c courtesy of B. S. C. Leadbeater, Birmingham; d from Kent: Manual of the Infusoria. Volume III. London 1882). Magn.: a 1500×, b 1800×, c 2500×, d 800×.

large (up to 2 mm) *Bursaria truncatella* was long believed to be member of the spirotrich ciliates.

The colpodid ciliates live predominantly in terrestrial or freshwater habitats. As food organisms, they take, depending on the size of the predator, bacteria, spores, other protozoa, and even rotifers.

Examples: *Bresslaua, Bursaria, Colpoda, Hausmanniella*

Phylum **Choanoflagellata**
Kent, 1880

The collar flagellates are uninucleated and small, seldom more than 10 μm in size. They occur both as sessile individual cells and as suspended colonies in marine or freshwater habitats. Their most important character, an anteriorly projecting basket-like collar made of dozens of slender microvilli, is interpreted as an apomorphic character not found in other protozoan groups (Fig. **141**). Only a single flagellum is present, the second (obviously formerly present) flagellum is reduced with the exception of its corresponding kinetosome. The flagellum projects above the brim of the collar and produces a water current by which food particles are transported to the outside of the microvilli. The sieved particles move down and are ingested at the base of the microvilli or at the apex of the cell body. Choanoflagellates are quantitatively important filter feeders of suspended bacteria in the world's oceans.

A number of sessile species live as solitary *(Monosiga)* or as colonial branch systems *(Codonocladium)*. Some live in stalked or stalkless loricae *(Salpingoeca)*. In marine species, the loricae are sometimes built from a framework of silicious rodlets (costae) (Fig. **141 b**).

The similarities in construction and function

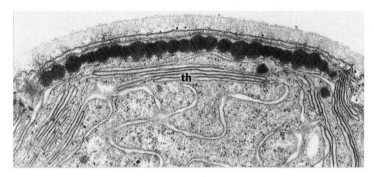

Fig. **142** Eyespot or stigma inside the chloroplast of *Chlorogonium*. th = thylakoids (courtesy of D. Fischer-Defoy, Wiesbaden). Magn.: 35 000×.

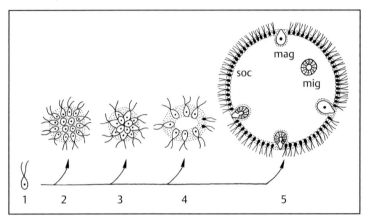

Fig. **143** Phytomonadea, Volvocida: Possible evolutionary pathway from monad flagellates such as *Chlamydomonas* (1) via *Gonium* (2), *Pandorina* (3) or *Eudorina* (4) to large colonies such as *Volvox* (5). Cells in *Eudorina* and *Volvox* drawn black represent somatic cells (soc). mag = macrogametes, mig = microgametes (adapted from Pickett-Heaps).

of the collars of choanoflagellates and choanocytes of sponges lead to the assumption of the presence of a synapomorphic character. Other arguments, used to support the theory of a common origin, are the abilities in both animal taxa to metabolize silica and to discharge osmotically invading water using contractile vacuoles. A recent "missing link" could be the occurrence of spherical colonies in *Sphaeroeca volvox* (diameter about 300–500 μm) or in *Proterospongia* (Fig. **141 d**). Such colonies would represent, if the proof of an epithelial arrangement of the cells could be supported, the Haeckelian blasteae, and therefore organisms forming the bridge between unicells and metazoans. Regarding the direction of the putative phylogenetic history, however, it should be mentioned that longer living fragments of true sponges (so-called reductia) have been described as choanoflagellates, so the evolution of recent choanoid flagellates could be the result of a retrograde development starting from metazoans. On the other hand, convergent development (analogy) could have occurred, caused by the evolutionary pressures of a sessile way of life and by similar strategies in filtration technique.

Examples: *Codosiga, Parvicorbicula, Salpingoeca*

Phylum   **Chlorophyta**
Pascher, 1914

Only the monadial (predominantly flagellated) representatives of this phylum, which as a whole comprise also the green algae and the embryophytes, are normally considered in a protozoological context. They differ from the other so-called phytoflagellates both by the secondary isokont and mostly flimmerless flagellation, and by the presence of chloroplasts with the same characters as in higher terrestrial plants (two covering membranes, chlorophylls a and b, thylakoids or thylakoidal stacks [= grana], and pyrenoids). The stigma, if present, is located in chloroplasts (Fig. **142**). If extracellular cell walls are formed, these consist predominantly of cellulose fibrils embedded in a polysaccharide matrix (pectin). Freshwater species without cell walls possess contractile vacuoles. Sexual reproduction is very common.

Order   **Volvocida**
Francé 1894 (within the class Phytomonadea Blochmann, 1895)

As the only permanently flagellated subtaxon
within the class, the order Volvocida must be
considered. Two, four, or (seldom) eight
amastigonemate flagella arise from the anterior
pole of the cell. The mostly cup-shaped chloroplast has the typical parietal position, and is responsible for the characteristic grass-green
coloration. The order comprises both unicellular
representatives *(Chlamydomonas, Chlorogonium),* and species with a conspicuous tendency to
multicellularity. Regarding the several types of
colonial societies, colonies with increasing numbers of cells would include *Gonium, Eudorina,
Pleodorina, Volvox* (Fig. **143**). The individual cells
are located inside a common gelatinous matrix,
and are frequently connected by cellular projections (plasmodesmata). Most of the species, occurring predominantly in freshwater, have a tendency toward mass development. Some species
are secondarily colorless and live as heterotrophic predators, such as *Polytoma* and *Hy-
alogonium.*
    Examples: *Chlamydomonas, Chlorogonium,
    Dunaliella, Pandorina, Volvox*

Class   **Prasinomonadea**
Christensen, 1962

In contrast to the Volvocida, the one to eight, but
mostly four isokont flagella of the prasinomonads arise from an apical invagination of the
cell body (Fig. **144**). The flagella are covered by
additional small and fine-sculptured scales, or
hairs of organic materials which are not comparable   with   mastigonemes   (Fig. **145**).
Moreover, the somatic cell surface also bears one
or more layers of such submicroscopic scales
that are produced or elaborated in the cisternae
of dictyosomes. In some cases, as in *Tetraselmis,*
the secreted scales produce a compact cover
(theca). Some species have extrusomes comparable to the ejectisomes of the cryptomonads.
Prasimonads are found mainly in marine environments; only a few species occur in freshwater. One species *(Tetraselmis convolutae)* lives
as a symbiont in the tissue of the marine platy-
helminth *Convoluta roscoffiensis,* which incorporates the symbiont in early stages of its ontogeny and which later develops a complete physiological dependency on its guest.
    Examples: *Mesostigma, Nephroselmis, Prasino-
    monas, Tetraselmis*

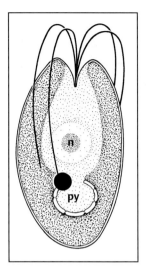

Fig. **144** Prasino-
monadea: *Tetraselmis
bichlora.* n = nucleus,
py = pyrenoid
(adapted from Ettl).
Magn.: 1700×.

METAKARYOTA INCERTAE SEDIS

(Groups which need to be added in the future to
other presently unknown metakaryotic taxa.)
    The groups, which are described in the follow-
ing chapters, have in general only one common
character which they share with the other
metakaryotes: They have mitochondria. This
character is probably a plesiomorphic character,
but these groups can not be subordinated with
certainty to one of the taxa already presented because they do not generally have everlasting
flagella, or flagellar root structures comparable
with those of the aforementioned taxa.
However, some groups could be affiliated with
the flagellate taxa because of the occurrence of
flagellate stages in their life cycle. This has already been done with some lower groups of the
former Sarcodina. The same could also be done
with the Foraminifera or the Heliozoa.
    The presence of pseudopodia, occurring in
the different subtaxa of the former Sarcodina
(see Figs. **22** and **146**), is increasingly seen as the
result of convergent evolution. It is also now regarded as a general eukaryotic property. We cannot use this as a principal character to recognize
monophyletic taxa other than in very small
groups.
    Doubts exist about the monophyletic origin of
the corresponding groups and they are designated with an asterisk (*).

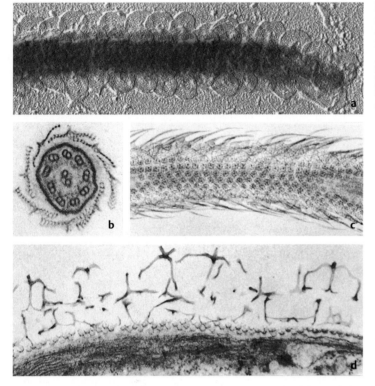

Fig. **145** Prasinomonadea: scales on flagellar (**a–c**) and somatic surface (**d**). a *Mantoniella squamata*; b *Mamiella gilva*; c and d *Pyramimonas tetrarhynchus* (courtesy of ø. Moestrup, Copenhagen). Magn.: a 45 000×, b 65 000×, c 65 000×, d 60 000×.

1. Amoebozoa*
Lühe, 1903

The protozoa combined in this polyphyletic group have pseudopodia which, so far investigated, never contain microtubules as stiffening or motility-mediating elements. Also, cytoplasmic microtubules seem to appear only in connection with the mitotic spindle apparatus. The pseudopodia are formed relatively quickly (seconds to minutes). The amoeboid locomotion is performed both by alternating new formation and retraction of pseudopodia, and by the continuous prolongation of the same pseudopodium. Flagella and centrioles are missing. Sexual reproduction is known in some cases. Multiplication takes place by mitotic binary fission, by multiplicative fission (schizogony), or by plasmotomy. Despite the conspicuous mutability of shape, morphologically recognizable types can be distinguished.

Lobosea*
Carpenter, 1861

The members of this group have lobate or tube-like and sometimes also pointed pseudopods (lobopodia) occurring singly, as in monopodial forms, or in greater numbers, as in polypodial forms. There are species both with and without shells (tests). Encysted stages seldom occur. Two subgroups can be separated from each other: Gymnamoebia (atestate genera) and Testacealobosia (testate genera).

Gymnamoebia*
Haeckel, 1862

Gymnamoebae represent naked cells without conspicuous extracellular structures (shells, scales). They inhabit all aquatic, terrestrial, and many endobiotic biotopes (Fig. **147**).

Locomotion is performed both by directed cytoplasmic streaming and by rolling motion. The size of the uni- or multinucleated cells ranges from only a few micrometers, as in small species of *Vannella*, to up to 5 millimeters, as in *Chaos carolinense*. The best known and well-investigated species is *Amoeba proteus*.

Some forms can be dangerous to humans, but normally only under specific nonhygienic circumstances. This is especially the case for the genus *Acanthamoeba*, which is present as harmless strains living as bacterivores in freshwater

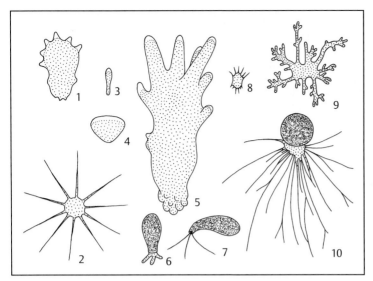

Fig. **146** Pseudopodial types in naked and testate (= dark-drawn) amoebae. 1 conical *(Mayorella)*; 2 radiating *(Actinophrys)*; 3 lobopodial, monopodial *(Saccamoeba)*; 4 lamellipodial *(Vannella)*; 5 digitate, polypodial *(Amoeba)*; 6 lobopodial *(Nebela)*; 7 filiform *(Cyphoderia)*; 8 filiform *(Nuclearia)*; 9 branching *(Stereomyxa)*; 10 netlike *(Allogromia)*.

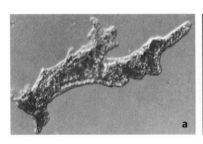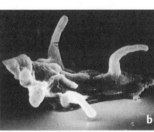

Fig. **147** Lobosea; *Amoeba proteus* by light (**a**) and by scanning electron microscopy (**b**). Magn.: 180×.

or moist soil, but also with strains appearing as opportunistic parasites exploiting deficiencies in the human immune system. They can evoke severe encephalitis (granulomatous amoebic encephalitis or acanthamoebiasis). The causative agent of amoebic dysentery, *Entamoeba histolytica,* is a common problem in tropical countries. The species lives as a harmless, so-called minuta form in the colon, but it is also able to transform itself into the malign magna form, to invade the intestinal tissues, and to cause severe ulcers there. The most conspicuous symptom is diarrhea, combined with fever and exhaustion. The infection is acquired by cysts ingested with feces-contaminated food.

Examples: *Acanthamoeba, Amoeba, Chaos, Entamoeba, Mayorella, Vannella*

Testacealobosia*
de Saedeleer, 1934

This group represents lobopodial Amoebozoa of freshwater, terrestrial, and marine habitats, whose cell bodies are partially covered with a shell or at least with a complex coating material, such as a layer of scales (Fig. **148a**).

The tests or envelopes, which consist of organic materials sometimes intermingled with inorganic matter and foreign bodies such as sand grains (Fig. **148b**) or diatom scales, bear a single aperture (= pseudostome), used for the projection of one or more lobose pseudopodia (Fig. **149**). A number of species live in moist soils or mosses. In such habitats, the shells function as a protection device against desiccation and mechanical stress.

Examples: *Arcella, Difflugia, Hyalosphenia*

Acarpomyxea
Page, 1976

The representatives of this taxon are uninucleated and organized as plasmodia. Some resemble slime molds in their outlines. They are always strongly branched and therefore extremely polymorphic (Fig. **146 (9)** and **150**). The formation of pseudopodia requires much time, so that only after several hours changes in body shape

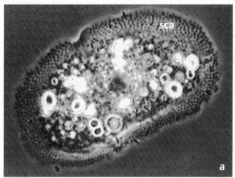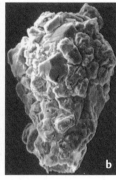

Fig. **148**   Lobosea:
a *Cochliopodium* with a layer of scales (sca); **b** shell of *Difflugia* (a courtesy of D. J. Patterson, Sydney). Magn.: a 800×, b 450×.

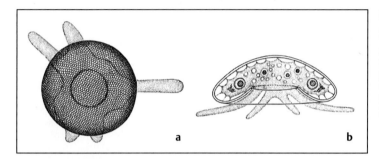

Fig. **149**   Lobosea: *Arcella vulgaris* in dorsal (**a**) and side view (**b**) (from Grell: Unterreich Protozoa, Einzeller oder Urtiere. In: Lehrbuch der Speziellen Zoologie, 4th edition, ed. by H.-E. Gruner, Fischer, Stuttgart 1980). Magn.: 220×.

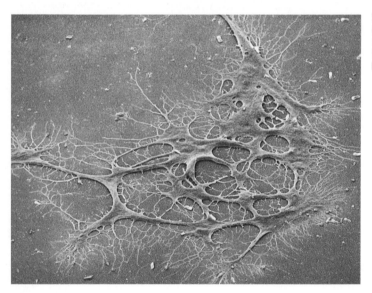

Fig. **150**   Acarpomyxea: Plasmodial network of *Thalassomyxa australis* (from Grell: Protistologica 21 [1985] 215). Magn.: 125×.

can be recorded. Neither shells, nor fruiting bodies, have been discovered. Knowledge of the biology of the Acarpomyxea is still fragmentary.

Examples: *Corallomyxa, Leptomyxa, Stereomyxa*

Filosea*
Leidy, 1879

Belonging to this group are both atestate and testate members, designated as Aconchulinida and Gromiida, respectively. A common, but apparently convergently derived character, is the presence of filiform pseudopodia (= filopodia)

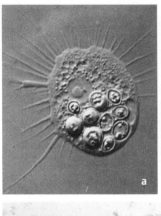

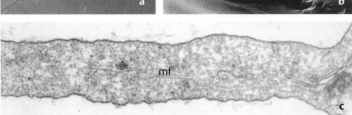

Fig. **151** Filosea: **a** *Nuclearia*; **b** shell of *Euglypha*; **c** filopodium of *Vampyrella lateritia* in longitudinal section. mf = microfilaments (a courtesy of D. J. Patterson, Sydney; b courtesy of K. Rauenbusch, Erlangen). Magn.: a 500×, b 750×, c 25 000×.

(Fig. **151**). They differ from the pseudopods of Lobosea and Acarpomyxea by their extremely small diameter, and a fibrillar axis made of 6 nm-filaments, possibly consisting of actin. Filopodia are normally formed and retract very rapidly, partially within seconds. With their motile activity, they draw the cell body forwards. Another morphological phenomenon, possibly homologous with the filopodia, are conspicuous, delicate, velum-like anterior pseudopodia. These so-called lamellipodia serve as bell-shaped projections for food ingestion or, as flat-spread projections with a ruffling frontal zone, for slow movements. In the latter case, they resemble the pseudopodial extensions of metazoan fibroblasts. In some groups, such as the Vampyrellidae, cyst formation occurs regularly. Filosea inhabit marine, freshwater, and terrestrial biotopes.

Examples: *Arachnula, Euglypha, Gromia, Hyalodiscus, Nuclearia, Vampyrella*

2. Granuloreticulosa*
de Saedeleer, 1934

The representatives of this putatively polyphyletic supergroup form a branching and anastomosing network of pseudopodia, with diameters decreasing continuously to less than 1 μm from the center to the periphery (Fig. **152**). As stiffening structural elements, microtubules occur both singly and arranged in loose clusters (see Fig. **23**). A particular type of motion is performed inside the pseudopods; a simultaneous bidirectional streaming normally including all cell organelles with the exception of the nuclei. The cells contain one to several, and sometimes numerous nuclei integrated into thicker centrally located protoplasmic strands or are found in regions covered by a shell. According to the presence of shells and their construction type, the following three subgroups are distinguished from each other.

Athalamea*
Haeckel, 1862

These Granuloreticulosa are atestate or athalam, but they sometimes have coverings interpretable as remnants of shells. Large species can occur, such as the marine species *Pontomyxa flava* (>10 mm), or the freshwater species *Reticulomyxa filosa* (several centimeters) The pseudopodia are multiple branched (Fig. **152**).

Examples: *Biomyxa, Pontomyxa, Reticulomyxa*

Monothalamea*
Haeckel, 1862

The main characteristic of this group is single-chambered organic shells, often with additional calcified material. Foreign materials can be included. An alternation of generations is not known. The group members occur in marine, freshwater, and terrestrial habitats (Fig. **153**).

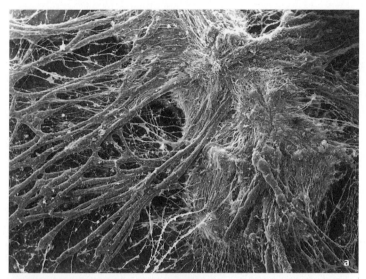

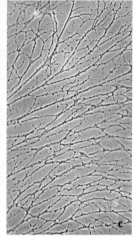

Fig. **152**   Granuloreticulosa, Athalamea: plasmodia of *Reticulomyxa filosa*. **a** spatial distribution of reticulopodia; **b** nucleated veins of the central region; **c** nuclear-free pseudopodia from the periphery. Magn.: a 70×, b 170×, c 300×.

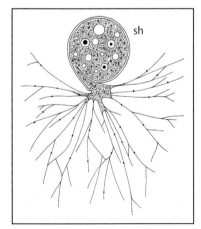

Fig. **153**   Granuloreticulosa, Monothalamea: *Lieberkühnia*. sh = shell. Magn.: 160×.

Examples: *Amphitrema, Lieberkühnia, Microgromia*

Foraminiferea
d'Orbigny, 1826

The foraminifers represent a large taxon. Their often calcified shells can be fossilized, so that presently more than 30000 ancient species are known. The number of recent marine species is about 4000. Most are benthic, some are planktonic. Individuals can reach an age of several months or even years.

The often finely perforated shells of foraminifers are single- or multichambered. The pseudopods emerge through the little pores and the larger pseudostome. The shells are often chambered, and additionally, the walls can ex-

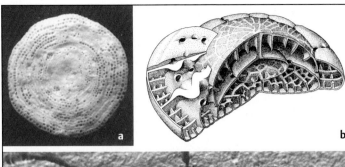

Fig. **154** Granuloreticulosa, Foraminiferea: **a** sightly scraped shell; **b** shell morphology of *Orbitolina*; **c** araldite cast of cavities in the shell of *Parrellina hispidula* (b from Hottinger, in R. H. Hedley, C. G. Adams: Foraminifera, vol. I–III. Academic Press, London 1974–1978, c from Hottinger: Schweizerische Paläont. Abh. 101 [1980] 115). Magn.: a 30×, c 200×.

hibit an intrinsic labyrinth of channels (Fig. **154**). Chambers and channels are filled with protoplasm.

The fossils of the foraminifers serve as very reliable indicators in stratigraphic investigations, such as in oil exploration. This explains the wealth of knowledge concerning taxonomic details. According to characters in the structure of shells, the Foraminiferea are divided into five orders: Allogromida, Textularida, Fusulida, Miliolida, and Rotaliida (Fig. **155**). The following fundamental construction features are distinguishable:

1. Organic, often single-chambered shells
2. Shells made of various inorganic and organic particles (e. g., skeletal needles of sponges) embedded in an organic matrix. To this type belong species up to several centimeters in size. Also, the Xenophyophorea (a little known taxon of fist-sized deep-sea organisms) can be grouped here
3. Calcified shells, normally with numerous chambers

The diversity of the shells is mostly based on differences regarding the chambers, which remain connected to each other by a foramen. In the *Nodosira* type they are ordered in a longitudinal row, and arranged helically in the *Rotalia* type. In the case of a plane spiral architecture, the shell is designated as planspiral; in a helical arrangement, the formation is called trochospiral. In the *Textularia* type, the chambers are ordered in a cable-stitch pattern, in two or three rows. The *Planorbula* type exhibits chambers which are ordered, from inside to outside, in more or less concentric circles.

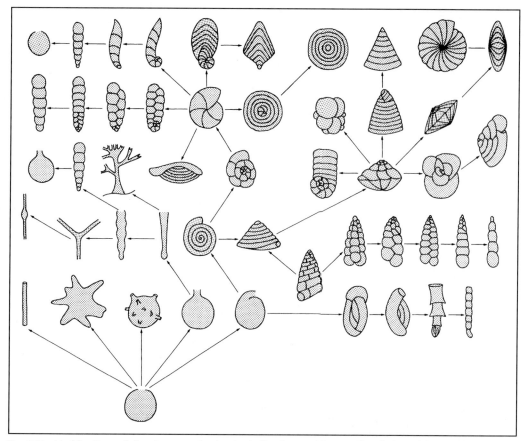

Fig. **155**   Modifications of chamber organization in shells of several foraminifers (adapted from Pokorny).

Symbiotic zoochlorellae or zooxanthellae occur frequently in both pseudopodia and the core cytoplasm. In such cases, the shells are normally transparent, and the organisms live near the surface in illuminated areas. The symbiont-free foraminifers feed on detritus, protozoans, and small metazoans. It is not clear whether they play an important role in the food chains of their habitats.

Sexual and asexual generations are characteristic of the life cycles of the better investigated species. Although proof is not available in all cases, the assumption is made that a heterophasic alternation of generations (Fig. **156**) is a general character of the foraminifers. The gametes are normally biflagellate. In some stages of the life cycle, a nuclear dualism is present.

Examples: *Allogromia, Hastigerina, Orbitulina, Rotaliella, Schwagerina, Textularia, Triloculina*

3. Actinopoda*
Calkins, 1902

The pseudopodia of all genera in this polyphyletic taxon are axially stiffened by bundles of highly ordered microtubules; they are called axopodia (Fig. **157**). The arrangement of the microtubules is specific for each subgroup, ranging from simple to geometrically complex patterns (Fig. **158**). Axopodia are formed relatively slowly. They are arranged in a radial symmetry and might bear extrusomes (e. g., kinetocysts). Axopodia serve both for locomotory or food capturing purposes and as projections which aid continued suspension in water for passive distribution.

With the exception of the heliozoans, which occur also in freshwater and in mosses *(Sphagnum)*, the actinopods are exclusively marine planktonic organisms. Important characters, both for systematic and determination, are inorganic scales, skeletons, and needles (Fig. **159**).

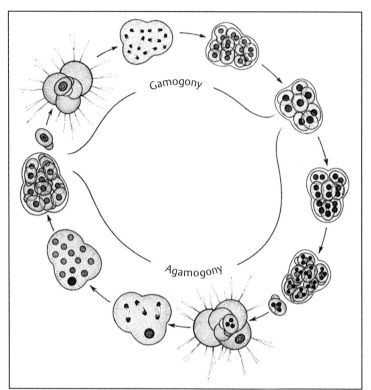

Fig. **156** Foraminiferan life cycle illustrated for *Rotaliella* (from Grell: Unterreich Protozoa, Einzeller oder Urtiere. In: Lehrbuch der Speziellen Zoologie, 5th edition, ed. by H.-E. Gruner, Fischer, Stuttgart 1993).

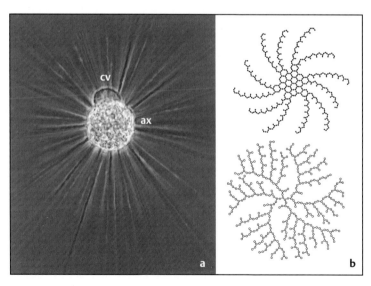

Fig. **157** Actinopoda: **a** *Actinophrys sol* with radiating axopodia (ax); cv = contractile vacuole. **b** taxon-specific patterns of microtubular aggregates in polycystines (b from Cachon and Balamuth in Hutner (ed): Proc. V. Int. Congr. Protozool. 1977. Magn.: 300×.

In the past, the Actinopoda were divided into the Heliozoa and the Radiolaria. However, new investigations on fine structural organization, development, and chemical composition of skeletal elements have shown a diversity which leads to rejection of these taxa. The name Radiolaria survives in the designation for the mineral radiolarite. This silicon composite has been produced since the Precambrian Era, mostly by Polycystinea (comp. Figs. **161** and **162**) and it appears in a diagenetically modified form as crystalline quartz.

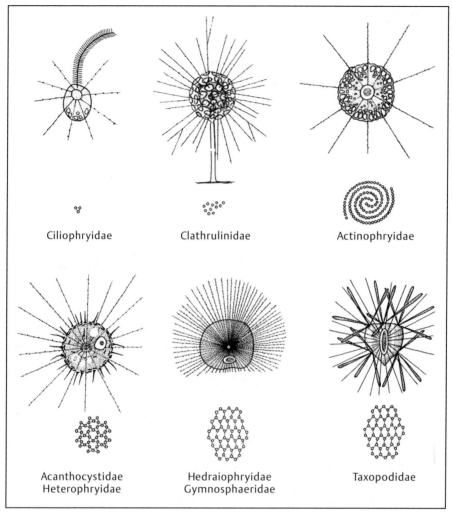

Ciliophryidae   Clathrulinidae   Actinophryidae

Acanthocystidae   Hedraiophryidae   Taxopodidae
Heterophryidae   Gymnosphaeridae

Fig. **158**   Arrangement and patterns of microtubules in different heliozoan taxa of family rank (courtesy of C. F. Bardele, Tübingen).

The subdivision of actinopods into four subgroups has no phylogenetic significance, but represents rather a pragmatic response to the lack of knowledge.

Acantharea
Haeckel, 1881

Acantharea (50 µm–1 mm) have 10 or 20 radially arranged spikes (spicula) made of strontium sulphate (celestite) (Fig. **160**). This mineral is soluble in seawater, so no acantharean fossils exist. The central endoplasm normally contains numerous nuclei and most of the other organelles. A perforated extracellular capsule of predominantly fibrillar material surrounds the endoplasm and the basal parts of spicula; the axopodia and some other cytoplasmic strands project to the outside through the pores. The cytoplasmic strands are connected with an ectoplasmic layer with many lacunae, which is covered by a periplasmic cortex, also perforated for the projection of axopods. Via contractile myophriscs (= myonemes), the extracellular cortex communicates with such regions of the spicula that are surrounded by cytoplasm. The axopodia are characterized by a hexagonally patterned microtubular axis. The endoplasm often contains zooxanthellae.

Fig. **159**   Skeletons of spumellarids and nasselarids in a sediment from Barbados island. Magn.: 200×.

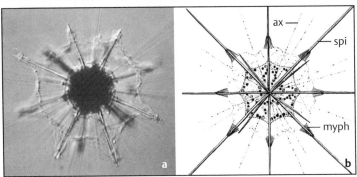

Fig. **160**   Acantharea: **a** living cell. **b** general organization. ax = axopodiaum, myph = myophrisk, spi = spiculum (a courtesy of M. Kage, Weißenstein; b from Grell: Unterreich Protozoa, Einzeller oder Urtiere. In: Lehrbuch der Speziellen Zoologie, 4th edition, ed. by H.-E. Gruner, Fischer, Stuttgart 1980). Magn.: a 40×.

Cysts and flagellated swarmers represent two stages of the life cycle, which are not completely known at present. The most favored subdivision into the four orders; Holocanthida (10 spicula), Symphyacanthida, Chaunacanthida, and Arthracanthida (each with 20 spicula) is predominantly based on the different patterns in the basal connections of spicula.

Examples: *Acanthocolla, Acantholithium, Acanthometra, Amphiacon*

Haeckel, Kunstformen der Natur.   Tafel 31 — *Cahegelas.*

Cyrtoidea. — Flaſchenſtrahlinge.   BRESLAUE

Fig. **161** Polycystinea: skeletons of Nasselerida. (from Haeckel: Kunstformen der Natur [table 31], Verlag Bibliogr. Inst., Leipzig and Wien 1899–1904).

Polycystinea
Ehrenberg, 1838

Solitary cells (30 μm–2 mm), as well as meter-large colonies, are characterized by elaborate silicious skeletons, made of needles or regularly perforated globes which fossilize easily. When occurring in vast numbers, the skeletons may be nested (Fig. **161**). The two cytoplasmic areas, the nucleated and optically dense endoplasm, and the peripheral, cleft, and often symbiotic algae-bearing ectoplasm, are separated by an intracellular and more or less distinct central capsule of mucoproteins. The capsule, not to be confused with the skeletons, consists of individual polygonal plates, which are fringed by cytoplasmic fissures through which endo- and ectoplasm communicate with each other. The plates of the central capsule exhibit complicated apertures (fusules), which bear the axoplasts or allow the traversing of the axopodial microtubules from endo- to ectoplasm. The microtubular bundles are linked crosswise, arranged in a blade-wheel pattern, or ordered in hexagonal six-partite and twelve-partite groups; often they reach deeply into the endoplasm, sometimes even into nuclear cavities. Besides axopodia, filopodia also occur.

Cysts and biflagellate swarmers appear during the life cycle of the better investigated polycystinea; sexual stages are unknown.

The figures of most representatives demonstrate exceedingly delicate skeletal elements. However, in living specimens, these skeletons

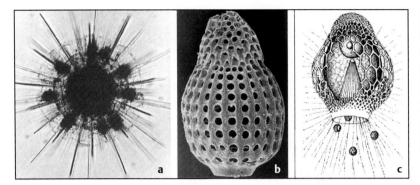

Fig. 162 Polycystinea: **a** living spumellarid. **b** skeleton of a nasselarid. **c** *Cyrtocalpis urceolus* (a courtesy of M. P. Kage, Weißenstein; c from K. G. Grell: Protozoology. Springer, Heidelberg 1973). Magn.: a 150×, b 300×, c 200×.

are covered from view by dense cytoplasm (Fig. **162 a**).

As much as possible, the classification is based on the ultrastructural data of axoplasts and central capsules. Since the era of Haeckel, the architecture of skeletons has played an important role. The order Spumellarida contains radially symmetrical forms with central capsules equally perforated on all sides. Also belonging to this taxon are the colonies of hundreds of individuals combined in a gelatinous matrix. The genera of the order Nasselarida have a conical basket-like skeleton with one main aperture (Fig. **162 b, c**).

Examples: *Collozoum, Eucoronis, Thallasicolla*

Phaeodarea
Haeckel, 1881

The mostly hollow skeletal needles and shell of the Phaeodarea consist of amorphous silica with the addition of organic components and traces of magnesium, calcium, and copper. They also have a central capsule; otherwise an envelope made of foreign materials. The biology of these organisms, occurring in deep-sea regions (bathypelagial), is poorly known.

The central capsule has three apertures: one astropyle, functioning as an oral apparatus, and two opposite parapyles, through which the axopodia project. In front of the astropyle, a yellow-brown mass of pigment is located, named the phaeodium (Fig. **163**), and which is possibly involved in silicon metabolism. The seven orders, Phaeogymnocellida, Phaeocystida, Phaeosphaerida, Phaeocalpida, Phaeogromida, and Phaeoconcida, were predominantly erected according to differences in skeleton morphology.

Examples: *Astracantha, Aulacantha, Coelodendrum, Phaeodina*

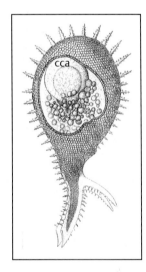

Fig. **163** Phaeodarea: *Challengeron wyvillei* with central capsule (cca) and phaeodium (from Grell: Protozoology, Springer, Heidelberg 1973). Magn.: 170×.

Heliozoea[*]
Haeckel, 1866

To this polyphyletic taxon belong actinopods which, because of their appearance, are often referred to as sun animalcules. At least five subgroups can be differentiated, each with their characteristic mode of constructing the axopodial microtubular bundles (see Fig. **158**) and with a characteristic number and location of the microtubules organizing centers (MTOC).

Actinophryida
Kühn, 1926

In cross section, the axopodial microtubules are arranged as two intermingled spirals (see Fig. **158**). The place of origin is always the nuclear envelope (Fig. **164**). Actinophryids are able to encyst. Within the cyst, the mother cell divides into two gametes, which undergo fusion

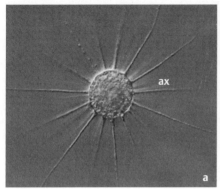

Fig. **164** Heliozoea: **a** *Actino-phrys sol* with axopodia (ax). **b** axopodial microtubules originating from the nuclear envelope (ne). (63). Magn.: a 280×, b 40 000×.

to a zygote. This kind of autogamy is called pedo-gamy (comp. Fig. **259**). The uni- or multinu-cleated cells are marine or freshwater. In peat mosses *(Sphagnum)* terrestrial species can be found.

Examples: *Actinophrys, Actinosphaerium, Camptonema*

Desmothoracida
Hertwig & Lesser, 1874

The cell body of these sessile heliozoans is sur-rounded by a perforated capsule. This envelope consists of organic materials or silica and is usu-ally stalked. The axopodial microtubules form an irregular bundle (see Fig. **158**). Besides axopodia, filopodia are also produced. During sexual repro-duction, uni- or biflagellate swarming cells ap-pear that move in an amoeboid manner and pro-duce, at the end of this stage, the new shell (Fig. **165**). The secretion of the material for the hollow or massive stalk takes place along a specially structured large pseudopodium con-taining hundreds of microtubules.

Examples: *Clathrulina, Hedriocystis, Orbu-linella*

Ciliophryida
Febvre-Chevalier, 1985

The naked cell bodies resemble those of the Acti-nophryida. The axopodia of the representatives of this order, however, have only very few micro-tubules (see Fig. **158**). During the life cycle, swarming stages with one to four flagella arise. In addition, the heliozoon stage is flagellated.

Examples: *Actinomonas, Ciliophrys, Pterido-monas*

Taxopodida
Fol, 1883

The bilaterally symmetrical cells of this marine heliozoon have conspicuous silicate spikes, ordered in rosettes over the cell surface (see Fig. **158**). The microtubules of the massive axopodia are ordered in a hexagonal pattern. The pseudopods swing in a paddle-like manner.

Example: *Sticholonche*

Centrohelida
Hartmann, 1913

The microtubules of the delicate and long axopodia arise from a single MTOC, the centro-plast (Fig. **166**), or from axoplasts. The nucleus is generally placed eccentrically, and in this ar-rangement has no morphological pecularities; in a number of cases, however, the nucleus is deeply tunnelled and surrounds the axoplast. As extrusomes, kinetocysts, and mucocysts are pres-ent; they serve for food capturing. The axopodial microtubules mostly form hexagonal or triangu-lar patterns (see Fig. **158**). A frequent phenome-non is the appearance of extracellular silicious spikes or scales.

Examples: *Acanthocystis, Gymnosphaera, He-draiophrys, Heterophrys*

4. Ascetospora
Sprague, 1978

Due to insufficient information, the obligatory parasitic Ascetospora are of questionable syste-matic position. In the past, they were often, due to the occurrence of spore stages, affiliated with the Apicomplexa; but apical complexes are ap-parently not present. The two subgroups, Haplo-sporea and Paramyxea, normally headed by the

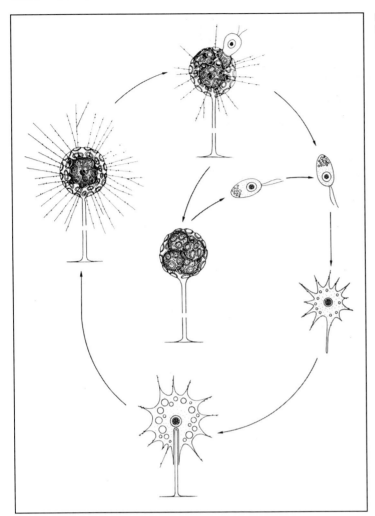

Fig. **165**   Desmothoracida:
life cycle of *Clathrulina elegans*
(adapted from Bardele).

Ascetospora, are also considered as independently evolved taxonomic units. There are two putatively synapomorphic characters: the appearance of so-called haplosporosomes or haplosporosome-like organelles, and the appearance of cell-in-cell formations, that develop ontogenetically in different ways. The general designation of the taxon refers to the delicacy of the conspicuous and complex ornamentation of the spores.

Haplosporea
Caulery & Mesnil, 1889

The representatives, named after the type-genus *Haplosporidium*, have unicellular spores. These are egested from their former hosts with feces and are ingested with food by a prospective invertebrate host. In the intestines, small amoeboid germlings are released from the spores; they invade connective tissues or epithelia and grow up to multinucleated plasmodia (= sporonts), later dividing into uninucleated sporoblasts. Two sporoblasts fuse to a zygote with an hourglass outline. The anucleate half, or episporoplasm, embraces the nucleate half, or sporoplasm, and separates from it by fission. This phenomenon of an autophagocytosis or self-engulfment leads to the situation of cellular nesting: A nucleate cell is located inside the vacuole of an anucleate cell. Before undergoing degeneration, the epispore secretes the material of the spore wall. During this process, both the specific ornamentations and a lid-like fold are formed at

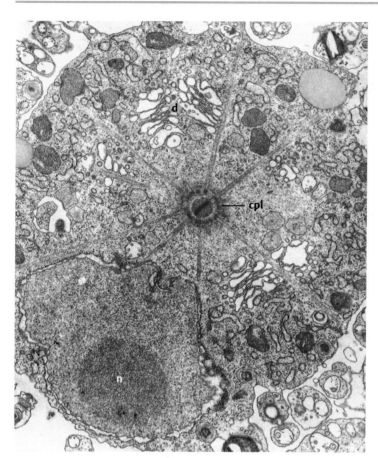

Fig. **166** Centrohelida: *Heterophrys marina*, axonemal microtubules arising from the centroplast (cpl). d = dictyosomes, n = nucleus (from Bardele: Cell Tiss. Res. 161 [1975] 85). Magn.: 40 000 ×.

the prospective spore aperture (Fig. **167**). More details of the life cycle are not known with certainty. It is suggested that intermediary hosts are involved.

Haplosporosomes can be found in the cytoplasm of sporonts and sporoblasts. They represent spherical, electron dense, and membrane-bound vesicles with a diameter of 70–250 nm. They contain an internal membranous substructure and, as formless constituents, also glycoproteins. Their functional role is unknown. With the exception of a group of putative Golgi derivatives (= spherule), the population of the other organelles bears no specialities.

Haplosporea occur as parasites in polychaetes, crustaceans, echinoderms, and especially molluscs. In oyster cultures, *Haplosporidium* species can cause important losses.

Examples: *Haplosporidium, Minchinia, Urosporidium*

Paramyxea
Levine, 1979

The spores of the representatives of the three genera comprising the taxon are always multicellular. They arise within a stem cell by endogenous budding. The result is a nesting of several generations of individual cells (Fig. **168**). The nesting principal is based on the phenomenon that, after karyokinesis, a portion of cytoplasm containing the smaller daughter nucleus separates from the cytoplasm of the stem cell by surrounding itself with merging cisternae of the endoplasmic reticulum. This process results in the establishment of an intracellular and intravacuolar daughter cell and is referred to as endogenous cleavage. In this way, formed daughter or secondary cells are able to perform two further cycles of identical reduplication, so that a maximum of four secondary cells appear, each inside an individual vacuole. During subsequent divisions by internal cleavage and with a con-

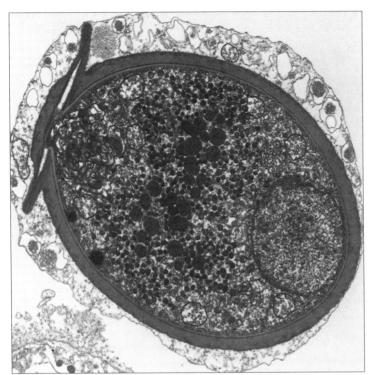

Fig. **167** Haplosporea: spore with operculum (upper right corner) of *Minchinia louisiana* (from F. O. Perkins, Gloucester Point). Magn.: 12 000×.

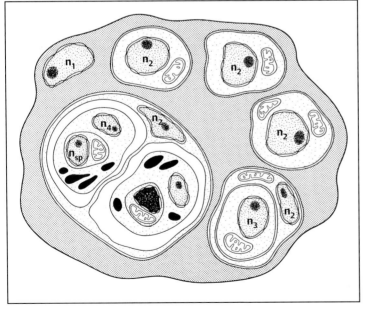

Fig. **168** Paramyxea: different stages of the life cycle of *Paramarteilia orchestiae*. n_1 (= nucleus of stem cell); n_2 and n_3 = nuclei of secondary and tertiary cells; n_4 = nucleus of a primary spore cell; n_{sp} = nucleus of sporoplasm (adapted from Desportes).

tinued growth of the stem cell, tertiary cells also develop inside the secondary cells. These divide into two identical primary (= outer) spore cells. From such outer spore cells arise, again by intracellular cleavage, the definitive reproductive units, the inner spore cells, called sporoplasm. This arrangement, illustrated for the genus *Paramarteilia*, represents the most simple situation;

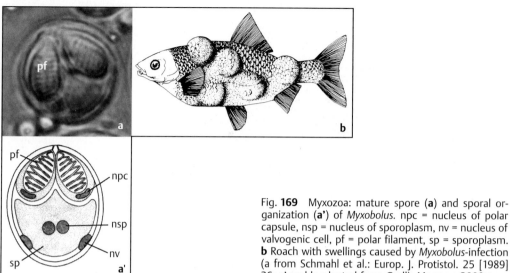

Fig. **169** Myxozoa: mature spore (**a**) and sporal organization (**a'**) of *Myxobolus*. npc = nucleus of polar capsule, nsp = nucleus of sporoplasm, nv = nucleus of valvogenic cell, pf = polar filament, sp = sporoplasm. **b** Roach with swellings caused by *Myxobolus*-infection (a from Schmahl et al.: Europ. J. Protistol. 25 [1989] 26; a' and b adapted from Grell). Magn.: a 2000×.

Fig. **170** Myxozoa: life cycle; infection stages (left) develop from multinucleated plasmodia (right). pf = polar filaments, sp = sporoplasm (courtesy of L. G. Mitchel, Ames).

with nine singular microtubules (9×1+0). By this characteristic alone, it must be assumed that this group had flagellated ancestors.

The Paramyxea live, as far as is known, as intracellular or tissue parasites of marine polychaetes, crustaceans, and commercially important molluscs. On the French coasts of the Atlantic, the population of oysters (*Ostrea edulis*) can be severely affected.

Examples: *Marteilia, Paramarteilia, Paramyxa*

5. Myxozoa
Grassé, 1970

This taxon comprises about 1200 species. The members are histozoic or cell parasites. Because of the formation of extrudible filaments, they were combined with the Microspora in the former taxon Cnidospora. However, the polar filaments of the Myxozoa (Fig. **169**) serve not for the transmission of the sporoplasm, but for anchoring the spores to the host tissue. Besides this, there are some other features justifying the erection of a specific taxon: the specific morphogenesis of the polar filaments, the appearance of true mitochondria and dictyosomes, and a conspicuous process of development, resembling that of the Paramyxea.

The life cycle of the Myxozoa is incompletely known, but the following are probably typical characters (Fig. **170**). The enigmatic construction of the spores, ingested with food by the host, contain either one binucleated or two uninucleated amoeboid germlings (sporoplasms). The haploid

in other genera there exist more complicated nesting principles, partly combined with meiotic divisions. As a conspicuous cytological character, there is the appearance of centrioles

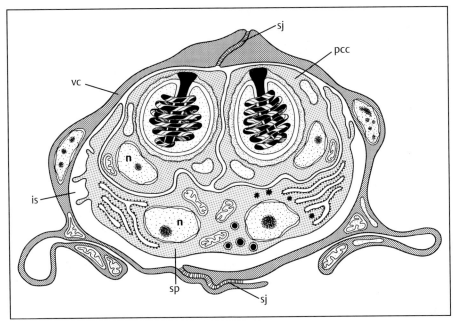

Fig. **171**   Myxozoa: diagram of a developing spore of *Leptotheca elongata*. is = intercellular space between sporoplasm (sp) and valvogenic cells (vc), n = nuclei, pcc = polar capsule cell, sj = septate junctions between valvogenic cells (vc) (adapted from Desportes-Livage and Nicolas). Magn.: 4000×.

germlings develop after a sexual fusion (autogamy) inside the spores or during release from it, to a multinucleated, diploid or polyploid, plasmodium. Such stages, which are able to multiply vegetatively by plasmotomy or by exogenous or endogenous budding, occur predominantly in the gallbladder, but also in connective or muscular tissues of invertebrates and cold-blooded vertebrates, especially bony fish. Inside the primarily motile plasmodia, the generative nuclei and their corresponding protoplasmic areas (= energides) separate by internal cleavage to individual cells. Each pair of these cells aggregates in such a way that one is engulfed by the other. In the resulting two-cell stage of the pansporoblast, the enveloping cell is designated the pericyte, the inner cell as the sporogonic cell. Whereas the pericyte, during the following development, degenerates to the enveloping layer, the sporogenic cell performs a number of additional divisions leading to the formation of two valvogenic (scale forming) and two capsulogenic (pole capsule forming) cells, as well as one or two sporoblasts. After meiotic divisions, the sporoblasts develop to haploid infectious amoeboid germlings. The valvogenic cells form a bipartite scale surrounding and protecting the germlings. Inside the capsulogenic cells, one

pole capsule differentiates, containing a coiled or stretched and extrudible polar filament.

In at least some cases, the assumption was made that the mature spores undergo, within another host, an comparable vegetative and sexual cycle, before, at the end of the complete life cycle, morphologically modified spores appear which are again infectious for the initial host. For the trout as host (parasite: *Myxobolus*) and *Tubifex* as host (parasite: *Triactinomyxon*), a corresponding alternation of hosts could be recognized. Thus the genus *Triactinomyxon* as a junior synonym of *Myxobolus*, should be rejected. Since both genera are representatives of different classes (Myxosporea and Actinosporea respectively), these subtaxa are possibly not reasonable and are therefore not described in this context.

From the high degree of cellular differentiation (Fig. **171**), with one generative and three somatic cell types, with the presence of quasi desmosomal cell contacts, and with the surprising similarities in construction and morphogenesis between the polar capsules and the nematocysts of Cnidaria, the question arises of whether the Myxozoa are really protozoa or modified and simplified metazoa. It is sometimes argued that a phyletic relation could exist between Myxozoa and the cnidarian taxon Narcomedusa, the life cycles of the latter also being characterized by cell-parasitic stages. Recent data of rRNA sequencing support a relation with the bilaterians.

In commercial fish farms, the Myxozoa can cause severe losses. *Myxobolus pfeifferi*, for example, is the causative agent of the bubonic disease in barbels (Cyprinoidae) (see Fig. **169**). The tumors reach diameters of 7 cm. In trout, *Myxobolus cerebralis* causes torsion disease. Even if both parasitic plagues lead to the death of fish only after a protracted period, the reduced growth and tissue ulcerations make affected fish hard to sell.

Examples: *Henneguya, Myxobolus, Sphaerosoma*

Part III: Selected Topics of General Protozoology

Comparative Morphology and Physiology of Protozoa

Skeletal Elements

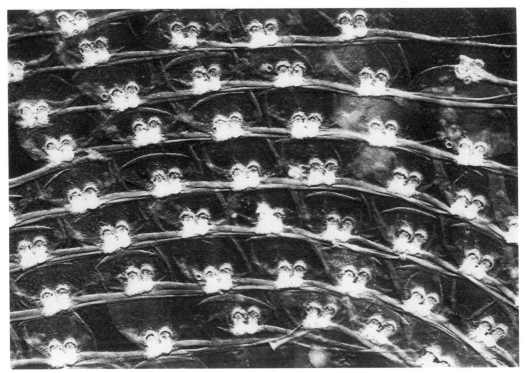

Fig. **172**   View from the inside of a shadow-cast pellicle fragment of *Paramecium*. Magn.: 12 000×.

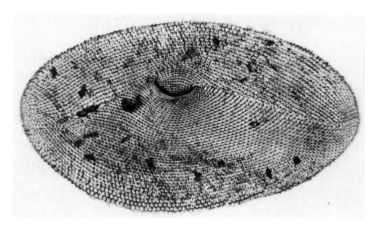

Fig. **173**   Silverline system of *Paramecium* (courtesy of W. Foissner, Salzburg). Magn.: 450×.

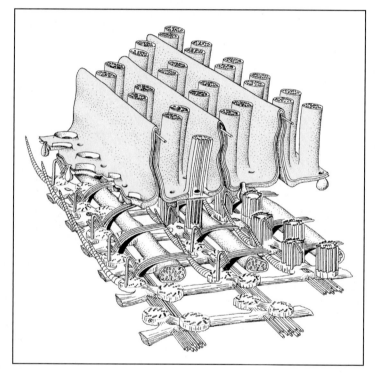

Fig. **174**   Three-dimensional reconstruction of cortical structures in the scuticociliate *Conchophthirius curtus* (adapted from Antipa).

The typical shape of the protozoan cell body is provided by intra- and extracellular skeletal elements which may be inorganic or organic. The organic elements are cell membranes, vacuoles, and bundles or other aggregates of microfilaments or microtubules (Fig. **172**). Other ill-defined filaments have also been detected to be part of the cytoskeleton.

Cortex

The cortical cytoplasm of most ciliates shows a complex organization which has been described in detail. This complexity can often be identified by light microscopy, especially if the organism has been stained by a method of silver impregnation (Fig **173**), a still essential technique for accurate systematic classification. Membranous and filamentous, as well as microtubular elements play an important role. However, most of the details of the ciliate cortex are observed only with electron microscopy (Fig. **174**).

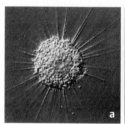
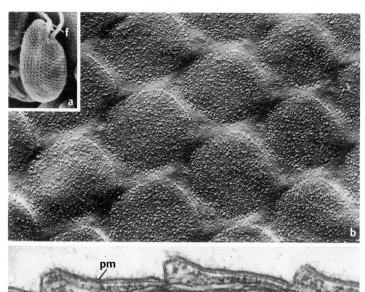

Fig. **175** Cortical structures of the cryptomonad *Rhodomonas* (= periplast) in dorsal view (**a**), freeze fracture (**b**) and ultrathin cross section (**c**). ch = chloroplast, f = flagellum, pm = plasma membrane (from Hausmann and Walz: Protoplasma 101 [1979] 349). Magn.: a 2500×, b 40 000×, c 70 000×.

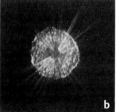
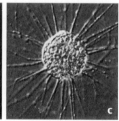

Fig. **176** Axopods of *Actinophrys sol* (**a**) exhibit birefringency due to the microtubular bundles (**b**); after application of mt-stabilizing drugs such as taxol the number of pseudopods increases dramatically (**c**). Magn.: 300×.

In addition to the ciliates, some flagellates such as the Euglenozoa (comp. Fig. **200**), the dinoflagellates (see Fig. **81**), and the cryptomonads (Fig. **175**) are known for their prominent cortical shape-forming components.

Intra- and Extracellular Skeletons

In addition to the cortex, or as an alternative, intracellular, or extracellular skeletal elements have been identified. These skeletons may be composed of inorganic or organic materials and in some cases, complexes of both. The systematic importance of skeletal elements is indicated in the taxonomic section. Inorganic, intracellular skeletons have very complicated geometries in the acanthareans and polycystineans.

The supporting role of organic skeletons of certain protists is easily demonstrated. For example, the axopodia of the heliozoa are stiffened by bundles of regularly arranged microtubules which can be degraded by abrupt cooling; this results in the depolymerization of the microtubules to tubulin subunits. When the organism is returned to room temperature, the kinetics of skeletal formation can be observed as the microtubules repolymerize and the axopodia form. If agents that inhibit (e. g., colchicine) or promote polymerization (e. g., taxol), are applied at the same time as the temperature shift to room temperature, disturbances in skeletal formation can be observed with the light microscope (Fig. **176**).

Scales and Loricae

Scales and loricae are external shaping and/or protective elements. Scales often occur in flagellates but they may also be found in some rhizopodial organisms. Scales originate from dictyo-

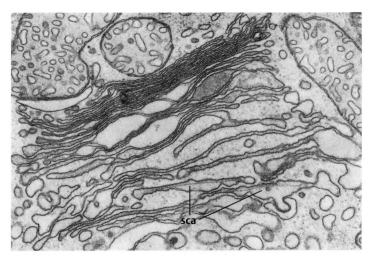

Fig. **177** Dictyosome of the prymnesiomonad *Pleurochrysis* with intracisternal scales (sca) (courtesy of W. Herth, Heidelberg). Magn.: 25 000×.

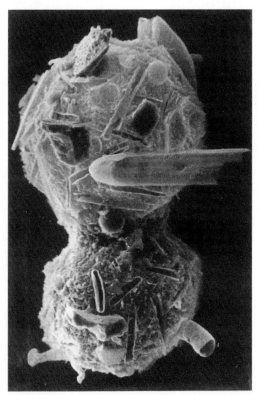

Fig. **178** Binary fission in the testate amoeba *Difflugia*; during mitosis the new shell (below) is formed by secretion of sediment grains into an amorphous matrix. Magn.: 400×.

somes (Fig. **177**) and are extruded to the surface by the process of exocytosis. They are often composed of polysaccharides such as chitin and cellulose, which may be encrusted with calcium. In some cases the scales have been found to be regularly organized at the cell surface by specific movements: the haptomonad *Pleurochrysis* rotates about its own axis during this process.

The lorica of flagellates, amoebae, and ciliates generally consists of an organic matrix to which the organism adds organic elements previously formed in the cytoplasm, or inorganic materials collected by the cell from the environment. The lorica may be distinctive and it is often species-specific. The mechanisms involved in the organization and formation of the lorica during the process of binary fission are still unknown (Fig. **178**). The results of experiments, such as those performed by Rumbler at the end of the 19th century, show some interesting similarities to the loricae of amoebae, but the mechanistic explanations do not appear to reflect the morphogenetic phenomena as they occur in nature.

Different stages of lorica formation have been identified for the marine ciliate *Eufolliculina uhligi* (Fig. **179**). The sessile stage of the ciliate lives in an organic lorica, which is primarily composed of chitin. In addition, there are proteinaceous components as well as pigments and inorganic traces including metals.

Cysts

Many protozoa form cysts (Fig. **180**). Cyst formation may be an obligatory step during the life cycle, e. g., digestive and reproductive cysts. Cysts may also be a consequence of sexual activity, as in *Actinophrys*, for example. Finally, cysts may be formed in response to unfavorable environmental conditions. Permanent resting cysts

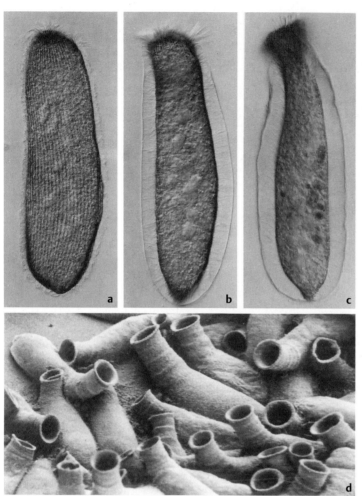

Fig. **179** The swarmer cell of *Eufolliculina* (**a**) during secretion of material for a new lorica (**b**, **c**); the lorica is ampulla-shaped and bears a curved neck (**d**) (from Mulisch and Hausmann: J. Protozool. 30 [1983] 97). Magn.: a–c 200×, d 120×.

are easily distributed by wind, spray, or migratory organisms such as birds, and since the encysted organisms may remain viable for 10–15 or even more years, the cyst is an important means of dispersal for those organisms that produce them.

Soil protozoa live in a biotope where the water content may change quickly. Those that form cysts (and most soil protozoa do) are generally able to encyst and excyst rapidly and the cyst walls are usually impermeable and capable of withstanding desiccation. The cysts of parasitic forms show similar adaptations. In contrast, other protists have less stable cysts which may require a moist environment to prevent them drying out. The palmella stage of euglenozoa and other photosynthetic flagellates may be regarded as cysts. In this case, the cells resorb their flagella and surround themselves with a thick mucus layer.

The chemical composition of some cyst walls has been investigated. Most contain chitin, which seems to be a basic feature of eukaryotic organisms. Less frequently, cellulose is a major component, and some forms have plates of silica. How cyst walls are formed is largely unknown, and little is known about the mechanisms that induce encystment. Some of the environmental factors which promote encystment have been identified. These include changes in temperature and pH, evaporation, low and high levels of oxygen, accumulation of the end products of metabolism, high population density, depletion, or excess of food. The effects of these external factors on the internal control of encystment are still unknown.

Excystment is also induced by a variety of different environmental factors, e. g., renewal of culture medium, addition of hypo- or hypertonic

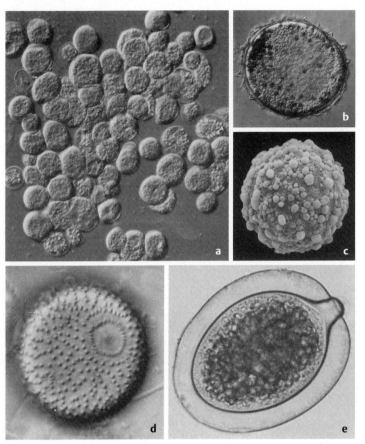

Fig. **180** Cyst formation in *Acanthamoeba* (**a**), *Climacostomum* (**b**), *Actinosphaerium* (**c**), in a sessile peritrich ciliate (**d**) and in *Frontonia* (**e**) (a and c courtesy of D. J. Patterson, Sydney; b courtesy of D. Fischer-Defoy, Freiburg; d from Hausmann and Rambow: Mikrokosmos 74 [1985] 208); e from Hausmann and Foissner: Mikrokosmos 75 [1986] 193). Magn.: a 500×, b 250×, c 700×, d 600×, e 400×.

medium, addition of certain organic substances, and addition of bacteria to the culture medium. Since excystment may be induced by external factors, it is clear that there must be a flow of information through the cyst wall.

During the process of excystment, the cyst wall disintegrates in response to enzymes produced by the encysted organism. In some cases, there is a preformed operculum in the cyst wall through which the organism can escape.

In some of the older literature, the process of encystment and excystment in *Paramecium* was described. It was reported to be a means of getting paramecia from a hay infusion. However, many recent experiments have failed to confirm the occurence of either process in *Paramecium*. The same holds true for *Amoeba proteus*.

Holdfast Organelles

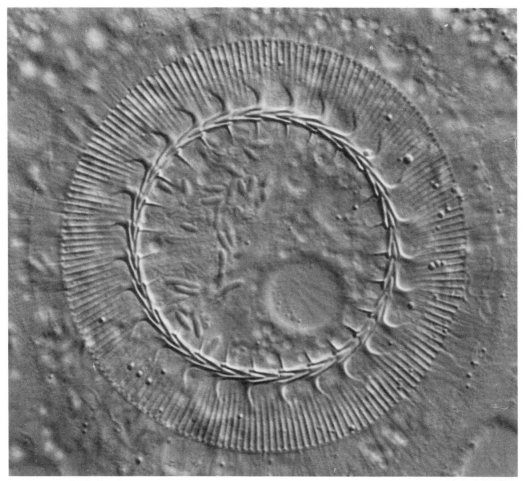

Fig. **181**   Adhesive disc of *Trichodina* (from Hausmann and Hausmann: J. Ultrastruct. Res. 74 [1981] 144). Magn.: 1700×.

In addition to those protozoa that are able to attach to a substrate via their lorica, a number of unicellular organisms can adhere either permanently or temporarily by means of a holdfast. In free-living forms, there are a variety of different methods of attachment. *Stentor* attaches by the secretion of mucus material, some other ciliates have intra- or extracellular stalks *(Vorticella,* *Carchesium, Tokophrya)* (see Figs. **133, 134, 136**), arm-like appendages of the cell (Fig. **182**), or even sucker-like organelles *(Trichodina)* (Fig. **181**). Some organisms that live as endobionts, e. g., diplomonads, have developed special suckers (see Fig. **44**), or specialized areas of the cell such as the epimerite of the gregarines (see Fig. **89**).

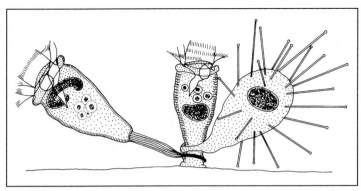

Fig. **182**   The peritrich ciliate *Epistylis lwoffi* (left) and the suctorian *Erastophrya chattoni* (right) embracing the peritrich *Apiosoma amoeba* (adapted from Matthes).

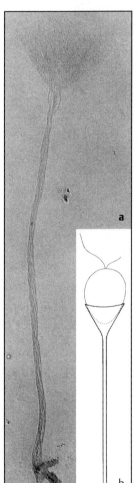

Fig. **183**   Shadow-cast lorica of *Poterioochromonas malhamensis* (Chrysomonadea) (**a**) and schematic organization (**b**) from Schnepf et al.: Planta 125 [1975] 45). Magn.: 4000×.
◁

Fig. **184**   **a** Cup and stalk development in *Poterioochromonas*. **b** negative staining by indian ink preparation. c = cell, cu = cup, stk = stalk (a adapted from Schnepf; b courtesy of G. Röderer, Stuttgart). Magn.: 1300×.   ▽

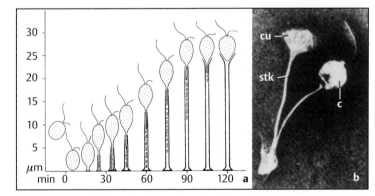

Stalks

Stalks are usually formed by the swarming stage of polymorphic unicellular organisms, although some species are known to be able to form a stalk at almost any moment of their life. The mechanisms of formation of the stalk differs between organisms. Some chrysomonads, e. g., *Poterioochromonas*, form a cytoplasmic tail, and the membrane of this tail becomes the origin of

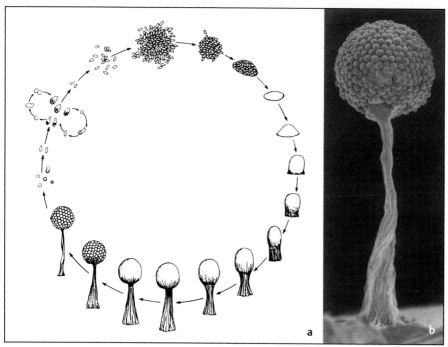

Fig. **185**   **a** Life cycle of the ciliate *Sorogena*. **b** sorocarp (123) (a adapted from Blanton and Olive; b from Olive and Blanton: J. Protozool. 27 [1980] 293). Magn.: 100×.

chitin fibrils. The resulting stalk widens at the top. After lorica construction is complete, the flagellate resorbs its cytoplasmic tail, and lives without special holdfast organelles in the wine-glass shaped lorica (Fig. **183**), which may later be abandoned in response to unfavorable conditions. The kinetics of stalk formation in this flagellate are easily observed (Fig. **184**).

Free-living swarmers of the sessile desmothoracidian heliozoans form a large pseudopodium with hundreds of microtubules that act as a template for the secretion of an organic fibrous sheath, which finally results in a hollow stalk. The organism lives at the top of the stalk, inside an extracellular, silicious lattice ball (see Fig. **165**).

Among the ciliates, the peritrichs and suctorians are in particular noted for their sessile life style and stalk formation. In the case of the peritrichs, the stalks are at least partially intracellular, while those of the suctoria are extracellular. Peritrich stalks may be highly contractile.

The peritrichs secrete the stalk material by means of an organelle called a scopula. This is a cup-shaped region of the cell that is covered by short, immotile cilia. Parasomal sacs are usually present in this region. In some organisms, the scopula may function directly as a holdfast organelle. When a stalk is secreted, the parasomal sacs may play an important role in stalk material secretion. It has been suggested that the assembly and disassembly of microtubules may be involved in the elongation of the peritrich stalk, and the microtubules may be responsible for the transport of secretory vesicles into the scopular region.

The stalk-secreting region of the suctorians resembles the scopula of the peritrichs. It does, however, appear to be different, and the extracellular stalks are probably not homologous to any peritrich stalk. Thus, this region has been referred to as a scopuloid to avoid confusion.

A special kind of stalk formation in ciliates has been described for members of the colpodid genus *Sorogena* (Fig. **185**). If food runs low, the cells become small and slender. Under proper conditions of illumination, usually in the early morning, they begin to aggregate. First, the aggregate forms a flat layer, which eventually becomes compact as the cells move closer together. Approximately three hours after the aggregation has begun, the aggregate is covered by a sheet. During the following 30–60 minutes, the sheet elongates and forms a stalk that rises above the surface of the water. A sphere forms at the top of the stalk and includes encysted ciliates. The whole structure, the sorocarp, resembles the fruiting bodies of lower fungi. When the cysts are released into the water, the ciliates excyst, and the cycle may start over again.

Attachment Devices

The attachment organelles of motile peritrichs and some endobiotic protozoa function in a sucker-like fashion. In this way, the attachment can be temporary and allow the organisms to make contact with or leave their substrate very quickly.

Microtubules are the main structural feature of the diplomonad sucker (see Fig. **44**). The complex adhesive disc of the urceolariids (motile peritrichs, e. g., *Trichodina*) consists of massive proteinaceous elements. There may be several hundreds of these elements arranged to form a circular, radially symmetrical holdfast (see Fig. **181**). The number and shape of the basic elements, as well as the general structure of the adhering disk, are important taxonomic features of the urceolariids.

Extrusomes

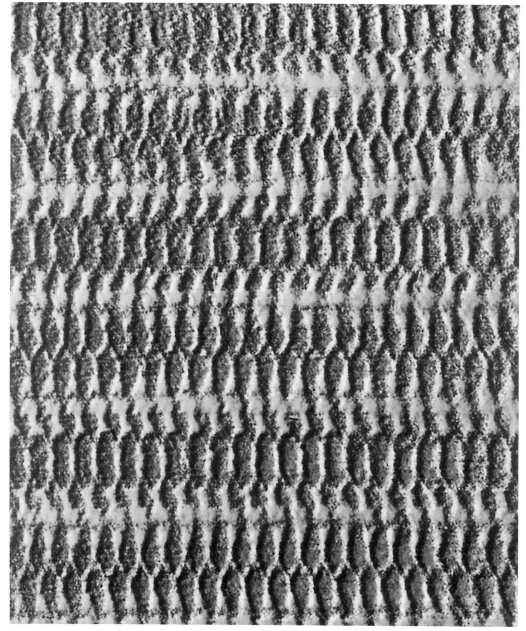

Fig. **186** Macromolecular network of protein filaments in *Paramecium* trichocyst (from Hausmann: Ann. Stat. Biol. Besse-en-Chandesse 7 [1973] 331). Magn.: 530 000 ×.

Table **6**   Characterization of extrusomes (in alphabetical order)

Name	Resting state	Ejected state	Mode of ejection
discobolocyst	spherical body with a disk-shaped ring at one pole	solid ring with the same dimensions as in the resting state furnished with a long tail	unknown (stretching?)
ejectisome	spirally wound ribbon, usually bipartite	tube-like organelle, smaller in diameter and longer than the resting state	sudden unrolling of the ribbon and formation of a tube by rerolling laterally
epixenosome	extracellular structure attached to the surface of certain hypotrich ciliates containing a tightly coiled band	as in ejectisome, a long tube-like element	sudden unrolling followed by a lateral rerolling
haptocyst	bottle-like organelle with a complex internal structure composed of several components	partly everted, excretion of poisonous (?) material	unknown
kinetocyst	compound organelle consisting of a central element enveloped by a ring-like jacket	central element lies in front of the open jacket	unknown
muciferous body	sac-like vesicle filled with un-ordered material	amorphous mucilage	secretion of the material through a pellicular pore
mucocyst	polyhedral paracrystalline body	polyhedral paracrystalline body whose diameter and length are multiples of those of the resting state	secretion lasting several seconds, unfolding of a preexisting network of filaments
nematocyst	spindle-shaped capsule with a coiled tube	capsule with an everted tube	eversion of the tube
rhabdocyst	stick-shaped organelle	tube-like structure with the same length and diameter as the resting state	telescopic expulsion
spindle trichocyst	spindle-shaped or rhomboid paracrystalline body, sometimes furnished with a specially constructed tip	thread shaped paracrystalline filament with the same diameter as the resting state, but several times longer	sudden (lasting milliseconds) unfolding of a preexisting three-dimensional network of protein filaments
toxicyst	inverted tube inside a capsule	capsule with an everted tube of the same width and length as in the resting state, secretion of poisonous material	sudden eversion and/or telescopic expulsion of a tube

Early microscopists observed a behavior that they considered to be unusual. Some protozoa extruded preformed structures without a change in their morphology. These events did not always appear to relate to the biological activities of the organism. All such extruded structures are now called extrusomes.

Extrusomes are vesicles which contain an organized substance which usually changes its form when the vesicle undergoes exocytosis, releasing its contents to the exterior of the cell. Extrusomes are usually located in the cortex of cells. The best known type of extrusomes are the spindle trichocysts of *Paramecium* (see Figs. **17**

and **186**). The different types of extrusive organelles have different morphologies and they may differ in their function. However, they all share one feature: mechanical, chemical, or electrical stimuli cause their extrusion, usually in much less than a second. During the process of exocytosis, the size and structure of the extruded material changes in a predictable and characteristic manner.

About a dozen different types of extrusomes are now known. They have been found among the flagellates, actinopods, and ciliates, although their frequency in the different groups is variable. It is always the case that all individuals of a

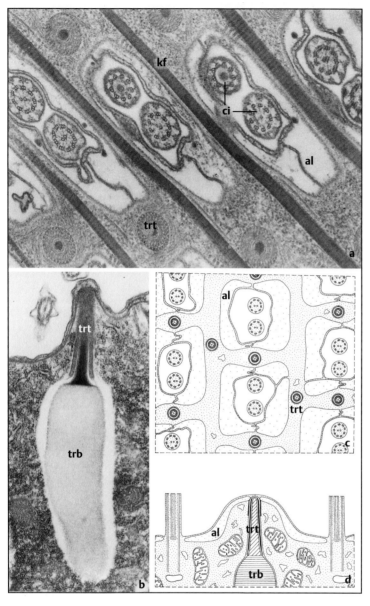

Fig. **187**   Trichocysts in *Paramecium*: resting stages in transverse section (**a**) and longitudinal section (**b**) and in corresponding schemes (**c, d**). al = alveolus, ci = cilium, kf = kinetodesmal fiber, trb = trichocyst body, trt = trichocyst tip (from Hausmann: Int. Rev. Cytol. 52 [1978] 197). Magn.: a 40 000×, b 18 000×.

species have the same kind(s) of extrusome(s). Extrusomes sensu stricto have not yet been detected among the Apicomplexa. However, the rhoptries of coccidia may be regarded as extrusome type vesicles. Additionally, there are a number of organelles found in protozoa that may be extrusomes. In most cases, further study is warranted.

Table **6** gives comparative data on extrusomes, which are distinguished by structural criteria and their mode of extrusion.

Spindle Trichocysts

The best known extrusomes are spindle trichocysts. Since they are easy to demonstrate by light microscopy and occur in large numbers, the spindle trichocysts of *Paramecium* are often demonstrated in introductory biology courses as being a special type of organelle of protozoa.

Spindle trichocysts originate in the cytoplasm, and in their resting states they are docked at predictable positions in the cortex (Fig. **187**).

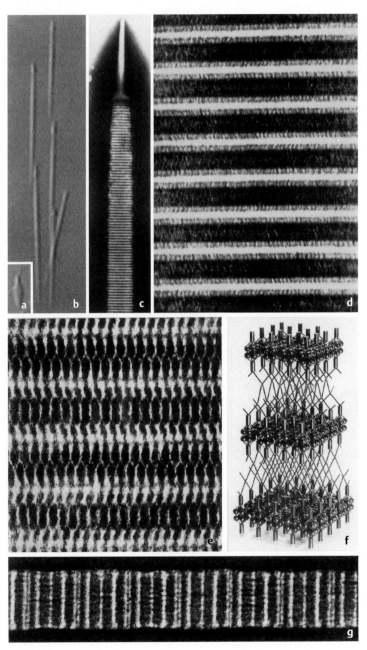

Fig. **188** Trichocysts in *Para-mecium* (**a–f**) and in the dino-flagellate *Oxyrrhis* (**g**): a rest-ing stage. **b–e** extruded forms with progressively in-creasing magnification **f** re-construction of the network of protein filaments visible in e. **g** discharged dinoflagellate trichocyst (from Hausmann: Int. Rev. Cytol. 52 [1978] 197). Magn.: a 800×, b 800×, c 10 000×, d 200 000×, e 300 000×, g 200 000×.

The tips are located in the space between adja-cent alveoli where they closely underly the plasma membrane. In this way, the trichocysts membrane may easily fuse with the plasma membrane to allow exocytosis. In *Paramecium*, the interior of mature trichocysts has a cross-striated periodicity of 7 nm. During extrusion, the membrane of the trichocyst fuses with the plasma membrane, and the extrusomal material is delivered to the outside of the cell in a matter of milliseconds.

Following this, the trichocyst membrane, re-maining in the cytoplasm, separates from the plasma membrane and disintegrates into small vesicles. This behavior of the trichocyst mem-brane also appears to occur following the dis-charge of other extrusomes and may represent a general mechanism in extrusomal discharge.

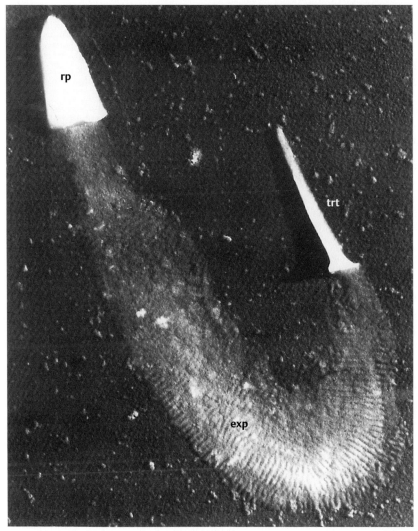

Fig. **189** Discharged but partially inhibited *Paramecium* trichocyst. exp. = expanded part, rp = resting part, trt = trichocyst tip (from Hausmann: Int. Rev. Cytol. 52 [1978] 197). Magn.: 25 000×.

The ejected trichocyst is about eight times longer than the resting trichocyst. In the extruded state, there are also periodic cross striations, but their periodicity is about 56 nm (Fig. **188**). This periodicity results from a regular three-dimensional arrangement of thin filaments which are connected in an orderly manner. A three-dimensional model of the spindle trichocyst at the macromolecular level has been produced, and based on this model, it has been demonstrated that the trichocyst shaft always has the same structure, although arrangements of the filamentous connections may change.

The driving forces for trichocyst extrusion, as well as the molecular events associated with this process, are still unknown. The examination of trichocysts during the process of extrusion suggests that there is a preformed network which expands during the process (Fig. **189**). The number of cross striations is the same in resting and extruded trichocysts, and the width of the cross striations is proportional to the increase in length. It is also known that trichocyst expansion is independent of ATP, but that it does require a certain concentration of calcium. Spindle trichocysts of flagellates are basically similar to those of *Paramecium* (see Fig. **188**).

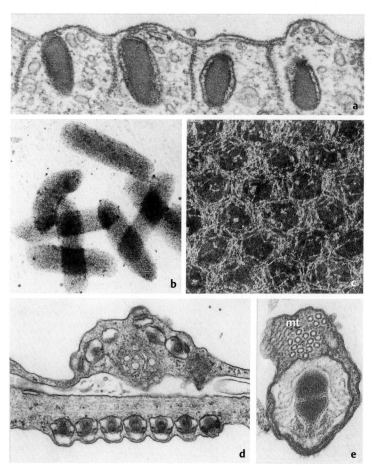

Fig. **190**   Mucocysts of the ciliate *Loxophyllum* in situ (**a**) and in extruded stages (**b, c**). The pattern of filaments at high magnification (**c**). **d** and **e** kinetocysts of the heliozoans *Clathrulina* (**d**) and *Acanthocystis* (**e**). mt = microtubules (a–c from Hausmann: Int. Rev. Cytol. 52 [1978] 197; d and e from Bardele: Z. Zellforsch. Mikrosk. Anat. 130 [1972] 219). Magn.: a 50 000×, b 6000×, c 110 000×, d 33 000×, e 65 000×.

According to their structure and behavior, spindle trichocysts are unique protein-containing organelles. Because of their superficial similarities, i. e., cross-striations, they have been compared to collagen. However, collagen does not have the dynamic properties of the trichocyst protein. Furthermore, both proteins differ substantially in the composition of their amino acids.

If all of the thousands of trichocysts in one *Paramecium* cell are experimentally removed, the cell can replace them all in about 5–8 hours. This indicates that there is an important function for these organelles, which is the repulsion of predators.

Mucocysts

In some respects, mucocysts resemble trichocysts. They are docked just below the plasma membrane. In the resting state there is a paracrystalline organization of filaments (Fig. **190**), and during extrusion, a preformed filamentous network expands. In contrast to the trichocysts, however, expansion occurs in three dimensions and continues for several seconds. Mucocysts appear to have a different chemical composition compared to trichocysts insofar as they probably also contain carbohydrates. They are found in flagellates, ciliates, and, in a modified form, in actinopoda. In ciliates (e. g., *Didinium*) they may be involved in the process of encystment, although there is some doubt that this is their main function. In the heliozoa, certain mucocysts are known as kinetocysts due to their characteristic movements. It seems plausible that the mucocysts of these protozoa are responsible for the sticky cell surface which is involved in the capture of food organisms.

Mucocysts may be the only type of extrusomes which correspond to similar organelles

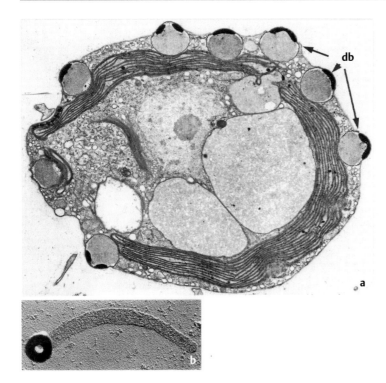

Fig. **191** Discobolocysts: **a** *Ochromonas tuberculatus* with resting discobolocysts at the cell periphery (db). **b** discharged stage with tail (from Hibberd: Brit. J. Phycol. 5 [1970] 119). Magn.: a 6200×, b 6500×.

found in multicellular organisms. They resemble organelles of the oomycetes, which are involved in primary cyst formation and the cortical granules of sea urchin eggs. Cortical granules are involved in the production of the fertilization membrane, which follows the fertilization event and aids in the prevention of polyspermy.

The muciferous bodies of the euglenozoa continuously secrete mucus material through pellicular pores. It has been suggested that this material might facilitate euglenoid movement.

Discobolocysts

Discobolocysts are found only in certain flagellates, particularly in chrysomonads. In the resting state, these organelles are nearly spherical. Their structure is polarized; the part adjacent to the plasma membrane contains a disc. Upon extrusion, the disc remains unaltered while the remainder of the organelle is transformed into long filamentous material (Fig. **191**), in a fashion resembling the extrusion mechanisms of trichocysts or mucocysts. The function of this organelle is unclear.

Toxicysts

There is no doubt about the functions of toxicysts. They serve as a defense against predators, and they are especially involved in the capture of prey. This type of extrusome is found primarily in carnivorous organisms, mainly ciliates, but they are also found in some phagotrophic flagellates.

Toxicyst vesicles contain a capsule. Within the capsule there is a long tube. Depending on the type of toxicysts, the inner tube may be either telescoped or everted during extrusion (Fig. **192**). In the latter case, the toxicyst resembles the function of the cnidarian cnidocysts, although the structures differ significantly in size.

During extrusion, the tubes of the toxicysts are forced into their prey organisms like hypodermic needles. The prey, usually unicellular organisms (or, exceptionally multicellular organisms, e. g., rotifers), is paralyzed or killed by the action of the toxicyst. Often the prey is mechanically disrupted by the toxicyst, but it is also apparent that the toxicyst contains a toxin which is injected into the prey during the process of extrusion.

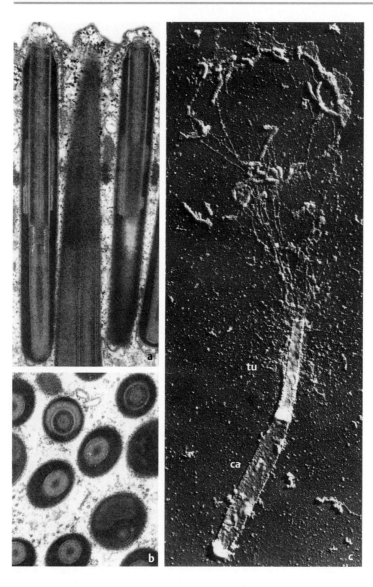

Fig. **192**   Toxicysts of the ciliates *Homalozoon* (**a**, **b**) and *Loxophyllum* (**c**): Resting stages in longitudinal (**a**) and cross section (**b**). **c** discharged stage. ca = capsule, tu = tubule (a and b from Kuhlmann and Hausmann: Protistologica 16 [1980] 125; c from Hausmann and Wohlfarth-Bottermann: Z. Zellforsch. Mikrosk. Anat. 140 [1973] 235). Magn.: a 35 000×, b 35 000×, c 15 000×.

Toxicysts are often concentrated in specific areas of the cell. In the haptorian ciliates, they occur primarily in the region of the oral apparatus. There may also be accumulations of toxicysts in bumps in the cell cortex. The presence of multiple toxicysts at one site probably has the purpose of providing a barrage of toxicysts that can be extruded simultaneously.

The toxicysts found in the knobs of suctorian tentacles are known as haptocysts. These are specialized extrusomes. The haptocysts are discharged on contact with the suctorian prey, and they provide a firm connection with the food organism. It is believed that they are mainly involved in the fusion of the plasma membranes of the two organisms (Fig. **193**).

Rhabdocysts

Rhabdocysts are rod-shaped extrusomes. As far as is currently known, they occur only in the karyorelictid ciliates, such as *Trachelonema sulcata* and *Tracheloraphis dogieli*. The mechanism of their ejection resembles that of those toxicysts having a telescopic discharge. A rhabdocyst is discharged like an arrow from a blowpipe.

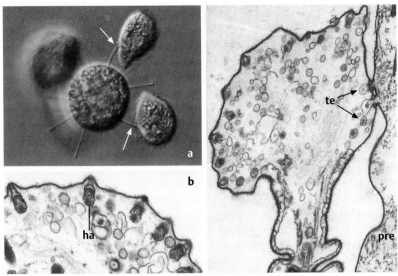

Fig. **193** Haptocysts in Suctoria: **a** the adhesive function of tentacles (arrows). **b** resting haptocysts (ha) at the apical tentacle knob. **c** extruded haptocysts (arrows) forming the interface between the arrested prey cell (pre) and the suctorian tentacle (te) (b and d from Bardele and Grell: Z. Zellforsch. Mikrosk. Anat. 80 [1967] 219). Magn.: a 250×, b 22 000×, c 22 000×.

Ejectisomes

Ejectisomes (Fig. **194**) are extrusomes that occur only in flagellates. They are found in crypto-monads and the prasimonads, and their function is unclear. In the resting state they are compact organelles which resemble tightly coiled ribbons. They usually consist of two parts. The extruded ejectisomes are tubes which may become extremely long as they are unrolled from the coiled extrusome. The ejection of these extrusomes results in an escape reaction in these flagellates.

Structures very similar to ejectisomes have been found in the kappa particles of *Paramecium*. Kappa particles are now known to be endosymbiotic bacteria. They contain R-bodies which are morphologically indistinguishable from ejectisomes. Under certain experimental conditions, the R-bodies have been observed to act like ejectisomes; they unroll and then fold up laterally to form a long tube.

It is difficult to imagine how convergent evolution may have produced such unusual and yet similar structures in widely different organisms.

The problem is even more profound when trying to imagine how a bacterial structure, the R-body, can be phylogenetically related to a flagellate organelle, the ejectisome.

Epixenosomes

From their ultrastructure and mode of function, the epixenosomes of certain hypotrich ciliates resemble the ejectisomes observed in flagellates (Fig. **195**). In these too, the tightly coiled band which is ejected is laterally unrolled to form a tube. However, in this case the situation is different insofar as these structures are not internal cytoplasmic organelles of the ciliates, but external elements that are tightly bound to the plasma membrane. In addition, the epixenosomes are reported to be eukaryotic in nature and thus, most probably, epibionts of unknown origin.

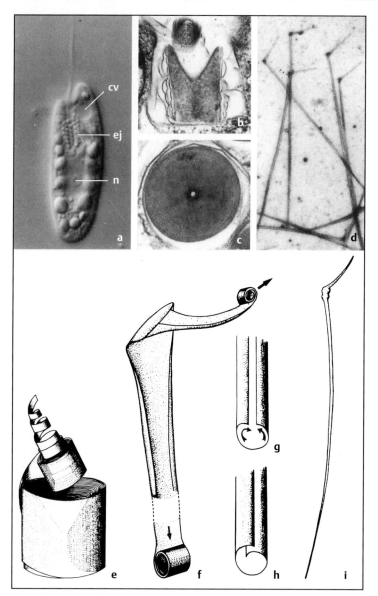

Fig. **194** Ejectisomes in *Cryptomonas paramecium*: **a** spatial relationships between the gullet-bordering ejectisomes (ej), contractile vacuole (cv) and nucleus (n) in a living specimen. **b** und **c** longitudinal and cross section of resting
ejectisomes. **d** discharged ejectisomes. e–i schematic representation of a resting ejectisome (e) and of the discharging process (f-i) from Hausmann: Int.
Rev. Cytol. 52 [1978] 197). Magn.: a 3500×, b 25 000×, c 40 000×, d 6500×.

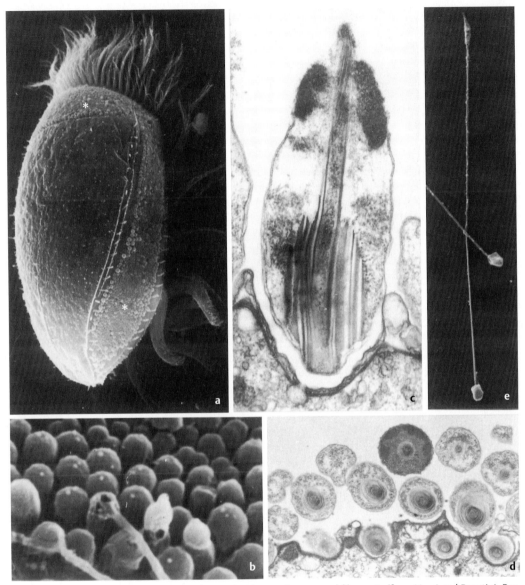

Fig. **195** Epixenosomes. **a** and **b** *Euplotidium itoi* with epixenosomes (*); ultrastructural organization in longitudinal (**c**) and cross section (**d**). **e** discharged epixenosomal filaments (from Verni and Rosati: J. Protozool. 37 [1990] 337). a 850×, b 8500×, c 28 000×, d 10 000×, e 2800×.

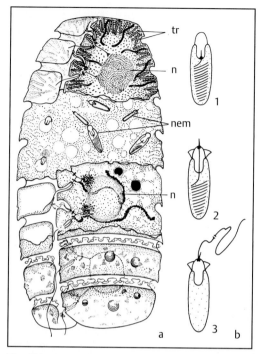

Fig. **196** Nematocysts: **a** different development stages in the dinoflagellate *Polykrikos*. n = nucleus, nem = nematocysts, tr = trichocysts. **b** tube discharging in a nematocyst (a from Chatton and Grassé: C. R. Sce. Soc. Biol. Ses Fil. 100 [1929] 281).

Nematocysts

Nematocysts have been found in certain dinoflagellates. As in the case of toxicysts, they resemble the cnidocysts of cnidarians. They are small capsules filled with a coiled tube that evaginates during discharge (Fig. **196**). The function of nematocysts is not as clear as it is for the toxicysts. Due to some structural specializations, as well as some features of their development, they are regarded as a separate, unique type of extrusomes. Possibly they do not represent a true organelle, but merely a highly adapted endobiont of so far unknown origin.

Contractile Vacuoles

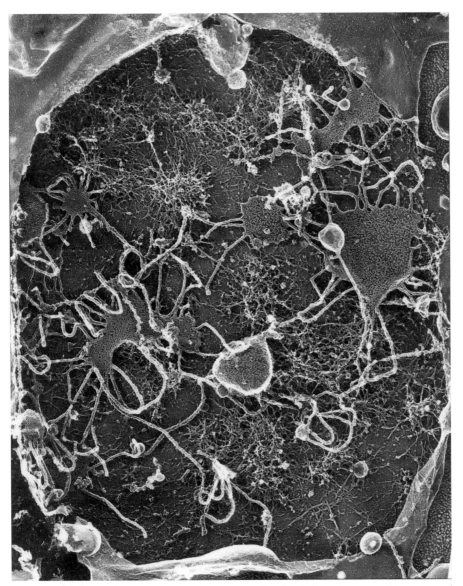

Fig. **197** View of the interior of a *Dictyostelium* amoeba showing an extensive, interconnected contractile vacuole system, the swollen cisterna being partially filled (from Heuser, Zhu and Clarke: J. Cell Biol. 121 [1993] 1311). Magn.: 20000×.

With the exception of some freshwater sponges, contractile vacuoles occur mainly in those protozoa which lack a cell wall (see Fig. **197**). Typically, the contractile vacuole slowly fills with fluid (diastole), which is periodically expelled to the surrounding medium by contractions of the vacuole (systole). The frequency of expulsion may vary, depending on the organism, between seconds and hours.

In most cases, a protozoon has only one contractile vacuole. However, there are a number of organisms which have two, e. g., *Paramecium* (Fig. **198**), or many more, such as the 15–20 seen in other ciliates such as *Dileptus* or *Homalozoon*. Some ciliates, e. g., *Loxophyllum* and *Spirostomum,* contain only one large vacuole, but it is connected to a long collecting channel that extends from the anterior to the posterior of the cell.

Contractile Vacuole Complex

Contractile vacuoles are easily identified in the light microscope, as they are transparent and contract regularly. In the electron microscope, it becomes apparent that the contractile vacuole itself is only a part of a larger system, which is called the contractile vacuole complex. The complex contains the contractile vacuole, the spongiome, and the contractile vacuole pore. It is through the pore that the fluid is expelled to the outside. In the ciliates, the complex appears to have its most complicated structure. Here, the vacuole may be connected with radially arranged ampullae which form long, narrow, collecting channels (Fig. **198**). The complex is supported and held in position by microtubular bundles that originate in the walls of the pore. Helically arranged microtubules are also found in the wall of the pore. The microtubules extend along each of the ampullae and pass to the very end of the collecting channels (Fig. **198**).

The contractile vacuoles of the chloromonads originate directly from the cisternae of dictyosomes. With the exception of a few unusual examples such as this, there is no evidence that the contractile vacuole complex is structurally continuous with other membranous organelles of the cell.

Spongiome

The spongiome consists of numerous small membranous vesicles or tubules. It is usually situated immediately adjacent to the contractile vacuole, and it has been suggested that these membranes are involved in the sequestration of fluid destined for export. The organization of the spongiome has been seen to differ in the different types of protozoa. The simplest type of spongiome is the vesicular type where many small vesicles surround a large vacuole (Fig. **199**). This type of spongiome is commonly found in those organisms which do not have a contractile vacuole in a fixed position within the cell, as, e. g., in some flagellates and many amoebae. The membrane of the spongiome of these organisms usually includes two distinctive types of vesicles: Some have a smooth membrane, while others are covered with a coating that resembles the clathrin coat of coated pits and coated vesicles (Fig. **199**).

It has been suggested that these coated vesicles are involved in the segregation of water from the cell, while the smooth-membraned vesicles are involved in reducing the membrane surface of the contractile vacuole. During diastole, the contractile vacuole originates from the fusion of smooth vesicles, and during systole, they fragment to individual vesicles. Since the fusing vesicles have a larger membrane area than the original vacuole, there must be a mechanism for rearrangement of the membrane, or, alternatively, not all vesicles are involved in each cycle of the vacuole.

Tubular spongiome (Fig. **199**) is found in some amoebae and flagellates, as well as in most ciliates. Here, the situation appears to be more complicated. The spongiome tubules, which are 60 nm in diameter, are permanent structures which are either directly or indirectly connected to the contractile vacuole. In the case of ciliates where there is a radial arrangement of collecting channels, e. g., *Paramecium*, the spongiome surrounds each of the collecting channels and is permanently connected to them (see Fig. **198**).

In the early 1960s, observations by electron microscopy suggested that the connections between the spongiome and the collecting channels were interrupted during each systole. This process was believed to prevent the flow of water back into the spongiome during contraction of the vacuole. According to this hypothesis, thousands of connections between the spongiome and collecting channels would be rapidly reformed following systole.

This hypothesis seemed plausible, and it was published in many textbooks until very recently. However, observations based on more advanced preparation techniques do not support this no-

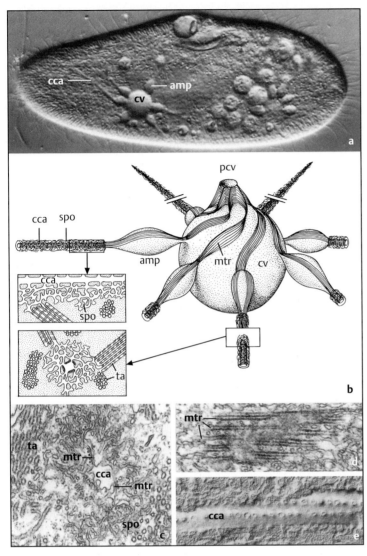

Fig. **198** The contractile vacuolar complex in *Paramecium*: **a** living specimen. **b** schematic diagram. **c–e** ultrastructural details. The complex consists mainly of ampullae (amp), collecting canal (cca), the contractile vacuole (cv), and a pore (pcv). Collecting canals are surrounded by and connected with irregularly arranged spongiomal tubules (spo). Tubular aggregates (ta) are located at a greater distance. The whole complex is stabilized by several microtubular ribbons (mtr) (a, c, d, e from Hausmann and Allen: Cytobiologie 15 [1977] 303). Magn.: a 500×, c–e 32000×.

tion. There are no indications at all that the spongiome is rhythmically connected with and disconnected from the collecting channels. Nor does it seem essential that such a mechanism should exist. The pressure that would be necessary to transport the water back into the spongiome would have to be extremely high, since the narrow tubules of the spongiome would present considerable resistance. The pressure, which lasts only a fraction of a second during the time of maximal filling of the contractile vacuole, and the fusion of the vacuolar membrane with the plasma membrane, is probably compensated for by the high hydrostatic pressure of the spongiome.

At the periphery of the spongiome, bundles of stiff, hexagonally packed tubules are occasionally observed. These tubules are approximately 50 nm in diameter and coated on their cytoplasmic face with mushroom-shaped, spirally arranged structures. They are directly connected with the smooth elements of the spongiome. It has been suggested that the fluid is segregated from the cytoplasm into these decorated tubules. From here it passes into the smooth elements of the spongiome and then into the ampullae. The ampullae are not directly connected with the spongiome. From the ampullae, the fluid is transported into the contractile vacuole.

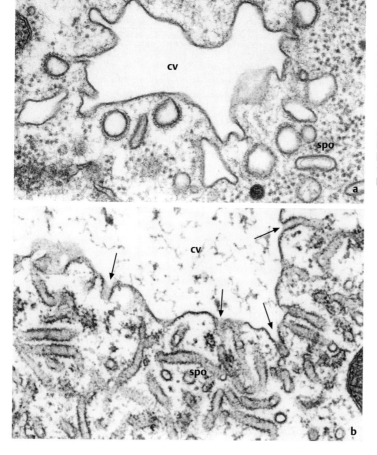

Fig. **199** Contractile vacuole (cv) and spongiome (spo) of the cryptomonad *Cyathomomas* (**a**) and the ciliate *Pseudomicrothorax* (**b**). In flagellates, normally a vesicular spongiome is found, while in ciliates, tubular systems dominate: the tubules are permanently connected with the lumen of the contractile vacuole (arrows) (a from Hausmann and Patterson: Exp. Cell Res. 135 [1981] 449; b from Hausmann: Arch Protistenk. 127 [1983] 319). Magn.: a 48 000×, b 55 000×.

Segregation of Fluids

It seems likely that the spongiome is involved in the process of segregation of fluids. Qualitative analyses of the cytoplasm, and the fluids of the contractile vacuole have shown that the fluid contains some ions, but far fewer than the cytoplasm. The molecular mechanisms of segregation of fluid is not yet understood in all details. The most likely mechanism is that the membranes of the spongiome contain pumps that transport protons into the lumen of the spongiome tubules. This view is supported by the presence of phosphatase activity within the contractile vacuole complex. Moreover, it has been shown that the mushroom-shaped elements represent those pumps. Similarly decorated membranous systems are found to be a part of the spongiome in almost all contractile vacuole complexes investigated in detail, as well as in metazoan cells like the apical membrane of toad bladder epithelium cells, in osteoclast's cell membranes, in regions where they secrete protons to dissolve bones, in the apical membranes of vertebrate kidney distal tubule intercalated cells, that pump protons into the urine.

Expulsion of Fluids

In those protozoa where the contractile vacuole has no fixed position, as, e. g., in most amoebae, the contractile vacuole fuses with the plasma membrane almost anywhere, although usually in the region of the uroid, and never in the region of an advancing pseudopod. This represents, then, a process of exocytosis. This is not always the case, however. In the chloromonad *Chlamydomonas,* expulsion of fluids is believed to occur without membrane fusion, and through hydrophilic channels located in the membrane. In the contractile vacuole of certain trypanosomes, only a small area of the contractile vacuole fuses with the plasmalemma, and thus small channels are produced for the expulsion of fluids.

Especially, in ciliates, but also in other protozoa, permanent, specially organized expulsion pores are present (see Fig. **198**). These preformed pores are necessary, as the presence of the complex cortical elements would preclude fusion with the plasma membrane.

Osmoregulatory Function

For many years, it has been suggested that the contractile vacuole complex is involved in osmoregulation. It is known that the osmotic pressure of the cytoplasm of a protozoon is higher than that of the surrounding medium. As a consequence, water flows into these cells. Water also enters the cell as a consequence of other activities such as feeding, e. g., via food vacuoles. In the absence of counteracting structures, such as a cell wall, or other mechanisms of controlling cell volume and internal pressure, the cell would burst from the influx of water. This may be easily demonstrated by increasing or decreasing the osmotic pressure of the surrounding medium, which results in a corresponding increase or decrease in the activity of the contractile vacuole.

In addition to the undisputed osmoregulatory function of the contractile vacuole complex, a number of other functions have been suggested. Cations such as sodium may be actively secreted via the contractile vacuole. The protozoan cell contains a relatively high concentration of potassium ions, but a low concentration of sodium. In contrast, the fluids secreted by the contractile vacuole are relatively high in sodium. However, an active mechanism for the transport of sodium has not yet been demonstrated for the contractile vacuole complex.

Mechanism for Regulating the Cell Volume

The contractile vacuole appears to represent only one mechanism, whereby a protozoon regulates its cell volume and responds to its environment. The response of the contractile vacuole seems to be involved in short-term adjustments. If a protozoon is placed in a hypertonic medium,

it immediately decreases in size, and its contractile vacuole stops cycling. If the osmotic shock has not been lethal, the cell eventually regains its normal size and the contractile vacuole begins to operate again. This adjustment is caused by an increase in the concentration of free amino acids, monovalent cations, and possibly sugars in the cytoplasm which act to offset the increased ionic strength of the medium.

Contractility of Contractile Vacuoles

Some scientists refer to contractile vacuoles as either pulsating vacuoles or water expulsion vesicles because the term contractile vacuole implies that these vacuoles actively contract. However, there has been no conclusive demonstration of the active contraction of these vacuoles, and it has been argued that the contractile vacuoles empty simply as a consequence of the cytoplasmic pressure within the cell.

One important argument in favor of contractility is (besides the somewhat superficial demonstration of actin in the periphery of the vacuole) the observation that the contractile vacuoles of different protozoa decrease slightly in diameter and become perfectly spherical just prior to expulsion of their contents. Furthermore, if the expulsion pore of a ciliate is mechanically blocked by the presence of a coverslip, the contractile vacuole increases in size and tentatively contracts with the same rhythm as if it was expelling its content. The simplest explanation for these observations is that at the moment of contraction, contractile elements act on all sides of the contractile vacuole.

Pulsating Cycle

The typical pulsating cycle is shown in figure **200** as it occurs in a primitively structured contractile vacuole complex. Following systole, no structures such as small vesicles are observed adjacent to the contractile vacuole. During diastole, small vesicles appear, grow in size, fuse with each other, and finally form the contractile vacuole. Just before the expulsion of its contents, the vacuole becomes spherical and is surrounded by small vesicles.

In addition to this simple process, there are more complicated ones, such as shown for *Paramecium caudatum* (Fig. **201**). Following diastole, the contractile vacuole is not visible. The ampullae are filled (diastole of the ampullae), and

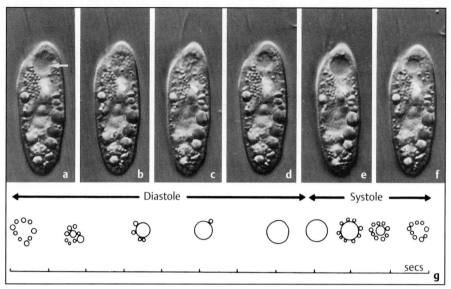

Fig. 200 Pulsation cycle of the contractile vacuole in the cryptomonad *Chilomonas* (**a–f**) and corresponding schematic representation (**g**); one pulsation cycle takes about 10 seconds (from Patterson and Hausmann: Br. phycol. J. 16 [1981] 429). Magn.: 2800×.

then the contractile vacuole appears and is filled from the ampullae (systole of the ampullae). Then the contractile vacuole becomes spherical and expels its contents through the pore (systole of the contractile vacuole) to the outside of the cell. In the meantime, the ampullae are filling once again. The fluid appears to flow continuously from the collecting channels into the ampullae. The ampullae may therefore act as a type of buffer, which allows the delivery of the vacuolar contents despite the continuous nature of the water segregation process.

Pusules

Although dinoflagellates do not have contractile vacuoles, it is suggested that the pusules, which are located at the bases of the flagella, perform a similar function. A pusule is a tube-shaped, branched invagination of the plasma membrane. It is closely associated with a system of intracellular vacuoles. Unlike contractile vacuoles, pusules are permanently open to the surrounding medium (Fig. **202**).

In some dinoflagellates, there are two types of pusules, i. e., collecting pusules and sac pusules. The latter consist of ampullae that are surrounded by vesicles and have a wide channel towards the outside. In contrast, collecting pusules are composed of one or two rather large vacuoles and a narrow channel toward the outside. The sac pusules are primarily found in armed, marine dinoflagellates.

Pusules may have an osmoregulatory function, but they only contract sporadically, if at all. There are doubts as to whether or not these structures are involved in osmoregulation at all, and convincing evidence remains to be presented.

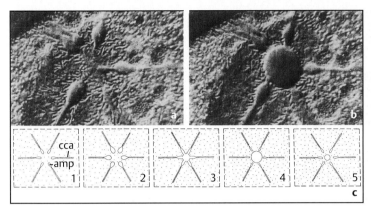

Fig. **201** Pulsation cycle of the contractile vacuole in *Paramecium*. **a** systole. **b** diastole. **c** schematic representation of some intermediate stages (1–5) of the cycle which in *P. caudatum* takes normally 8 seconds. amp = ampulla, cca = collecting canal. Magn.: and a b 750 ×.

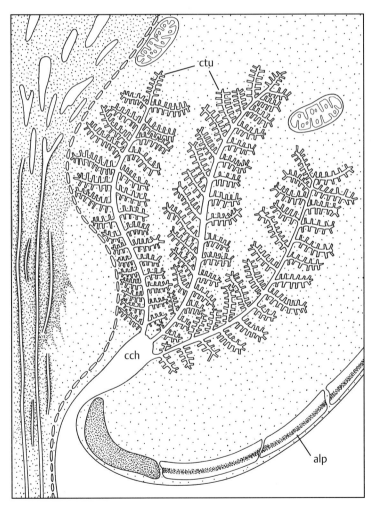

Fig. **202** Canal system of the pusule in the dinoflagellate *Oodinium*. alp = alveolar plates, cch = collecting chamber, ctu = collecting tubules (adapted from Cachon).

Motility

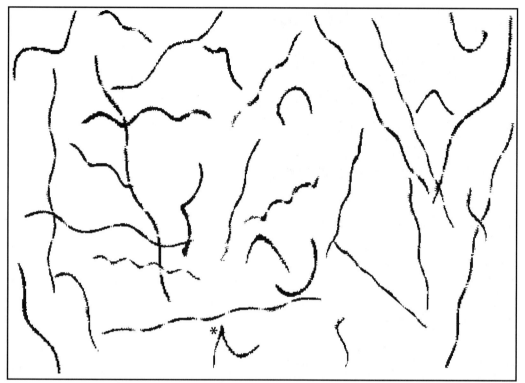

Fig. **203** Traces and velocities of swimming *Paramecium* cells in a stimuli-deprived environment (video analysis for 2.5 seconds). The traces are left-winding helices and correspond to a maximum distance of about 4 mm. * = swimming reversal (courtesy of H. Machemer and R. Bräucker, Bochum).

All protozoa are motile (Fig. **203**). Motility includes locomotion, the creation of water currents, and intracellular movements. Most types of motility are the result of the activity of cilia or flagella (swimming, streaming of fluids, etc.). While the mechanisms underlying protozoan motility are known in broad outline, detailed explanations of the fundamental mechanisms are either unavailable or largely speculative. This is certainly the case for specialized movements such as gliding and certain changes in cell shape.

Cilia and Flagella

Cilia and flagella are thread-like structures with a diameter of 0.2 μm. They originate from basal bodies, which are usually located in the cortex. The ciliary or flagellar shaft (the axoneme) is covered by the plasma membrane (Fig. **204**). The cilia and flagella of eukaryotes are, therefore, intracellular organelles in contrast to prokaryote flagella, which are extracellular. Much of our knowledge about structure and function of the axoneme comes from studies on flagellated and ciliated protozoa.

The beat of cilia and flagella is a striking phenomenon which has been observed since the

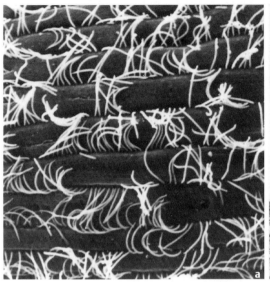

Fig. **204**   Cilia of ciliates in front view (**a**, *Homalozoon*) and in longitudinal section (**b**, *Paramecium*). axn = axoneme, ks = kinetosome, pm = plasma membrane. Magn.: a 1700×, b 45000×.

dawn of protozoological studies of ciliates and flagellates. The first impression is that the movement of cilia seems to differ from that of flagella. Flagella appear to beat in a three-dimensional wave pattern whereas many cilia seem to beat in a two-dimensional plane (Fig. **205**). However, more detailed investigation reveals that cilia also beat in a helical or threedimensional manner, and some flagella move only in a single plane. Cross sections of cilia and flagella show the axoneme with its characteristic arrangement of microtubules, the so-called 9×2+2 pattern, which is surrounded by the plasma membrane (Fig. **206**). There are nine peripheral double microtubules (= doublets) and two individual central microtubules (= singlets). The central tubules are surrounded by a sheath made of clasps. The doublets each consist of a complete microtubule with 13 protofilaments (= A-tubule), and an attached microtubule (= B-tubule) which has only ten protofilaments of its own, but shares three protofilaments with the A-tubule.

Each A-tubule is the origin of pairs of arm-like structures (dynein arms) which are directed toward the B-tubule of the adjacent doublet. The doublets are interconnected by nexin links which also attach to the A-tubules. The central microtubules are surrounded by a central sheath material from which radial spokes are directed towards, and connected with, the A-tubules of each doublet. None of these structures are continuous along the length of the axoneme, and each has its own periodicity of occurrence.

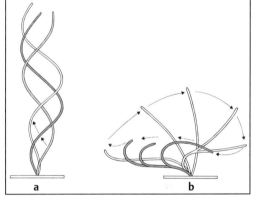

Abb. **205**   Diagram illustrating undulatory flagellar (a) and ciliary movements (b) (adapted from Satir).

The organization of the basal body differs from that of the axoneme. The central tubules of the cilium terminate at an axial granule; the peripheral doublets continue into the basal body where an additional C-tubule is added. Like the B-tubule, the C-tubule is an incomplete microtubule, in this case, sharing protofilaments with the B-tubule (Fig. **207**). In the ciliates, there are often nonciliated (sometimes referred to as barren) basal bodies which may be paired with ciliated ones. The basal bodies of flagellates and ciliates are associated with microtubular and microfilamentous structures which are arranged in a species-specific manner. The number, site of insertion, and organization of these elements may

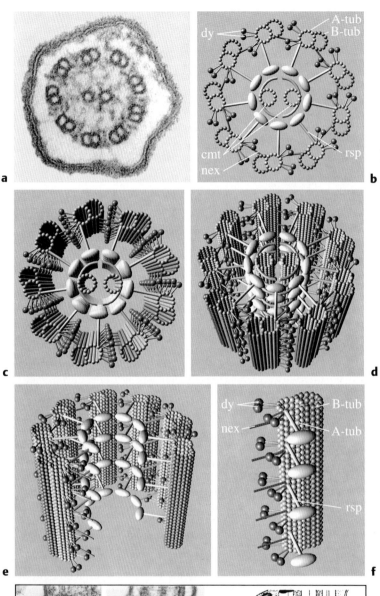

Fig. **206** Cilium in transverse section (**a**) and in corresponding schematic 9 × 2 + 2-representation (**b**). c–f three-dimensional reconstruction of the axoneme (**c–e**) and a microtubular doublet (**f**). A-tub = A-microtubule, B-tub = B-microtubule, cmt = central pair of microtubules, dy = dynein arm, nex = nexin (interdoublet link), rsp = radial spoke (b–f adapted from Hausmann and Gradias). Magn.: a 150 000×.

Fig. **207** Ciliary base: **a** transverse section at the transitional region between axoneme and kinetosome; nine axonemal doublets (upper part) and three kinetosomal triplets (lower part) are visible. **b** longitudinal section of axoneme and kinetosome. c schematic representation (c from Dentler: Int. Rev. Cytol. 72 [1982] 1). Magn.: a and b 60 000×.

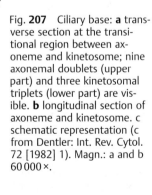

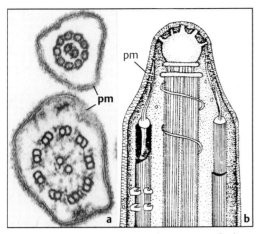

Fig. **208** Ciliary apex: **a** transverse section of ciliary tips exhibits only A-microtubules. **b** anchoring of central pair microtubules at the tip. pm = plasma membrane (b from Dentler: Int. Rev. Cytol. 72 [1982] 1). Magn.: 85 000×.

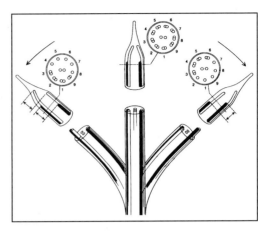

Fig. **209** Model of the sliding mechanism between outer doublet microtubules (adapted from Satir).

be used to determine phylogenetic relationship between different groups of ciliates and flagellates.

The tips of cilia and flagella are organized in a specific manner. The B-tubules are shorter than the A-tubules, and the peripheral doublet microtubules are shorter than the central pair of microtubules. Thus, a section through the very tip of a flagellum or cilium may show two central microtubules or two central microtubules and nine peripheral A-tubules (Fig. **208**). The central tubules may be attached to the plasma membrane in a variety of different ways. This is a species-specific phenomenon.

The 9×2+2 pattern is highly conserved from one organism to another. The diameter of the axoneme is always 0.2 µm, and the arrangement of the additional elements such as the dynein arms, nexin links, and radial spokes always shows the same specific periodicity along the axoneme. However, several variant forms have been found. The flagella of certain gregarine gametes may have a 9×2+5, 9×2+0, 6×2+0, or 3×2+0 pattern. These modified forms are still motile, but are much less active than is the case in protozoa with the normal axonemal pattern.

The asymmetry of microtubules at the tip of the cilium or flagellum stimulated the hypothesis that the activity of cilia and flagella might be propagated by the sliding of peripheral doublets against one another (Fig. **209**). The evidence for this came from the ultrastructural examination of cilia bent in different directions. The interconnection of peripheral doublets and the central pair of microtubules by the nexin links and radial spokes convert the sliding movements of the peripheral doublets into a bending motion.

It is now established that the dynein arms have an ATPase activity; they are the force-generating elements of ciliary or flagellar motion (Fig. **210**). In the resting state, dynein arms point diagonally towards the cell surface. The addition of ATP and activation causes them to contact the adjacent B-tubule. They then move upwards to a horizontal position pushing the doublets relative to one another. The splitting of ATP to ADP + P_i results in detachment of the dynein arms from the B-tubule. The dynein arms regain their attachment and upward position as they are charged with another ATP molecule, and the cycle starts again. The cyclic attachment, translocation. cleavage, and recharging of the dynein arms may be repeated at a rapid rate approximately 50 times per second. The plasma membrane which covers the cilium guarantees an equal distribution of ATP at all levels of the axoneme. Dynein arms are known to contain at least five subunits, and the beat of the movement of the arms is a consequence of the three-dimensional rearrangement of the subunits as their protein conformation changes due to the presence or absence of ATP, ADP + P_i, and Ca^{2+}.

The remarkably smooth movement of the cilium or flagellum must be a result of the activity of different arms on different peripheral doublets acting in a specific sequence. For example,

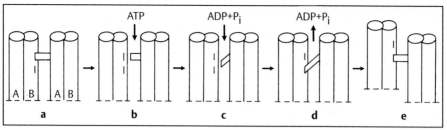

Fig. **210** Possible scheme for the dynein crossbridge cycle and ATP hydrolysis. **a** in the absence of ATP, the dynein arm of A-tubule binds to the adjacent B-tubule. **b** binding of ATP to dynein causes the arm to detach and to shorten. **c** and **d** splitting of ATP coincides with shifting of the arm to an angle of about 40° and reattaching the B-tubule at the next possible position. **e** reorientation of the dynein arm to the original position causes sliding of the adjacent doublet (adapted from Satir).

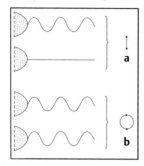

Fig. **211** Flagellar beat patterns and waveforms in frontal and side view. a uniplanar beating. b helicoidal beating.

if the effective stroke of a cilium is directed towards the left, dynein arms of the upper doublets (in the cross sectional view) must be active, while doublets of the lower half are inactive. The opposite situation would be true for the recovery stroke; that is, the lower doublets would be the active ones.

It has been suggested that the central tubules may have a regulatory function. At least in some cilia with a three-dimensional beat, as in *Paramecium*, it has been demonstrated that the central pair of microtubules rotates within the central sheath. Others suggest that rotation may be a passive response, and that central microtubules provide the axoneme with extra strength.

The ciliary beat may be divided into two phases: the effective stroke and the return stroke (see Fig. **205**). During the effective stroke, the cilium remains largely straight; it bends primarily at its base. During the return stroke, the cilium bends along its length in a wave which is propagated from base to tip; during this course, the cilium remains at a relatively short distance from the cell surface. During the effective stroke, the cilium moves the surrounding water, whereas during the return stroke, there is little net movement of the water. In this way, ciliary activity either moves the organism in the opposite direction of the effective stroke or moves water in the direction of the effective stroke.

The mode of flagellar beat may be different from this. In extreme cases, we see either a uniplanar stroke or a helical stroke (Fig. **211**), and there are many variations between these two extremes. The sinusoidal wave of flagellar motion generates a water current which moves the organism forward. In the case of flagella which originate at the posterior end of the cell (recurrent flagella), the sinusoidal wave is directed from the basal body toward the posterior end of the organism. If the flagellum originates at the anterior end of the organism, the waves are often propagated from the flagellar tip towards the basal body. In both cases, the organism swims forward. In the case of those organisms which have flagella covered with mastigonemes such as *Ochromonas* (Fig. **212**), the tip-to-base wave can move the organism in the direction of the tip of the flagellum. This is because the paddle-like mastigonemes reverse the hydrodynamic forces (comp. Fig. **231**). There are many different patterns of ciliary and flagellar activity, and some cilia and flagella can perform different types of beat, often associated with different cellular activities. For instance, some cilia have multiple functions, being used for locomotion in one type of movement, and for feeding in another type of activity. This raises the question of how flagellar and ciliary activities are regulated.

Cilia and flagella can change the direction of their effective stroke. This is best seen in the classic example of the avoiding reaction of *Paramecium* (Fig. **213**). When a *Paramecium* collides

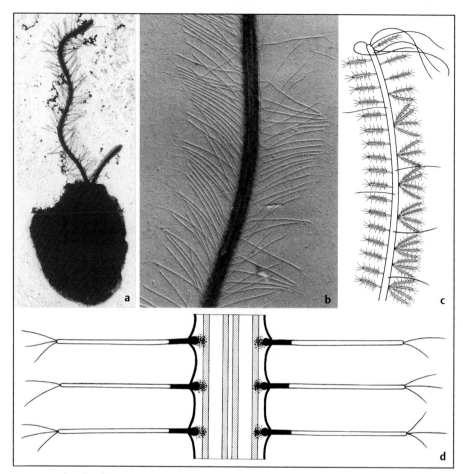

Fig. **212** Flagellar hairs or mastigonemes in chrysomonads. **a** heterokont flagella in *Poterio-ochromonas*; **b** flagellar mastigonemes in *Paraphysomonas cylicophora*; **c** orientation of several types of mastigonemes at the longer flagellum of *Ochromonas danica*; **d** schematic representation of the tripartition of mastigonemes or stramenopili (b courtesy of B. S. C. Leadbeater, Birmingham, c adapted from Bouck). Magn.: a 5000×, b 16 000×.

with a barrier, the organism suddenly swims backwards, tumbles, and swims off in a new direction. This behavior is explained as follows: Contact with the barrier causes a depolarization of the plasma membrane resulting in a change in calcium permeability of ciliary membranes, and a sudden influx of calcium. The increase in ciliary calcium concentration results in a reorientation of the effective stroke, and the cell swims backward: Immediately following mechanical contact, the calcium is pumped out of the cytoplasm, which results in a gradual reorientation as the level reaches the point where there is neither active backward swimming nor active forward swimming. Finally, the cell starts swimming forward again when the intracytoplasmic calcium level becomes normal.

It was suggested that intramembranous particle aggregations, which are located at the base of the cilium, might represent the calcium channels. These presumed channels are located in such a manner as to be aligned with the peripheral doublets (see Fig. **207 c**). However, detailed observations reveal that membrane aggregations found in other ciliates are arranged in a different manner to those of *Paramecium*, and do not often occur with nine-fold symmetry. There have been many studies which document the role of calcium in the regulation of ciliary and flagellar activity, so it is likely that there are locations within the cell where calcium can be sequestered. One suggestion is that the alveoli below the cell surface provide such a location. If

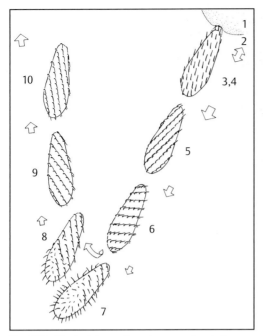

Fig. **213** Bioelectric and motor events during an avoiding reaction of *Paramecium* induced by a local stimulus. Swimming velocity is illustrated by size of open arrows. Cilia are represented at the end of their power strokes and as part of a metachronal pattern (exceptions: initial synchronous ciliary reversal (4), and inactivated cilia (7, 8). During tumbling at the end of the backward swimming period (4–7) reorientation is arbitrary and results from transient ciliary inactivity (7, 8) (from Machemer and de Peyer: Verh. Dtsch. Zool. Ges. 70 [1977] 86).

this is the case, the alveoli would represent a repository for calcium and would be analogous to the sarcoplasmic reticulum of vertebrate skeletal muscle.

Not only is the beat of a single cilium or flagellum regulated, there is also a coordination of the activities of these organelles wherever they occur in close associations. This is best seen in the metachronal wave patterns in the cilia of *Paramecium* and flagella of *Opalina*. The cilia parallel to the beat direction of *Paramecium* are all in the same phase of beat, and their state of bending is shifted in phase with respect to the adjacent cilia. This results in a characteristic wave pattern (Fig. **214**). In *Paramecium*, the beat orientation is inclined with respect to the longitudinal axis of the organism. This results in a helical motion of the organism as it swims. During forward locomotion, the metachronal waves move

from the posterior-left to the anterior-right of the organism. Currently, the opinion is that the coordination of ciliary beat is a consequence of hydrodynamic effects with are influenced by the positioning of the cilia. Cilia which are closely associated with one another, such as the paroral membrane (= undulating membrane) of oligohymenophoreans or the cirri of hypotrichs, move in a highly coordinated manner, or they are synchronized, giving the impression of being a single morphological unit. Such compound ciliary organelles were considered in the older literature to be mechanically coupled to one another. There is no evidence of this from electron microscopy. Careful observations demonstrate that cirri and membranelles are composed of individual cilia in tight groups, and only in a few single cases has there been any demonstration of connections between cilia.

Another ancient hypothesis is that ciliary activity is coordinated by a neuromotorium brain and neurofibrillar elements (the equivalent to subcellular neurons). These ideas were based on the results of silver impregnation which elucidated some of the fiber systems of the infraciliature of ciliates. The fact that many of the early silver impregnating procedures were originally devised to stain neurons in vertebrates lent weight to this notion. Experiments were eventually performed in which some of these connections were destroyed without consequential loss of coordination. Results from electrophysiological studies also contradict this hypothesis. A strong biological argument against coordination by a neurofibrillar network is based on observations of *Mixotricha paradoxa*. This organism is a flagellate which lives in the gut of cockroaches. Its cell surface is covered with commensal spirochetes. The spirochetes are not connected to one another, but they nevertheless move in a metachronal wave pattern reminiscent of the flagella of *Opalina*.

There are different types of metachronism (Fig. **215**). If the effective stroke goes in the direction of the phase shift, this is called symplectic metachrony, and is commonly found in endocommensals such as *Opalina* and *Isotricha*. If the effective stroke goes at right angles to the right of the phase shift, this is called dexioplectic metachronism, and is found in most free-living ciliates, such as *Paramecium*. The type of metachronism of an organism is coupled to the sense of counterclockwise ciliary gyration, but an increase in the viscosity of the surrounding medium can modify it. For instance, in the presence of the viscous methyl cellulose, *Para-*

Fig. **214** Metachronal waves in *Paramecium* (from Hausmann: Mikrokosmos 63 [1974] 165). Magn.: 2000×.

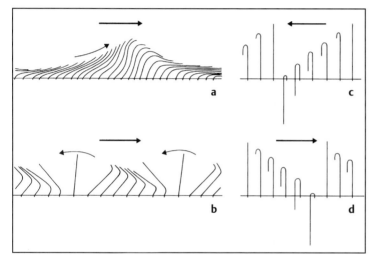

Fig. **215** Coordination types and waveforms of cilia. **a** symplexy. **b** antiplexy. **c** laeoplexy. **d** dexioplexy (adapted from Sleigh).

mecium may convert metachronism to a symplectic form.

In addition to the structures of the axoneme described above, certain cilia and flagella may have accessory structures. For example, the flagella of euglenoids, kinetoplastids, and the transverse flagellum of dinoflagellates contain a paraxial rod which lies parallel to the axoneme and may act to stiffen the flagellum (see Fig. **24**). The trailing flagellum of dinoflagellates may include contractile filament bundles which allow these flagellates to quickly retract this flagellum into the cytoplasm.

Many flagellates have extracellular accessory structures associated with the plasma membrane such as the mastigonemes (see Fig. **24** and **212**) and scales (see Fig. **145**). Undulating membranes, seen for instance in the trypanosomes (see Fig. **56**), appear to be especially effective for locomotion in relatively viscous media. A special modification of a flagellum has been described in the kinetoplastid *Cryptobia*. This organism lives as a symbiont in the spermatheca of the land snails *Triadopsis* and *Helix*. The membrane of the basal 5–10 μm section of the flagellum forms finger-like projections which wind around the microvilli of the spermathecal epithelium. In this way, these flagellates attach to their host (Fig. **216**).

Axostyle and Costa

Axostyles are massive bundles of closely packed sheets of microtubules (see Figs. **21** and **45**). They occur in symbiotic flagellates such as oxymonads, trichomonads, and hypermatigids. In some of these organisms, the axostyles are motile and are responsible for a winding, twisting movement of the cell. Additionally, they may rotate around their own axis and contribute to the motility of the organism. Among other proteins, a dynein with ATPase activity has been isolated from axostyles. Thus, it seems reasonable

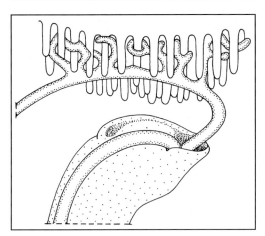

Fig. **216**   Flagellar branching as holdfast device in the trypanosomatid *Cryptobia* (adapted from Current).

to assume that axostyle motility is based on the same fundamental mechanism which exists in cilia and flagella; that microtubules move past one another due to the activity of dynein molecules, and at the expense of ATP. Interconnections between microtubules of the axostyle have been observed. Presumably these represent dynein molecules.

A number of the flagellates that have an axostyle also contain a rod-shaped organelle called the costa. The proteinaceous costa originates at the flagellar basal body and extends towards the posterior end of the cell. The costa appears as a striated fiber in the electron microscope. In addition to a skeletal function, it may also be responsible for some bending movements of these organisms.

Haptonema

The haptonema is an organelle which is found only in members of the prymnesiids (formerly called Haptophyceae). It is situated between the two flagella (see Fig. **68**), and this extension of the cell surface has approximately the same diameter as the adjacent flagella. Haptonemes may be of very different lengths. In cross section, a haptonema is seen to include 6–7 microtubules and an ER cisterna within the limiting plasma membrane. A haptonema is glutinaceous and, acting as a holdfast, allows an organism to attach to a substrate. Haptonemes show pronounced movements and can coil up in 1/50th to

1/100th of a second. The process of uncoiling lasts 2–10 seconds. The motive force associated with this activity is still unknown. In at least one case, the biological significance of the haptonema has been elucidated convincingly: It is involved in food capture (comp. Fig. **234**).

Amoeboid Movement

While amoeboid movement is normally associated with the locomotion of certain protozoa by means of pseudopodial extension, the phenomenon and general mechanism may be relevant to other forms of cytoplasmic streaming. As in other eukaryotic cells, cytoplasmic streaming (e. g., cyclosis) is a natural activity of protists. However, cytoplasmic movements do not usually result in whole cell translocation or change in cell shape.

Amoeboid motion has also been examined in the amoebae of *Dictyostelium*; most intensively in the lobose rhizopods *Amoeba proteus* and *Chaos chaos* (the mechanism of locomotion in the many other different types of amoeba, with different types of pseudopodia, is largely unknown). In *Amoeba proteus*, there is a hyaline cap or clear region at the tip of an extending lobopodium. Just posterior to the tip, the hyaline cap narrows to a thin hyaline layer which surrounds the rest of the cell (Fig. **217**). During pseudopodial extension, the cytoplasm in the central part of the pseudopodium streams forward as a granular liquid (the sol state of the endoplasm). In the cortical region, however, the cytoplasm is stationary (gel state of the ectoplasm). When the sol endoplasm reaches the hyaline cap, it flows like a fountain into the adjacent cortical area where it changes to the gel state, and becomes part of the ectoplasm. At the posterior end of the amoeba, the uroid, there is a continuous transformation of ectoplasm into endoplasm (Fig. **217**).

The proteins actin and myosin have both been isolated from amoebae. In metazoan cells, these proteins are known to be responsible for the forces involved in muscle contraction. There is no doubt that this is the role that they play in amoebae, providing the motive force responsible for cytoplasmic streaming. There is, however, controversy as to how this is accomplished. At present there are two major hypotheses: the hydraulic-pressure hypothesis, and the frontal-zone contraction hypothesis. In both hypotheses, it is assumed that motion is produced by the transformation of ectoplasm to endoplasm or endoplasm to ectoplasm.

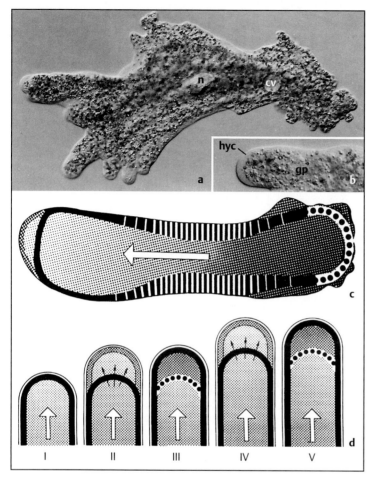

Fig. **217** Amoeboid movement: **a** locomotory form of *Amoeba proteus*. n = nucleus, cv = contractile vacuole. **b** in frontal regions of pseudopodia of *A. proteus* the hyaline cytoplasm (hyc) and granuloplasm (gp) are clearly discernible. **c** model of cytoplasmic streaming (arrow); lightly stippled = sol cytoplasm; darkly stippled = gel cytoplasm; solid back layer = newly formed microfilament system; interrupted black layer = contracted microfilament system; black dotted layer = disassembling microfilament system. **d** formation of pseudopodia by cyclic local relaxation of the ectoplasmic cylinder by separation of the cortical layer from the plasma membrane (II, IV), formation of a new layer at the pseudopodial tip (III, V), and destruction of the cortical layer at the hyalogranuloplasmic border (III, V) and alternate solation (lightly stippled) and gelation of the hyaloplasm between the layer and the cytoplasm (darkly stippled) (from Stockem and Klopocka: Int. Rev. Cytol. 112 [1988] 137). Magn.: a 350×, b 400×.

The hydraulic-pressure hypothesis holds that there is an actin/myosin network located beneath the plasma membrane of the entire cell which contracts in the intermediate–posterior region and generates an anteriorly–directed pressure; this drives the cytoplasm towards the advancing pseudopodia. This suggestion is supported by observations of the acellular slime mold *Physarum*, in which the phenomenon known as shuttle streaming occurs. In this case, there can be no doubt that the streaming to the cytoplasm is caused by a posterior pressure. The extension of this hypothesis of other lobopodial amoebae is supported by the observation that the bulk of the actin filaments are located in the region of the uroid. At the uroid, the plasma membrane is extensively folded, an observation which suggests that a contractile process is at work.

The frontal-zone contraction hypothesis favors the idea that the motive force for cytoplasmic streaming is produced at the front of the advancing pseudopodium. It is thought that, at the tip of the pseudopodium, the sol-like cytoplasm changes to the gel state by a process of contraction, and that during this process, more sol-like cytoplasm is carried along. The ectoplasm remains in the gel (contracted) state until it reaches the uroid, where it is again transformed to the sol state. This hypothesis is supported by two observations. Firstly, streaming often appears to begin at the front edge of the cell during pseudopodial formation. If one experimentally generates a low pressure within the uroid cytoplasm, there is no influence on cytoplasmic streaming. Secondly, the observation that demembranated cytoplasm can, under certain circumstances, show streaming phenomena

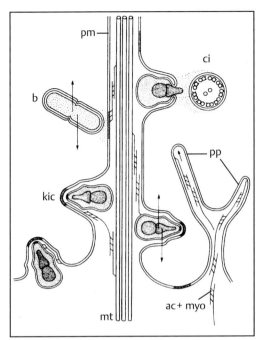

Fig. **218** Scheme of axopodial motion phenomena and pseudopodial movements in a heliozoan. ac+myo = actomyosin filaments, b = bacterium affixed to the axopodial surface, ci = cilium of a prey organism, kic = kinetocyst, mt = microtubules, pm = plasma membrane, pp = pseudopodial projections (adapted from Bardele).

like those in natural pseudopodia seems to support this hypothesis.

As of yet, there is no experimental observation which seems to unambiguously support either of these current hypotheses. However, recent measurements indicate that a high pressure is produced, not only in the most posterior region of the cell, but also in the middle region. Forces arising from actin polymerization and myosin activitiy (including myosin I as well as myosin II) drive a number of different actin-based cell movements in metazoan cells, including lamellipodial protrusion. These mechanisms are also discussed to be of importance, e. g., for the locomotion of small amoebae.

Extensive data concerning the organization and contractile activities of microfilaments, and by this the physiology of amoeboid movement, come from studies on *Physarum* shuttle streaming. Due to the large variety of life cycle stages and experimentally derived stages of *Physarum*, microfilaments can attain mainfold manifestations ranging from a rather primitive actomyosin

cortex beneath the plasma membrane in flagellated amoeboid stages, to an extremely complex arrangement of actomyosin fibrils extending through the entire cytoplasmic matrix in vegetative plasmodia. The cortical microfilament system shows a common and permanent cytoskeletal differentiation which serves two main functions: together with the spectrin-like membrane skeleton, the actomyosin cortex stabilizes the cell surface, and participates in morphogenetic events; the actomyosin cortex delivers motive force by regular contractile activities which are transformed via the plasma membrane into hydraulic pressure gradients, and finally result in protoplasmic streaming via gelsol transformations. Besides actin and myosin, a set of different actin-binding proteins has been shown to participate in the assembly and disassembly of the microfilament system in *Physarum*, and to control its numerous functions. Internal Ca^{2+} plays a crucial role in the regulation of microfilament assembly and contractile activities by controlling the myosin ATPase. Moreover, a Ca^{2+} / cAMP loop seems to function as the oscillator triggering the typical protoplasmic shuttle streaming activity with a periodicity in the minute range.

Actin and myosin are certainly involved in other types of contractility such as the filamentous contractile ring involved in the division furrow of dividing ciliates. Furthermore, it has been suggested (although there are no reliable results to date) that cyclosis of food vacuoles and other organelles is driven by actin/myosin complexes. Actin has been demonstrated to be an important component of the cytopharyngeal basket of ciliates, and a suggestion as to how actin and myosin might be involved in the positioning of kinetocysts within the axopodia of Actinopoda and the growth of their pseudopodia is presented in figure **218**.

Euglenoid Movement

The movement of euglenoids, which resembles a flexing, worm-like movement, is often referred to as metaboly or metabolic movement. It is characterized by wave-like peristaltic changes in cell shape along the longitudinal axis of the cell (Fig. **219**). At least in the species *Distigma*, two phases of movement can be distinguished: an expansion or dilation of the cell body which travels as a wave from the posterior part of the organism towards the anterior; followed by an extension of the anterior part of the body, which is

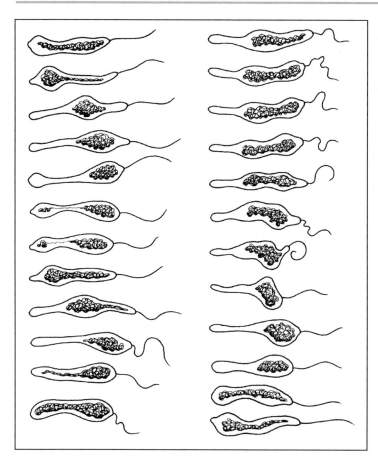

Fig. **219** Typical euglenoid (or metabolic) motility in *Distigma*. Time interval between figures about 2 seconds (from Hausmann and Hülsmann: PhotoMed' 4 [1991] 253).

accompanied by a posterior-directed streaming of the cytoplasm. At the conclusion of these two phases, the organism regains its original form. While euglenoid movement is often accompanied by movements on a substrate, it can also result in motility in the absence of contact with a surface.

The mechanism of this movement is still unclear. The cortex of euglenoids is distinguished by pellicular strips which are usually arranged in a spiral fashion (Fig. **220**). These strips correspond to the location of pellicular proteinaceous plates. Microtubules are found below the pellicular plates, and the arrangement and numbers of microtubules differ according to species. In some species, the plates are rigidly interconnected; such organisms do not exhibit euglenoid movement. In the case of *Distigma*, it has been postulated that the expansion of the cell caused by the activity of the subpellicular microtubules and cytoplasmic streaming in the posterior direc-

tion is a consequence of sol-gel transformations of actin/myosin.

Cell Contraction

A number of heterotrich ciliates are able to contract rapidly. *Stentor*, for example, can contract to one third of its length in a matter of milliseconds. Cell elongation following contraction requires 5–10 seconds.

The elements responsible for the contraction and elongation or relaxation are located in the cortex. The cilia of these heterotrichs are organized in longitudinal kineties, each kinetid composed of a pair of basal bodies. One basal body of each pair (usually the posterior one) has a cilium. Postciliary microtubules are associated with each ciliated basal body. The postciliary microtubules are in the form of a ribbon of microtubules, and they are directed toward the poste-

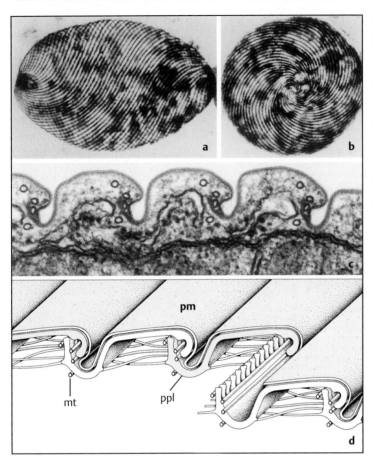

Fig. **220** Pellicular structures in euglenids. **a** and **b** silver-line system of *Euglena*. **c** the helicoidal structures derive from subpellicular protein plates *(Euglena gracilis)*.
d schematic reconstruction of *Astasia longa* pellicle: mt = microtubules, pm = plasma membrane, ppl = protein plates (a and b from Foissner: Acta biol. Acad. Sci. hung. 28 [1977] 157, c courtesy of T. Suzaki, Hiroshima, d adapted from Suzaki and Williamson). Magn.: a and b 1000×, c 90 000×.

rior of the cell. Since they are longer than the distance between kinetids, they overlap and can interact with one another (Fig. **221**). Therefore, each kinetal row is accompanied by a band of overlapping postciliary microtubules which extend from the anterior to the posterior of the cell.

Below each microtubular band, there is a band of 4 nm filaments, called myoneme, located in the cytoplasm. This band is always surrounded by cisternae of the endoplasmic reticulum.

Several different observations and experiments support the idea that the rapid contractions are caused by the activity of the myonemes. During contraction, these filaments change their conformation to become 10–12 nm thick, tubular, and shorter. This conformational change does not require ATP, but it is influenced by calcium. This suggests that the ER cisternae adjacent to the myonemes serve as a site of calcium sequestration, and are analogous to the sarcoplasmic reticulum of vertebrate skeletal muscle. Presumably, they release calcium in response to stimulation, and later actively pump calcium back into their space, thus acting to regulate the contractile state of the myonemes.

The microtubular ribbons are thought to play an active role in the extension of the cell. During contraction, they passively slide past one another. During extension or relaxation, they may actively slide against one another, possibly by means of dynein arms. In this way, they elongate the cell much as one would extend a telescope. This idea is supported by the observation of bridge-like elements seen between adjacent microtubular ribbons. During contraction, the bridges do not have contact with the adjacent microtubular ribbon, but they interconnect the ribbons during elongation.

This system of contraction and elongation is based on two antagonistic mechanisms, and it appears to be widely distributed among members of the Heterotrichida.

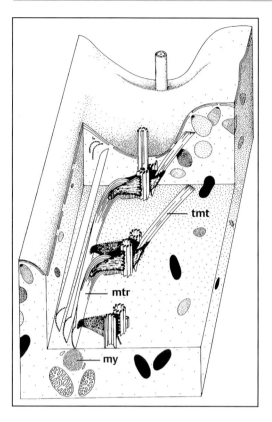

Fig. **221**   Cortical organization of the heterotrich ciliate *Eufolliculina*. The myonemes (my) are responsible for rapid contractions of the cell, the microtubular ribbons (mtr) for reextension of the cell. tmt = transverse microtubules (from Mulisch et al.: Protistologica 17 [1981] 285).

Contractions of Stalks

Many peritrich ciliates are attached to their substrate by means of a stalk. The stalks of some species have the capacity for contraction. In the relaxed state, the stalks can be quite long and and rather straight. Mechanical stimulation will cause them to contract rapidly. When this occurs, the stalk coils up, thus shortening its effective length (Fig. **222**). The stalk itself is bounded by a membrane-like material and contains a membrane-bounded, tube-like projection of the cell which includes a bundle of 2–3 nm filaments, the spasmoneme (= myoneme), narrow tubes of ER which originate in the cell body, and possibly mitochondria. In *Vorticella*, the spasmoneme has a diameter of approximately 1 µm, while the stalk has a diameter of 5–7 µm. Opposite the spasmoneme, there are a number of

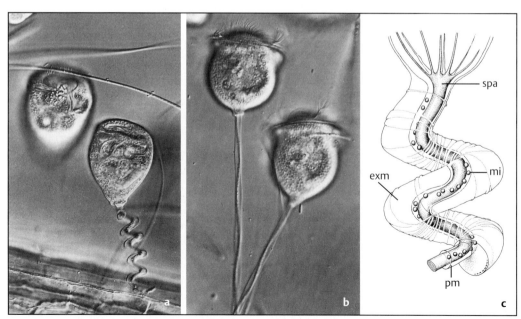

Fig. **222**   Rapid contraction (**a**) and slow reextension of *Vorticella* stalk (**b**); stalk with spasmoneme (spa) flanked by numerous mitochondria (mi) (**c**). exm = extracellular stalk matrix (stalk sheath), pm = plasma membrane (a and b from Kleinig and Sitte: Zellbiologie, Stuttgart 1992; c adapted from Amos). Magn.: a and b 220×.

Fig. **223**   Contractile behavior of the apical region of the haptorian ciliate *Lacrymaria olor* (**a–c**) (from Hausmann: Mikrokosmos 76 [1987] 176). Magn.: a and b 170×.

stiffening rods which are closely associated with the wall of the stalk. The remainder of the stalk is filled with a loosely defined fibrillar material.

It has been demonstrated that stalk contraction is independent of ATP but sensitive to calcium. This suggests a mechanism which is similar to contraction of heterotrich ciliates. The thin ER tubes inside the spasmoneme may store calcium. The extracellular stalk material of the peritrichs is probably very elastic and a likely antagonist to the contractility of the spasmoneme. Activity of the oral ciliature is not required for stalk elongation. The stiffening rods appear to be responsible for the coiling which occurs upon contraction.

Body Bending

Many ciliates are able to bend, fold, or stretch their cell body (Fig. **223**). This may rely on a mechanism similar to that of cell contraction. Ciliates which show this capability have a filamentous sheet located between the cortex and the endoplasm. The cytoplasmic face of this sheet is covered with mitochondria, and the cortical side is associated with numerous small vesicles (Fig. **224**).

This filamentous system may also contract in response to calcium that could be released from the small vesicles. However, to explain the kind of body bending which occurs, the filamentous layer must be capable of being activated at selected sites. Since an antagonist system has not been observed in these organisms, it is

possible to assume that natural turgor pressure will return the cell to its original shape following the contraction phase.

Other Phenomena of Contraction

In addition to the types of contractility discussed above, there are a number of other systems which are involved in movements. In most cases, their mode of function is either unclear or unknown.

The rootlet system, which originates at the basal bodies of flagellates and extends toward the posterior of the cell or interconnects the basal bodies, is one such system. These are bundles of filaments that appear in the electron microscope as striated fibers (Fig. **225**). The corresponding protein has been called centrin. The addition of calcium can cause these fibers to contract. The degree of contraction corresponds directly to changes observed in the periodicity of their striations. In addition to their role as supporting structures for the basal bodies, it has been suggested that they may also influence the direction of the effective stroke of the flagellum by changing the orientation of the basal body with respect to the cell surface.

Filament bundles with a similar morphological appearance have been observed in some ciliates. Here, however, they do not originate from basal bodies, but are found at the filamentous layer between the cortex and the endoplasm (Fig. **225**). They may be involved in the bending of the cell.

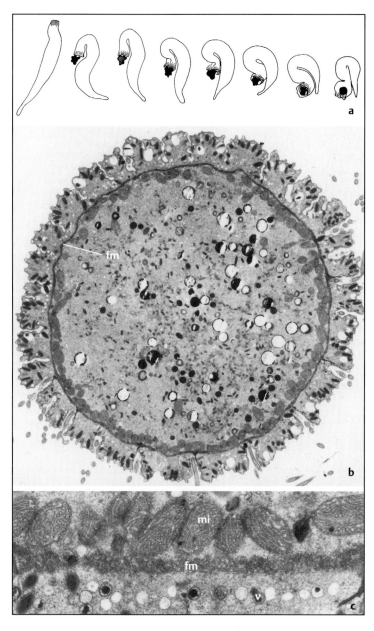

Fig. **224** Twisting cell contractions during food uptake in the haptorid ciliate *Homalozoon* (**a**). **b, c** transverse sections of *Homalozoon* revealing a continuous filament sheath or filamentous mantle (fm) which separates the inner mitochondria-containing cytoplasm from the cortical region. mi = mitochondria, v = vesicle (from Kuhlmann et al.: Protistologica 16 [1980] 39). Magn.: b 5000×, c 27 000×.

Kinetodesmal fibers originate from the basal bodies of ciliates. Although they also show a striated appearance in the electron microscope, there is no evidence that they are contractile. In contrast, the striated myonemes (myophrisks) of the Acantharea are obviously contractile. They originate from spines, and their contractions and relaxations result in the movements of the spines (Fig. **226**). On the structural and functional basis, they resemble the myonemes which are found in the tentacle of the dinoflagellate *Noctiluca scintillans*.

A number of filamentous systems observed in some protozoa have unknown functions. This is true for the cortical filaments of *Paramecium* (see Fig. **107**). Such systems are often described as contractile in the absence of experimental evidence. The filamentous cords associated with the adhesive disc of *Trichodina* are often implicated in the activity of adhesion, yet their actual role has not been documented.

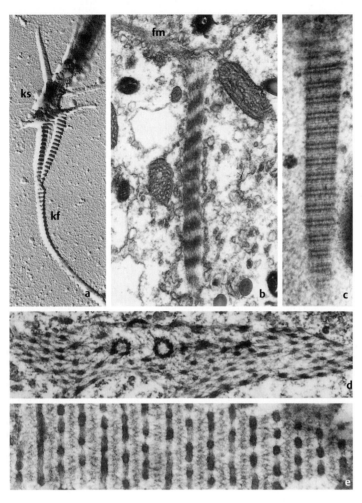

Fig. **225** Structural components possibly involved in motive force generation for bending movement: **a** kinetodesmal fiber (kf) deriving from the kinetosome (ks) of an isolated kinetid. **b** cross-striated filament bundle affixed to the filament mantle (fm) of the haptorian ciliate *Loxophyllum*. **c** periodically striated flagellar rootlet in *Chilomonas*. **d, e** filament systems with regular patterns in the peritrich ciliate *Trichodina*. Magn.: a 25 000×, b 20 000×, c 60 000×, d 30 000×, e 65 000×.

Gliding

There is also a type of movement which is best described as gliding. Gliding is especially found in the trophic stages of gregarines and in the sporozoites of coccidians. These organisms can move along a substrate without any visible sign of contraction or organelle activity that might be responsible for providing the motive force.

It has long been suggested that the movement of the gregarines is based on a directed secretion of mucus, a sort of jet propulsion system. Ultrastructural observations have shown that the cell surface of the gregarines is highly sculptured by many, parallel, longitudinal folds of the cell surface (Fig. **227**). These folds contain microfilaments which are located above microtubules. It was thought that these folds were capable of subtle undulations that could not be seen in the light microscope. Thus they generated the necessary force, yet were below the limit of resolution of the light microscope. This hypothesis was supported by electron microscopy, which showed that the orientation of the folds was not constant.

It is now possible to observe these folds in living material in the light microscope, and no movements of the folds have yet been observed. For this reason, some authors still believe in the older hypothesis of mucus secretion. The folds are thought to guide the expanding mucus towards the posterior of the cell, thereby providing for the forward movement of the cell. However, at present newly designed experiments indicate that an actomyosin system may be implicated in gliding movements of gregaines, and intriguing evidences accumulate in favor for this opinion.

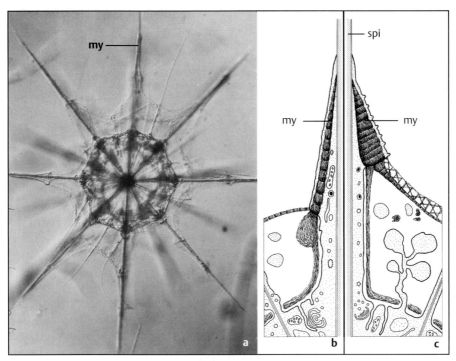

Fig. **226** Acantharian myophrisks (my), surrounding each spine (spi) (**a**), in relaxed (**b**) and contracted state (**c**) (b and c adapted from Febvre and Febvre-Chevalier). Magn.: a 200×.

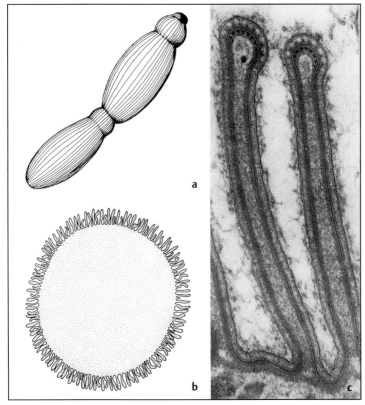

Fig. **227**   Pellicle folds of gregarine syzygy in frontal view (**a**) and in cross-section (**b**). **c** Higher magnification showing pellicle folds of a trophozoite of *Gonospora beloneides* (c courtesy of I. Desportes, Paris). Magn.: 65 000×.

A different type of gliding is suggested by the sporozoites of the coccidia. These infectious stages are able to carry out different kinds of movements, yet they do not appear to have any organelles that could be responsible for movement.

Movements that have been observed do not necessarily need to involve contact with a substrate. It has been suggested that there are binding sites in the plasma membrane which make contact with the substrate. These binding sites are thought to move in a predetermined spiral pathway from anterior to posterior by the action of a contractile filament system. As a consequence, the cell moves. A rather similar model has been proposed to explain the gliding movement of gregarines.

Some bacteria and green algae are also known to move by gliding. In these, the cells are surrounded by a cell wall. In some cases, it is known that gliding is based on the directed secretion of mucus. In most cases, however, the mechanism of gliding is still unknown.

Ingestion, Digestion, and Defecation

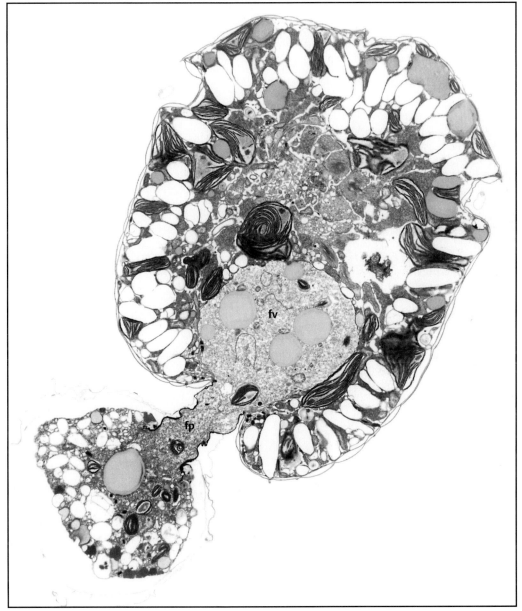

Fig. **228** The mixotrophic dinoflagellate *Amphidinium cryophilum* phagocytozing another dinoflagellate. The prey cytoplasm is ingested by a specialized pseu-dopodium. fv = food vacuole, fp = feeding pseudopodium (from Wilcox and Wedemayer: J. Phycol. 27 [1991] 600). Magn.: 4000×.

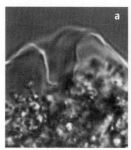 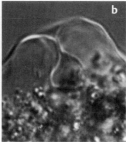 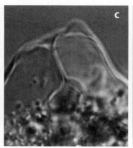 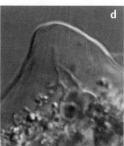

Fig. 229 Endocytosis in *Amoeba proteus*: sequences of pinocytosis induced by 0.5% ovalbumin. The interval between invagination of the cell surface (**a**) and vesiculation (**d**) is about 3 minutes (from Stockem and Klein: Europ. J. Protistol. 23 [1988] 317). Magn.: 1500×.

The protozoa demonstrate a variety of ways of assimilating the substances necessary for life. They include heterotrophs, autotrophs, and organisms that can be either. The autotrophs use exogenous sources of energy, e. g., sunlight to produce organic compounds from inorganic molecules (CO_2) and ions. This is the case for those protozoa which live primarily by photosynthesis. Heterotrophic unicellular organisms, on the other hand, require preformed organic compounds. These organic compounds are derived directly or indirectly from autotrophic organisms. Autotrophic organisms, which also take up or ingest organic compounds or organic particles are referred to as mixotrophic organisms (Fig. 228); and those which can be either autotrophic or heterotrophic, depending upon the prevailing conditions are referred to as amphitrophic. Many parasitic protists rely on the diffusive transport of essential compounds across the plasma membrane. Such organisms are referred to as osmotrophs.

Pinocytosis and Phagocytosis

Nutrient uptake can be divided into two general categories: pinocytosis and phagocytosis. Pinocytosis ("cell drinking") is the uptake of nutrients in solution. The process may also involve some selection and concentration of specific ions and other nutrients. Phagocytosis ("cell feeding") is the ingestion of particulate matter. This division between the two types of feeding is not clear cut since depending upon the organism, there may be transitions between pino- and phagocytosis. However, there is evidence that receptors exist in the plasma membrane by which either pinocytosis or phagocytosis is initiated. The distinction between particulate and nonparticulate food is unclear, since it depends on the degree of optical resolution. The distinction is usually based on observations by light microscopy, and an electron microscope often detects particles which are not detectable by light microscopy.

There are cases where clear distinctions can be made. In *Amoeba proteus*, the production of pinocytic channels and pinosomes is clearly an phenomenon which is distinct from that of phagocytosis (Fig. 229). Pinocytosis in amoebae appears to happen by the contraction of a membrane-associated network of filaments associated with the formation and retraction of pseudopodia. It has been suggested that an external stimulus induces a local influx of calcium ions. This may cause contraction of the membrane-associated filaments. At the same time, the pseudopodium extends. This causes the formation and elongation of the pinocytic channel. Further contraction of the cortical filaments results in the pinching off of pinosomes. The details of this process are only documented for *Amoeba proteus*.

Although phagocytosis occurs in many unicellular organisms, pinocytosis has been observed only in a relatively small number of protists. Pinocytosis appears to occur e. g., at the micropores of the sporozoite stages of the Apicomplexa (see Fig. 84), the parasomal sacs of ciliates (Fig. 230), and in a number of flagellates. In flagellates, pinocytosis is especially evident in the region of the flagellar pocket (e. g., in euglenoids) or at the flagellar groove (e. g., in the Retortamonadea and Diplomonadea).

Selection of Food

Unicellular organisms are able to differentiate between different kinds of food particles.

However, they usually cannot recognize whether the food is digestable or not. *Paramecium* will ingest bacteria as readily as particles of carmine, latex beads, or iron filings. The speed with which particles are ingested is often variable. Yeast cells, for instance, are often rapidly ingested, yet they may not be digested and are often egested in an undigested, intact, viable state.

It is not clear what factors are required for the activation of phagocytosis. The surface properties of foods have been experimentally modified, and this has shown that the physical and chemical surface properties of the food are important for phagocytosis. These properties may be recognized by the protozoa, most probably by the glycocalyx at the plasma membrane. As a consequence, phagocytosis may or may not be activated. After ingestion of the food, internal control mechanisms inform the cells whether the ingested food is digestable. In *Climacostomum*, latex beads are egested much more rapidly than the indigestable residues of bacteria and algae. Predatory ciliates can obviously distinguish between different prey organisms, and since we do not often see examples of cannibalism, they must also be able to distinguish between prey and other members of the same species. However, cannibalism does occur under certain circumstances.

Acquisition of Food

Protozoa acquire their food as particles from the surrounding medium by a variety of means.

Pelagic (= floating) organisms such as *Actinophrys* come in contact with their food organisms more or less by chance. The prey becomes enmeshed in a sticky substance or mucus that is secreted by the axopodia. Electron microscopy has demonstrated that the plasma membrane of such organisms is elaborated into a highly convoluted surface. This enhances the interaction between the axopod and the prey organism by increasing the surface area and increasing the area of contact and the amount of mucus available for contact.

Filter feeders use cilia or flagella to create a current of water which carries the food particles toward the mouth (Figs. **231, 232, 233**). Selection of food particles may occur in the oral area. The haptonema of the Prymnesiomonada (see Fig. **68**) might be, at least in some species, a means for collection of food particles (Fig. **234**).

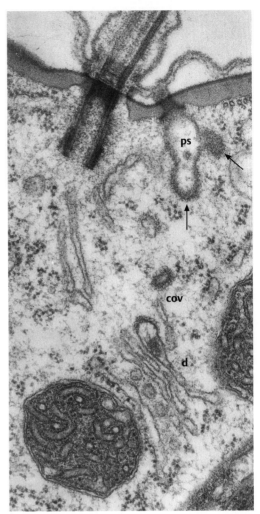

Fig. **230** Pinocytosis in the ciliate *Pseudomicrothorax*: pinching off of coated vesicles (cov) with clathrin-like layer (arrow) at the parasomal sac (ps). d = dictyosome. Magn.: 40 000×.

Gulpers are usually, although not always, slow organisms which are not able to actively hunt their prey. These organisms often have toxicysts which paralyze or kill swimming prey organisms. The immobilized prey can then be ingested without difficulty. The oral area of gulpers can usually be greatly extended (Fig. **235**).

The most sophisticated organelles involved in ingestion are the cytopharyngeal basket of flagellates and ciliates (Fig. **236**), and the tentacles of suctoria (see Fig. **128**). Suctorian tentacles are morphologically and functionally reminiscent of cytopharyngeal baskets that consist of hundreds

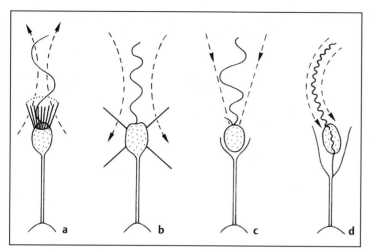

Fig. **231**   Schematic representation of water currents in sessile heterotrophic flagellates with specialized flagellation types: **a** *Codonosiga*. **b** *Actinomonas*. **c** *Ochromonas*. **d** *Bicosoeca*. The naked flagellum (**a**) causes a different streaming compared to flagella with mastigonemes (**b–d**) (adapted from Sleigh).

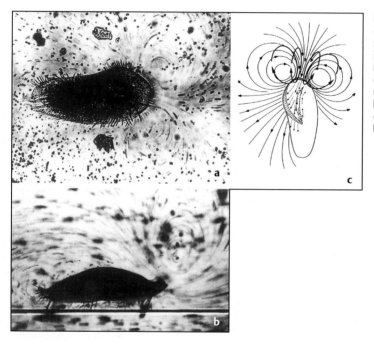

Fig. **232**   Water currents caused by *Stylonychia* in dorsal view (**a**), side view (**b**) and in summarizing schematic representation (**c**); visualization of streaming phenomena by use of coal particles under long-time exposure (courtesy of H. Machemer, Bochum). Magn.: a and b 100×.

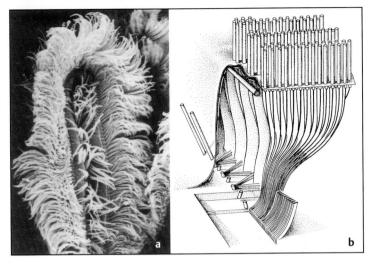

Fig. **233**   Regularly arranged membranelles of *Eufolliculia* at the periphery (**a**) and inside the peristomal wings (**b**) cause water currents under which food particles are transported to the oral region (from Mulisch and Hausmann: Profistologica 20 [1984] 415). Magn.: 1000×.

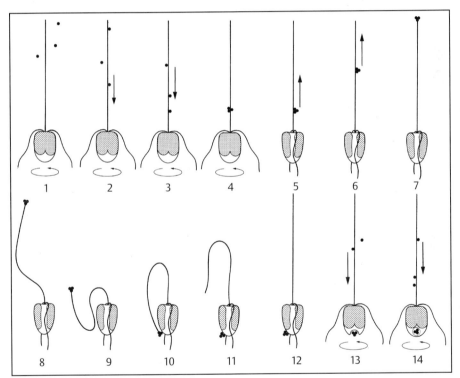

Fig. **234**   Capture (1–2), basal-directed transportation (3), accumulation (4), apical-directed transportation (5–7) and phagocytosis (8–14) of food particles by the haptonema of *Chrysochromulina hirta*; circular arrows indicate rotation phases of the cell (adapted from Kawachi et al.).

sary for the digestion of the cell wall of these organisms (Figs. **237** and **238**). The motive force needed for phagocytosis is created within the cytopharyngeal basket/tentacles. A central role in this process is played by some of the microtubular lamellae. In *Pseudomicrothorax dubius*, ATPase has been localized within these lamellae. However, the microtubules have not been identified with the production of the motive forces, but rather a filamentous system seems responsible for this. At least some of these filaments are actin. It appears that the microtubules

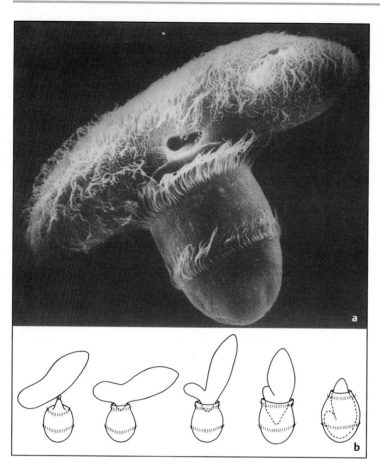

Fig. **235** Ingestion of *Paramecium* by *Didinium nasutum* (a courtesy of G. Antipa, San Francisco). Magn.: 400×.

serve a skeletal role. At the molecular level, the motive force may be created by a gliding mechanism which actively moves the food vacuole into the cell. From this point of view, it is incorrect to describe the ingestion by suctoria as sucking.

Mycophagous ciliates such as *Grossglockneria* and *Pseudoplatyophrya* feed exclusively on fungi and yeast. They have a feeding tube that is somewhat similar to a cytopharyngeal basket, as radially arranged microtubular lamellae are important structural elements here also. During food uptake, the ciliate produces a small hole in the cell wall of its food organisms through which the prey cytoplasm is endocytotically ingested. The mechanism of breaking the fungal cell wall is unknown, but it is conceivable that the perforations are caused enzymatically.

Myzocytosis is a mechanism for food uptake which is found in many ectoparasitic dinoflagellates such as *Gymnodinium* and *Paulsenella*. Here, the prey is pierced by means of an extensible, tube-like peduncle (see Fig. **228**) and the prey cytoplasm is ingested including both dissolved and particulate organic substances.

Some members of the Filosea show a different mode of specialization. Although no permanent oral apparatus for food ingestion is known, a transient circular ingestion area arises. This is indicated by the round holes in the cell walls of food algae created by e. g., vampyrellids. The organisms are able to punch circular holes through the cell walls of their food (green algae such as *Oedogonium* and *Spirogyra*). This may occur by a localized secretion of enzymes. Through this hole, a large feeding pseudopodium invades the cell and ingests the entire protoplast (Fig. **239**). Moreover, circular feeding structures, called amoebostomes (Fig. **240**), can be induced at least experimentally in certain schizopyrenid sarcodines, which normally feed on small bacteria. The origin of these structures and their biological significance are still obscure.

Many armored dinoflagellates digest enclosed large prey cells outside the cell body

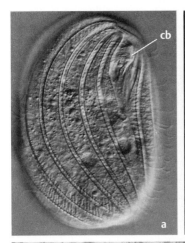
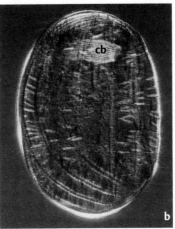

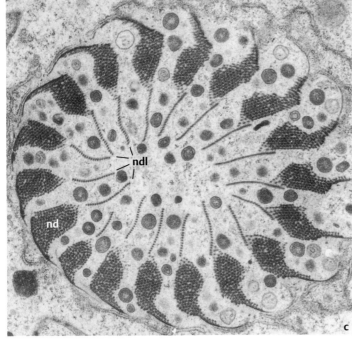

Fig. **236** Cytopharyngeal basket (cb) of *Pseudo-microthorax* using Nomarski optics (**a**), polarized light (**b**) and in cross section (**c**). The birefringence is due to the presence of numerous micro-tubules forming the nema-todesmata (nd) and nemades-mal lamellae (ndl) (from Hausmann and Peck: Differen-tiation 11 [1978] 157). Magn.: a and b 800×, c 15 000×.

proper since the sulcal opening is too narrow to internalize large particles. In *Protoperidinium conicum*, an unusual kind of extracellular diges-tion has been documented. These protozoa ex-trude a pseudopodial feeding veil through the flagellar aperture that surrounds the prey (Fig. **241**). The prey, preferentially diatom cells, are gradually digested within 20–30 minutes but are not drawn inside the theca. The veil is re-tracted after digestion has been completed and the dinoflagellate resumes swimming.

Food Vacuole Formation

In those protists which are surrounded by a single plasma membrane, food vacuole forma-tion usually occurs by an invagination of the plasma membrane that encloses the food. This process is probably driven by an actomyosin sys-tem. This type of food vacuole formation occurs in amoeboid and amoeboid-like (e. g., *Trichonym-pha*) protozoa.

Food vacuole formation is more complicated in those organisms with a constant body shape

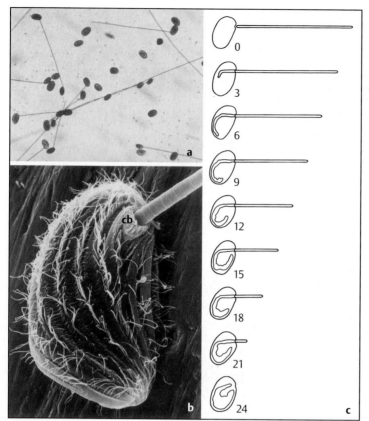

Fig. **237** **a** Feeding *Pseudomicrothorax* cells; **b** ingestion of an *Oscillatoria* filament with aid of the cytopharyngeal basket (cb); **c** schematic representation of feeding velocity: numbers correspond to elapsed time (in seconds). The diameter of the cyanobacterium is about 5 μm (from Hausmann and Peck: Differentiation 14 [1979] 147). Magn.: a 40×, b 1000×.

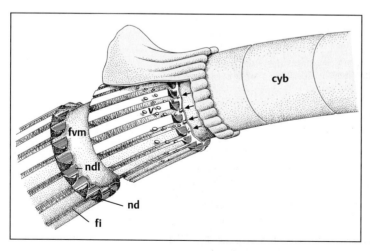

Fig. **238** Cytopharyngeal basket of *Pseudomicrothorax* during ingestion of prey: the basket consists of nematodesmata (nd) and nemadesmal lamellae (ndl) and is stabilized by filaments (fi). In the upper region of the basket, small vesicles (v) enter the lumen by passing through the nematodesmata (arrows) and fuse with the membrane of the food vacuole (fvm). cyb = cyanobacterium.

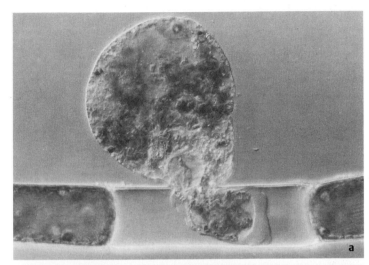
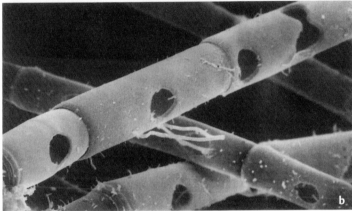

Fig. **239** Food uptake in vampyrellids. **a** *Hyalodiscus* during ingestion of the cytoplasmic content of an *Oedogonium* cell. **b** penetration holes in the cell walls of vanished *Oedogonium* filaments caused by *Gobiella*. Magn.: 550×.

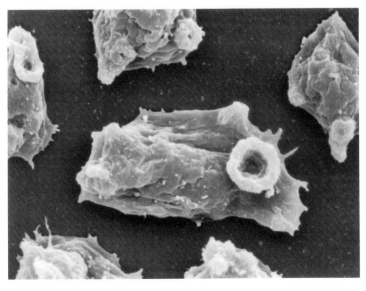

Fig. **240** *Naegleria fowleri* with bell-shaped ingestion pseudopods (amoebostomes) (from John et al.: J. Protozool. 32 [1985] 12). Magn.: 2500×.

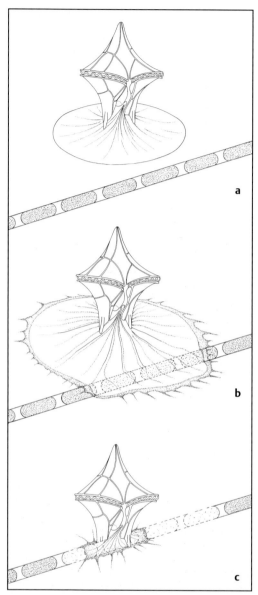

Fig. 241 Approach to (**a**) and digestion (**b, c**) of the cytoplasm of colonial diatoms by a velum-like feeding pseudopodium of a heterotrophic dinoflagellate of the genus *Peridinium* (adapted from Jacobsen and Taylor). Magn.: 400×.

means that this area must lack alveoli, epiplasm, and somatic cortical fiber systems, such as the filamentous network and microtubules. Thus, in the region of the cytostome, the plasma membrane is directly in contact with the cytoplasm.

Because there is no extra plasma membrane available for food vacuole formation, a mechanism is required to produce the membrane for the food vacuole. A transport system brings preformed membrane from the interior of the cell to the site of food vacuole formation. This is accomplished by small vesicles that are often visible by electron microscopy. In some cases, these vesicles are known to be transported along ribbons of microtubules toward the cytostome, although the actual mechanism of transport is unknown. At the cytostome, the vesicles fuse with the growing food vacuole membrane (Fig. **242**). All phagotrophic ciliates examined so far show this mode of growth of the food vacuole membrane.

Growth of the food vacuole by this means requires hundreds or even thousands of vesicles. It has been demonstrated that these vesicles do not need to be synthesized during feeding but are already present in the cytoplasm. This means that ciliates are able to feed only until all vesicles are consumed. In some cases, the vesicles are synthesised by the Golgi apparatus and may contain digestive enzymes. How they are segregated within the cytoplasm and how they are moved over large distances toward the cytostome is unclear.

The transport of these vesicles may be observed by light microscopy within the cytopharyngeal basket of certain ciliates. In these cases, one observes a stream of fine granules moving in the opposite direction to the cyanobacterial filament being ingested. The granules move to the outer region of the basket, to the area between the microtubular bundles; a region where the granules have access to the lumen of the basket (see Fig. **236c**). The granules must reach the food vacuole membrane through the spaces between the microtubular bundles. When they fuse with the food vacuole membrane, they contribute to its growth.

Vesicle transport, incorporation of the vesicles into the food vacuole, and inward transport of the food vacuole itself occur in a specific sequence. Each event depends upon the other. At the maximal velocity of ingestion (15 μm of cyanobacterial filament per second), there is no interference between these events.

In these ciliates, the fusion of vesicles with the growing food vacuole also initiates diges-

and a well developed and structured cortex. This is especially the case for ciliates (see Figs. **98, 99, 107, 174**). In these organisms, there must be a specific area where the normal somatic cortical organization is absent, and where the oral apparatus allows for food vacuole formation. This

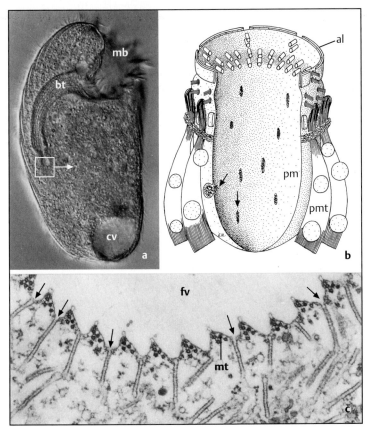

Fig. **242** Growth of food vacuoles in ciliates: **a, b** in *Climacostomum* the food particles are carried by membranelles (mb) into a buccal tube and are ingested periodically at the end of this tube (bt). The plasma membrane (pm) of the growing food vacuole enlarges by incorporation of small flattened vesicles (arrows). al = alveolus, cv = contractile vacuole, pmt = postciliary microtubles. **c** at the cytostomal region of *Trichodina*, flattened vesicles (arrows) fuse with the food vacuole (fv). mt = microtubules (b from Fischer-Defoy and Hausmann: Differentation 20 [1981] 141; c from Hausmann and Hausmann: J. Ultrastruct. Res. 74 [1981] 131). Magn.: a 500×, c 55 000×.

tion, as the small vesicles contain the hydrolytic enzymes necessary for digestion. In *Pseudomicrothorax dubius*, these enzymes digest the wall of the cyanobacterial prey in a matter of seconds. Once this is accomplished, the cyanobacterial filament becomes flexible with the naked but intact protoplast making its way into the growing food vacuole. In this way, large numbers of filaments can be ingested. The degree of specialization found for *Pseudomicrothorax* has not been found for all similar ciliates. There are other cyanobacteria ingesting ciliates which are less specific; they often require several hours to accomplish digestion of a single filament.

In contrast to the simple naked amoebae, floating organisms with axopods produce food vacuoles in a much more complicated manner. *Actinophrys sol* can ingest free-swimming ciliates (Fig. **243**). The first step in this process involves the adhesion of prey to the axopod. This is brought about by the action of kinetocysts. In the second phase, which occurs over approximately 15 minutes, a very large food vacuole is formed. The membrane of this food vacuole does not originate from either the plasma membrane or simple vesicular membrane precursors. Rather, it originates from the rapid transformation of specific, membrane-bound, electron-dense granules that are abundant beneath the plasma membrane. Following stimulation, these granules are transported along the membrane of the site of the food vavuole being formed. Here, they expand as their contents become decondensed, and the resulting vesicles fuse with one another to supply membrane for the developing food vacuole.

In general, little is known about the factors which initiate food vacuole formation in the protozoa. In some cases, particulate substances in the environment appear to be required. The plasma membrane of organisms with an oral apparatus must make contact with these particles several times in order to induce food vacuole formation. In addition, food particles must have certain (unknown) surface properties.

The time required for food vacuole formation

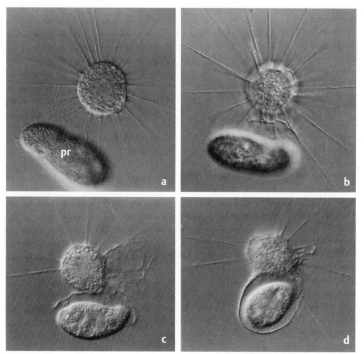

Fig. **243** Capture (**a**) and ingestion of the ciliate *Colpidium* (**b–d**) by the heliozoan *Actinophrys sol*. Within about 15 minutes the prey (pr) is engulfed by a funnel-like pseudopodium (from Hausmann and Patterson: Cell Motil. 2 [1982] 9). Magn.: 300×.

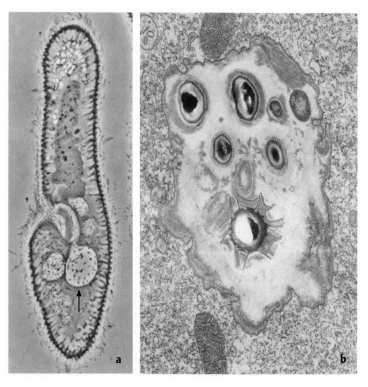

Fig. **244** Separation of a food vacuole (**a**, arrow) and a bacteria-containing food vacuole (**b**) in *Paramecium* (a courtesy of R. Stiemerling, Bonn). Magn.: a 500×, b 18 000×.

is influenced by physicochemical conditions such as temperature, pH, and ionic composition. These factors also influence the properties of the plasma membrane and the fusion of vesicles with the plasma membrane.

Digestion

After the food vacuole separates from the plasma membrane and/or cytostome, it is referred to as the digestion vacuole (Fig. **244**). One of the first events in digestion is a drop in pH of the vacuolar contents. This results in an environment conductive to digestion as the hydrolytic enzymes involved have an acidic pH optimum. Following enzymatic digestion of the food, usable compounds are distributed throughout the cell, and the undigestible residues are defecated. The entire process may require from 20 minutes to several hours, depending upon the protozoon involved. Each phase is accompanied by morphological changes of the digestion vacuole and its contents.

The different steps of digestion may be separated into several defined phases (Fig. **245**), the details of which are known only for *Paramecium*. After the food vacuole is pinched off from the cytostome, its diameter is reduced. This vacuole is then called the digestive vacuole I (DV-I). It has an average diameter of 15 μm which decreases to approximately 6 μm. The perimeter of DV-I is irregular; it shows numerous tubular extensions. The tubes have a diameter of 0.1 μm and they are 0.4–0.8 μm long. Vesicles pinch off from the food vacuole. This results in a reduced surface area and a decrease in the vacuolar diameter. However, this event alone does not account for the drastic reduction in volume. Due to the small size of the vesicles, their surface area is much larger than their volume. This suggests that most of the liquid passes through the vacuolar membrane. The details of this event are still not understood but it is known that membranes are generally permeable to water, and water might pass into the cytoplasm by osmosis.

Prior to and during the pinching off of the tubular vesicles, there is also a fusion of irregular vesicles with the DV-I. These vesicles have a size of up to 1 μm. They are incorporated into the membrane of the DV-I and are associated with the drop in pH of the vacuolar contents. These vesicles are known as acidosomes.

When the DV-I has shrunk to a diameter of 6 μm, the second or DV-II phase begins. DV-III is characterized by a sudden increase in diameter,

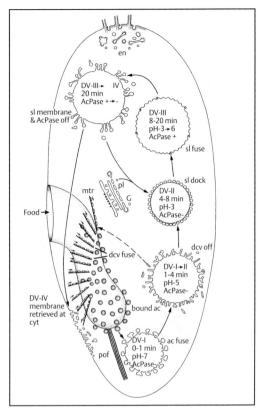

Fig. 245 Map of the flow of membrane through the phagosome-lysosome system in *Paramecium*. ac = acidosomes; AcPase = acid phosphatase; cyt = cytoproct; cyx = cytopharynx; dcv = discoidal vesicle; DV-I to DV-IV = digestive vacuole stages; en = endosomes with clathrin coat, pinched off from parasomal sacs; G = Golgi complex; mtr = microtubular ribbons; ndv = nascent digestive vacuole; pl = primary lysosome; pof = postoral fibre; sl = secondary lysosome (from Fok and Allen: Membrane flow in the digestive cycle of *Paramecium*, in H. Plattner (ed): Membrane traffic in protozoa. JAI, Greenwich 1993).

an increase which continues until the DV-III reaches the original size of DV-I. This increase is a consequence of the fusion of primary lysosomes with the DV-II. The oval shaped lysosomes have a size of approximately 0.2×0.4 μm. The same point raised above with respect to the vesicles involved in reduction of the DV-I volume also occurs here: the increase in volume accompanying the fusion of the primary lysosomes cannot be explained simply as an increase due to membrane added by fusion. There must be another (unknown) mechanism at work by which water and diffusible molecules enter the DV-II from the cytoplasm.

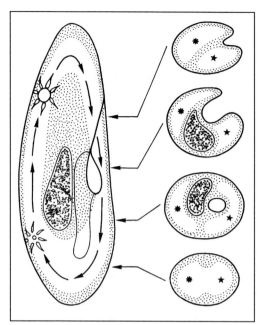

Fig. **246** Cyclosis of digestion vacuoles in *Paramecium* along a distinct tubelike pathway; stationary areas of the cell are dotted; streaming rate of the cytoplasm (marked by arrows or asterisks) is about 2–3 μm per second (adapted from Sikora).

It is difficult to understand the reason for the reduction and then increase in volume of the digestive vacuole. It has been suggested that the food particles are concentrated before being exposed to the digestive enzymes, which results in more efficient digestion. However, since DV-III is nearly the same size as the original food vacuole, there are probably some other activities involved. As has been shown by calculation, only part of the additional volume can be attributed to lysosomal contents, therefore, it is probably the case that it is not a reduction in volume which is required, but rather a change in the composition of the digestive vacuole membrane. Some membranes, such as the tube like vesicles, may be removed from the vacuole and transported back to the cytostome where they can support growth of another food vacuole.

The DV-III stage is characterized by digestion of the food. During this process, there is continuous activity at the periphery of the digestive vacuole. This may be due to fusion of additional lysosomes with the vacuole, and/or the removal of substances from the vacuolar lumen. The latter are enclosed in small vesicles which pinch off from the vacuole. The undigestible residues re-

main inside DV-III, which now becomes a residual vacuole, DV-IV. During defecation, the residues are released to the outside by exocytosis,

While the events described above are best understood for *Paramecium*, they may also be generally true for other protozoa.

Cyclosis

The majority of events associated with digestion do not occur in a specific region of the cell, but rather during transport of the digestive vacuole through the cell. In *Paramecium*, the movement of the digestive vacuole follows an elliptical path (Fig. **246**). This movement is called cyclosis; unidirectional cytoplasmic streaming with a velocity of about 3 μm per second.

The mechanism of cyclosis remains a matter of speculation. Due to the rheological properties of this streaming, the movement can not be caused by shear forces between stationary and moving cytoplasm. Therefore it has been suggested that there is an actomyosin system operating within the streaming cytoplasm. This is presumed to consist of many small subunits which are randomly distributed within the entire stream of the cyclosis. The subunits are thought to undergo energy-dependent transformations from the sol to the gel state and vice versa. These events might be accompanied by contraction and relaxation of the actomyosin system. If so, they should be controlled by free calcium ions.

Defecation

Defecation vacuoles contain the undigestible residue, and they no longer show enzymatic activity; their pH has returned to the original, near neutral value. These vacuoles are now known as residual vacuoles, and their content is expelled (Fig. **247**). In the simplest case, as in the rhizopods, defecation occurs as the reverse of phagocytosis. The residual vacuole fuses with the plasma membrane and its contents are extruded to the outside. The membrane of the vacuole is initially incorporated into the plasma membrane but it may be retransported to the cytoplasm.

The process of defecation by exocytosis is hindered in ciliates by the presence of cortical structures, so exocytosis takes place at a special site within the cell, an area that is specially modified for this purpose. This area is called the cytoproct (= cytopyge), and it is often a ridge-shaped

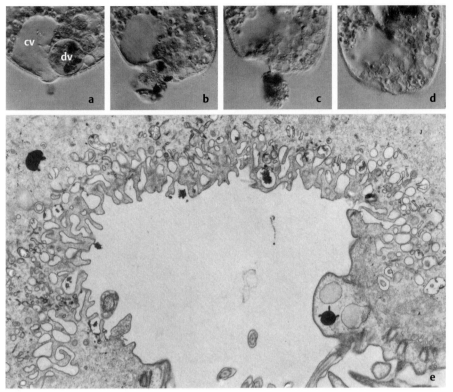

Fig. **247**   Defecation in *Climacostomum* (**a–d**) and structural organization of a collapsed defecation vacuole with peripheral vesiculation processes (**e**). cv = contractile vacuole, dv = defecation vacuole (from Fischer-Defoy and Hausmann: Zoomorph. 100 [1982] 121). Magn.: a–d 400×, e 11 000×.

convolution of the pellicle (Fig. **248**). The alveoli, epiplasm, and filamentous networks terminate adjacent to the ridge, enabling the residual vacuole to come into direct contact with the plasma membrane.

Individual microtubules and ribbons of microtubules are directed into the cytoplasm from the region of the cytoproct and from the unciliated basal bodies lying adjacent to the cytoproct. These microtubules may guide the residual vacuoles toward the cytoproct, or they may even actively transport them to the cytoproct.

At the cytoproct, the membrane of the residual vacuole fuses with the plasma membrane. During extrusion of the vacuolar contents, the membrane divides into small vesicles and vesicular subunits (Fig. **248**). In this way, the membrane materials remain inside the cell. In some cases, these vesicles may serve as a source of membrane material for the growing food vacuole. In other cases, it is more likely that the membrane material first undergoes transforma-

tions which involve passing through the Golgi apparatus before they may be used again in the digestive process.

Crystals

Many protozoa contain crystals which, because of their birefringent properties, are clearly visible in the polarization microscope. They vary in shape (Fig. **249**), but they are always surrounded by a membrane. In heterotrophic organisms, their occurrence and shape appear to be correlated with the nutrition of the organism. Paramecia fed on bacteria contain small particle-like crystals; after being fed on protein or meat extracts, they contain numerous large crystals. Foraminifers contain crystals after feeding on copepods or ciliates, but no crystals are found after a diet restricted to diatoms. However, direct relationship between digestion of the food and the formation of crystals has yet to be found.

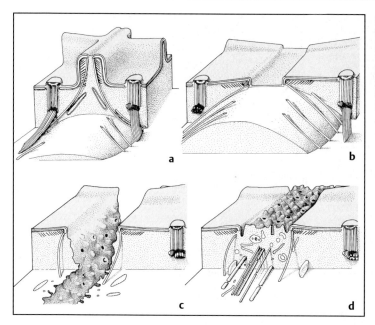

Fig. **248** Function of the cytoproct of *Paramecium:* approach of a defecation vacuole to the cytoproct (**a**, **b**) and vesiculation of the former vacuolar membrane after fusion with the cell surface (**c**, **d**) (adapted from Allen and Wolf).

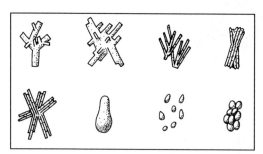

Fig. **249** Crystal shapes in *Paramecium* (adapted from Wichterman).

The crystals are composed mainly of salts of calcium and phosphorus. In addition, small amounts of magnesium and chloride have been detected. Organic components are occasionally detected. The best example of an organic crystal is the lithosome, which is a spherical deposit produced from concentric layers of organic and inorganic material (Fig. **250**).

In the past, crystals were considered to represent metabolic by-products (excretion grains) which were later extruded. Today, two alternate hypotheses are discussed: 1) the crystals are deposits or reservoirs of ions which are absolutely necessary for some metabolic activities; 2) the crystals are involved in the regulation of levels of certain ions in the cell. Both hypotheses are supported by observations. In certain amoebae, crystals are always present, and when removed experimentally, they are rapidly replaced by new ones. Many protists typically form crystals when they are grown under optimal conditions.

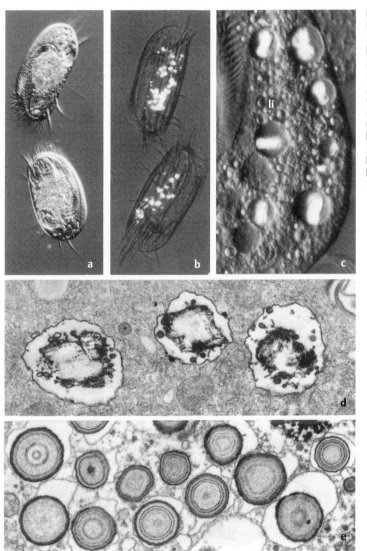

Fig. **250**   *Euplotes* (**a**) with birefringent crystals (**b**) which are membrane bound (**c**); li = lithosomes. Crystal vacuoles in an ultrathin section of *Paramecium* (**d**). Lithosomes of *Euplotes* are characterized by concentric layers of organic and inorganic material (**e**) (a–d from Hausmann: Mikrokosmos 71 [1982] 33; e from Hausmann and Walz: Protoplasma 99 [1979] 67). Magn.: a and b 280×, c 1800×, d 20 000×, e 10 000×.

Nuclei and Sexual Reproduction

(by Maria Mulisch, Köln)

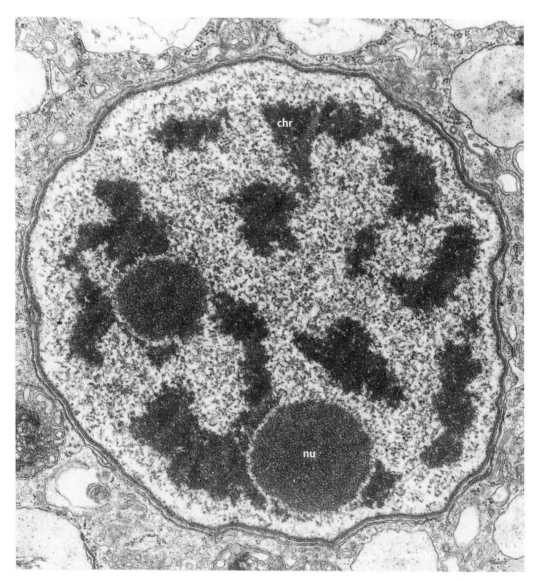

Fig. **251**   Condensed chromosomes (chr) and nucleolus (nu) in a nucleus of *Lateromyxa gallica* (Vampyrellidae). Magn.: 30 000 ×.

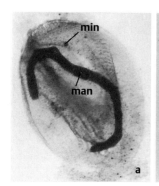

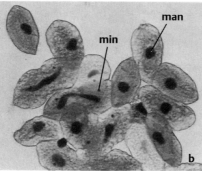

Fig. **252**   Stained nuclei of *Euplotes* (**a**) and *Tetrahymena* (**b**). man = macronucleus, min = micronucleus. Magn.: a 500×, b 200×.

Structure and Function of Nuclei

The nuclei of protozoa are more diverse with respect to their shapes, morphologies, and modes of division, than the nuclei of metazoa. Most protozoa have only a single nucleus, although in many genera there are species with several or many identical nuclei. These are referred to as homokaryotic and include *Chaos, Lateromyxa, Pelomyxa, Actinosphaerium,* and *Opalina* (Fig. **251**). In some protozoa with a complex sexual cycle (e. g., Apicomplexa), uninucleated and multinucleated stages occur. This is the case in coccidians, where the sporozoites and the merozoites have only one nucleus, while schizonts and macrogamonts contain many nuclei. The life cycles of a number of foraminifers include uninucleated gamonts as well as multinucleated agamonts.

Protozoa are called heterokaryotic if different types of nuclei occur simultaneously in one cell. There may be two types of nuclei specialized for different functions (nuclear dualism). A true nuclear dualism with one or several (usually small) micronuclei, and one or several (usually larger) macronuclei (Fig. **252**), is a feature of all ciliates and certain stages (agamonts) of some foraminifers (e. g., *Rotaliella heterokaryotica*; see Fig. **156**). The micronucleus is the germline nucleus. It shows little or no gene expression. The macronucleus is the somatic nucleus that determines the phenotype of the cell. It does not contribute to the sexual progeny of the cell. In contrast, the different nuclei in some other organisms (e. g., *Peridinium balticum*) is not a true nuclear dualism with both based on the same genome. Rather, one nucleus is of an endosymbiont and thus of foreign origin.

Most nuclei of protozoa are rounded or oval shaped. Their diameters vary from less than 1 μm to more than 100 μm. The most complex shapes may be observed in the macronuclei of ciliates. This is especially true for macronuclei of large ciliate species, which are often lobed or segmented. The enlargement of the surface area of the macronucleus provides a better exchange between nucleus and cytoplasm.

The nucleus (see Figs. **8, 18, 251**) is surrounded by a nuclear envelope. This consists of a cisterna that is derived from the rough endoplasmic reticulum (rER) and is penetrated by pores as in all eukaryotes. The inner or the outer membrane of the nuclear envelope may be supported by filamentous layers. For example, the inner nuclear membrane in *Amoeba proteus* is underlain by a thick, characteristically structured layer (= honeycomb layer) (Fig. **253**), the function of which is still unknown. The nuclear lamina inside the nuclear envelope of some heliozoans (e. g., *Actinophrys)* supports the axopodial microtubules.

The karyoplasm usually contains chromatin, which ultrastructurally has a fine fibrillar appearance, and one or several electron-dense aggregates that are nucleoli (see Fig. **18, 251**). Nucleoli are sites of synthesis of the ribosomal subunits. They are absent from the micronuclei of ciliates and from some gametic nuclei.

The degree of condensation of the chromatin varies during interphase. Compact, densely packed chromatin that fills the entire nucleus is characteristic for the nuclei of microgametes of most Apicomplexa and for the micronuclei of many ciliates. In many euglenoid flagellates, the chromatin in the nondividing nucleus forms dense rods whose ultrastructure is similar to that of mitotic chromosomes. The chromosomes of many hypermastigids (e. g., *Barbulanympha, Trichonympha*) are permanently visible and attached by their kinetochores to the inner side of the nuclear envelope. The small nuclei of most dinoflagellates always contain compact chromo-

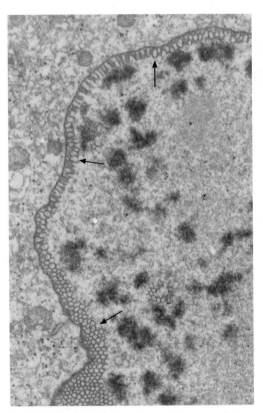

Fig. 253 Honeycomb layer (arrows) beneath the nuclear envelope of *Amoeba proteus* (courtesy of W. Stockem, Bonn). Magn.: 12 000×.

A number of protozoa (volvocid flagellates, many dinoflagellates, some hypermastigids and oxymonads, Myxogastra, Apicomplexa) have haploid nuclei; most others (e. g., *Actinophrys, Noctiluca*) have diploid nuclei during most of their life cycle. In haploid organisms, meiosis occurs in the zygote, this usually being the only diploid stage. Double chromosomes that have been equated to diploidy have occasionally been observed in some flagellated protozoa, for example, in *Chlamydomonas*. In diploid organisms generally only the gametes are haploid. There are also protozoa (e. g., foraminifers) with alternating diploid and haploid phases.

Large nuclei are often polyploid. The micronuclei of the ciliates studied to date are diploid, whereas the macronucleus contains many copies of the genetic material. The macronuclear DNA content of *Tetrahymena thermophila* is 10- to 20-fold greater than the micronuclear DNA content. The macronucleus of hypotrich ciliates contains only about 5% of the entire genome of the micronucleus. However, the macronuclear DNA content is much greater because the copy number of a few genes in the macronucleus is dramatically increased. The macronuclear DNA of hypotrichs is present as millions of short, linear, gene-sized fragments. The number of copies of each fragment is directly proportional to the cell size. The macronuclei of karyorelictid ciliates differ from those of higher ciliates by their much lower ploidy (they are almost diploid) and by their inability to divide. They must be differentiated from the diploid micronucleus at every cell cycle.

somes. These differ from those of other protozoa by their ordered fibrillar structure. It is proposed that each consists of a long, circular DNA molecule twisted into a super helix. As the structural organization of the dinoflagellate nucleus is very different from all other eukaryotes, it is regarded as an independent type of nucleus and referred to as a dinokaryon.

As in metazoa, the DNA in the nuclei of protozoa is usually organized into chromatin by association with histones. All five histones (H1–H5) found in metazoa have been identified in the macronucleus of *Tetrahymena*. The histones of ciliates differ from those of metazoa in some parts of the amino acid sequence. In contrast, the histone composition in the transcriptionally inactive micronucleus of ciliates differs substantially from that of the macronucleus. The nuclei of most dinoflagellates lack histones but contain small, variable amounts of other basic proteins.

Roles of Micronucleus and Macronucleus in the Functioning of Ciliates

Experimental analyses have led to the widespread opinion that the only role of micronuclei is the storage and recombination of the genetic material, but that micronuclei are not involved in maintenance and regulation of cellular functions. This is supported by the existence in nature of species without micronuclei. Recent transplantation experiments, especially on *Paramecium* and hypotrich ciliates, indicate that this idea should be revised.

Paramecium cells from which the micronuclei have been experimentally removed, develop an abnormal oral apparatus in the next generation. The observations suggest that the micronucleus is involved in oral development (= stomatogenesis). During vegetative reproduction the macronucleus may replace the functions of the lost mi-

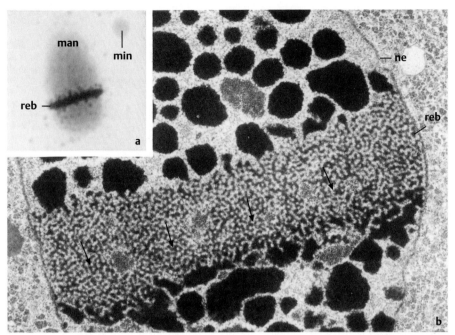

Fig. **254**   Replication band (reb) within a macronucleus (man) of *Stylonychia* in a light- (**a**) and of *Euplotes* in an electronmicrograph (**b**). ne = nuclear envelope, min = micronucleus. Arrows indicate directions of movement of the replication band (a from Steinbrück: Europ. J. Protistol. 26 [1991] 2; b courtesy of K. Eisler, Tübingen). Magn.: a 800×, b 16 000×.

cronucleus, because the oral apparatus becomes normal after some generations.

During sexual reproduction (conjugation), the macronucleus breaks down. The micronucleus is essential for regular stomatogenesis following this process, although gene products of the old (degraded) macronucleus may influence oral development until ten generations after conjugation.

The Nucleus during the Cell Cycle

The eukaryotic cell cycle is traditionally divided into several distinct phases (see Fig. **274**). During interphase the cell grows continuously; during M phase it divides. The interphase includes a phase auf DNA replication (the S phase) and usually two gaps, called G_1 phase and G_2 phase. The M phase includes the segregation of the replicated chromosomes into two daughter nuclei (by mitosis) and the division of the cell (by cytokinesis).

Replication of the Genetic Material

DNA synthesis may occur during different stages in the life cycle; during the vegetative growth of the cell, or directly before or after nuclear division (= karyokinesis). In many multinucleated, homokaryotic protozoa, the replication of genetic material and karyokinesis occurs independently in each nucleus, and may not be linked directly to the cell division cycle. The micronuclei of many ciliates with multiple micronuclei also enter the S-phase independently of each other and independently of the macronucleus. Synthesis of DNA and of histones in the macronuclei of hypotrich ciliates occurs synchronously and is locally restricted to one or two so-called replication bands (Fig. **254**). The appearance of chromatin structures changes drastically at certain sites near the nuclear envelope. These changed areas develop into a band (= replication band) and then move (from one or from several sites) through the whole macronucleus. At the site of the replication band, the amount of DNA and histones is duplicated. The replication band disappears just before the onset of nuclear and cellular division. In most other ciliates, DNA

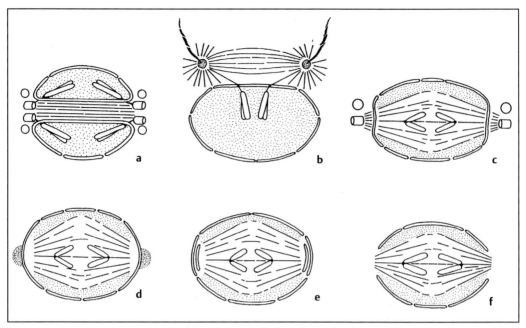

Fig. **255** Types of mitosis in different taxa (anaphase): **a** microtubules passing through the nucleus in channels (dinoflagellates); **b** cytoplasmic spindle microtubules interacting with chromosomes (hypermastigids); **c** and **d** intranuclear spindle and ex- tranuclear spindle-organizing centers (foraminifers); **e** spindle apparatus within closed nuclear envelope (ciliates); **f** nuclear envelope perforated at the nuclear poles (after Grell).

synthesis in the macronucleus is not detectable ultrastructurally.

Mitosis

During nuclear division, the duplicated genetic material becomes evenly distributed between the two daughter nuclei. This is achieved by a process called mitosis.

Mitosis is usually divided into five stages (prophase, prometaphase, metaphase, anaphase, and telophase) that generally include the condensation of chromosomes and the separation of the sister chromatids by a bipolar microtubular spindle. The spindle microtubules nucleate at microtubule organizing centers (= MTOCs), which are often associated with centrioles or basal bodies, respectively.

In contrast to the higher eukaryotes, which vary little in their modes of nuclear division, protozoa show different types of mitosis (Fig. **255**). These differ in the fate of the nuclear envelope as well as in the site, symmetry, and morphology of the spindles and their organizing centers.

As in metazoa, the nuclear envelope may disintegrate totally (open mitosis), or only at the poles (semiopen mitosis). The open mitosis is found in chrysomonads, cryptomonads, haptomonads, prasinomonads, pedinellids, some amoebae, labyrinthulids, some gregarine, and some Myxogastra. Chloromonads, some gregarines, some heliozoans, and some Myxogastra show a semiopen mitosis. The spindle develops inside a closed nuclear envelope (= closed mitosis or cryptomitosis) in kinetoplastid, oxymonad and euglenoid flagellates, some amoebae, some gregarines, foraminifers, radiolarians, in ciliates (micronucleus), and in opalines.

In groups with total or partial degradation of the nuclear envelope, the MTOCs for the spindle microtubules are situated in the cytoplasm, and the microtubules gain direct access to the chromosomes through the perforations in the nuclear envelope. In protozoa which undergo cryptomitosis, the spindle organizing centers may be placed inside (e. g., in ciliates) or outside the nuclear envelope (e. g., in foraminifers). In trichomonads, hypermastigids, and dinoflagellates, the spindle organizing centers are placed outside the closed nuclear envelope; the spindle develops laterally to the nucleus (trichomonads, hypermastigids) or inside a channel through the nucleus (dinoflagellates). The microtubules of the extranuclear spindle interact via the nuclear envelope with the intranuclear chromosomes, as in Polymonadida and some dinoflagellates. In other dinoflagellates, the centromere regions of the chromosomes are linked to the nuclear envelope, but not directly to the extranuclear spindle.

The new envelope for the daughter nuclei is made from parts or fragments of the old nuclear envelope. In some cases (e. g., *Actinophrys*), it becomes visible inside the old nuclear envelope before the latter is degraded.

In many protozoa, individual chromosomes become visible during mitosis (e. g., five in *Tetrahymena*, more than one hundred in *Stylonychia*, about five hundred in *Amoeba proteus*). Other genera (e. g., the ciliates *Homalozoon* and *Nyctotherus*) form aggregates of chromosomes that later break apart into smaller subunits.

The great diversity of the modes of nuclear division suggests that mitosis has diversified during the evolution of the protozoa. Ancestral features may be conserved in many recent taxa with primitive or with derived characters. During division in prokaryotes, the chromosomes are attached to the plasma membrane and become separated by the addition of membrane between the attachment sites. This superficially resembles the type of mitosis (dinomitosis) in dinoflagellates, which is why some authors regarded the dinomitosis as primitive. However, molecular studies suggest that dinoflagellates represent a highly derived group. Dinomitosis differs substantially from division in prokaryotes, for example by the involvement of microtubules. The evolutionary sequence of the diverse types of mitosis in protozoa is still unknown.

Division of the Ciliate Macronucleus

The macronucleus of ciliates (except the karyorelictids) divides during cytokinesis. Before the division furrow is formed, the macronucleus usually migrates in the division plane. It elongates longitudinally between the daughter cells and constricts (apparently passively) in the cell division plane during cytokinesis. The halves separate just before or during separation of the daughter cells. The envelope of the macronucleus remains intact during this process.

Intranuclear and cytoplasmic filamentous and microtubular systems are involved in the movement, stretching, and shaping of the nuclei.

Experiments on *Stentor* have shown that the position and elongation of the macronucleus are controlled by cortical elements. This may also be true for the macronuclei of other ciliates (e. g., *Bursaria, Paramecium*) which have been observed to be attached to the cell cortex.

The mode of division of the ciliate macronucleus is often termed "amitosis," as no typical spindle apparatus and usually no chromosomes appear. On the other hand, it is unlikely that the genetic material is distributed randomly between the daughter cells. Branched or segmented nuclei condense into a single, compact mass before division. This occurs also with the number of small macronuclei in, e. g., *Urostyla grandis*. The genetic material may be mixed by rotation of the macronuclear content. This has been observed, for example, in *Homalozoon*. Intramacronuclear microtubules appear to be involved in the separation of elements resembling chromosomes in the macronuclei of *Protocruzia*. They have also been observed in other ciliates. Differences in the DNA content between the daughter nuclei may be adjusted by selective synthesis or removal of genetic material.

Meiosis

Sexual reproduction in eukaryotes is linked to the reduction of the number of chromosomes. This is achieved by a meiotic division of the nucleus. Meiosis occurs in diploid cells, and yields haploid products that each contain a complete set of chromosomes with one of each kind present. Recombination of genetic material is an important potential benefit of meiosis.

Meiosis may take place at different phases of the cell cycle. In haploid organisms, meiosis occurs just after fusion of the gametic nuclei (e. g., in volvocal flagellates, many dinoflagellates, *Dictyostelium, Trichonympha*, Apicomplexa). In contrast, meiosis in diploid protozoa takes place during the development of gametes (e. g., *Opalina*, some heliozoa). This is also true for ciliates which, however, do not produce gametes but gamete nuclei that are exchanged during conjugation. In foraminifers, meiosis of the diploid asexual stages (agamonts) leads to haploid cells (gamonts) that give rise to multiple gametes after mitotic divisions. Meiosis in foraminifers is called intermediate.

Meiosis in most eukaryotes consists of two subsequent nuclear divisions that may be attended by cell divisions. If there are no interphase chromosomes, chromosomes will con-

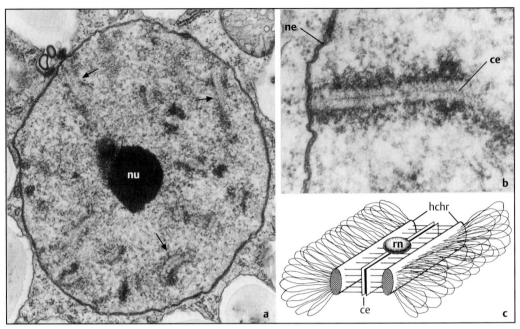

Fig. 256   Synaptonemal complexes in *Lateromyxa gallica* (Vampyrellidae): **a** nucleus during zygotene with nucleolus (nu) and several synaptonemal complexes (arrows); **b** attachment of a synaptonemal complex at the nuclear envelope (ne); **c** schematic reconstruction of pairing of homologues in a synaptonemal complex. hchr = homologue chromosomes, rn = recombination nodulus, ce = central element (a and b from Röpstorf et al.: Europ. J. Protistol. 29 [1993] 302; c after Krstić). Magn.: a 32 000×, b 75 000×.

dense during the prolonged prophase I of the first meiotic division. Homologous, already duplicated chromosomes pair to create bivalents and are connected by a synaptonemal complex (Fig. **256**). The breakage and reciprocal reunion of the homologous DNA helices involves crossing over between nonsister chromatids (genetic recombination). The bivalents line up in the equatorial plane on the spindle during metaphase I. The homologue chromosomes separate and move toward opposite poles of the cell (anaphase I). As the maternal and paternal homologues are randomly distributed between the daughter nuclei, the genes are reassorted, and their number is halved during this process. The second meiotic division resembles mitosis. The four haploid nuclei produced from each diploid nucleus by these two meiotic divisions are genetically different.

As far as has been observed, meiosis in most protozoa appears to follow this general scheme. In many protozoa, however, precise details are not known.

A different chromosome behavior, which probably only superficially resembles meiosis, is reported from oxymonads, some hypermastigids, and Apicomplexa. In some of these organisms, chromosomes seem to exist permanently in a kind of double state. Replication of these double structures results in pairs of dyadlike structures that become visible in all cell divisions. This kind of chromosome behavior was equated to the first meiotic division. A simplified two-step meiosis has been reported from other hypermastigids. Here, following the kind of first meiotic division just described, a second type of division (by some authors equated to a second meiotic division) can occur without replication, reducing the state from double to single. In these protozoa, the doubled chromosomes are not the result of the association of homologue chromosomes from separate nuclei (an essential ingredient of meiosis).

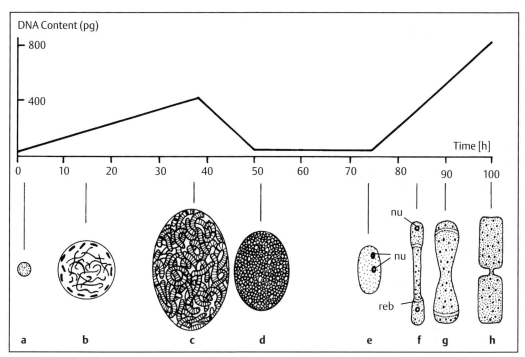

Fig. **257** Development of macronuclear anlage in *Stylonychia lemnae*. **a** directly after separation of exconjugants (t = 0) the size and DNA content of a macronuclear anlage correspond to those of a micronucleus; **b–c** increase of DNA content and formation of giant chromosomes (c); **d** fragmentation of giant chromosomes and decrease of DNA content; **e** formation of nucleoli (nu); **f** and **g** replication bands (reb) appear and the DNA content increases; **h** after 100 hours, the macronucleus exhibits its characteristic features (after Ammermann).

Macronuclear Development in Ciliates

The genome of the macronucleus develops from the genome of the micronucleus. In the Karyorelictea the macronuclei do not divide. They arise from the micronucleus after every division cycle. The karyorelictids, therefore, have been regarded as a very primitive ciliate group. This view is supported by recent findings on the phylogeny of ciliates based on ultrastructural and molecular data.

Most ciliates develop new macronuclei only after sexual reproduction (conjugation). The macronuclei derive from the newly recombinant genome of the micronucleus. The process of differentiation is very complicated and not yet fully understood. In most ciliates so far analyzed, at least three types of events occur in the DNA rearrangement process: chromosomal breakage, internal sequence elimination, and DNA amplification. The final degree of amplification varies from one species to another, but it generally appears fairly constant within a single species.

Extensive DNA replication leads to the development of giant chromosomes in the macronuclear primordia of most hypotrich ciliates (*Stylonychia, Euplotes, Oxytricha*), but also of some members of other groups (e. g., of *Chilodonella cucullus, Loxophyllum meleagris, Nyctotherus cordiformis*). The giant chromosomes resemble the polytene chromosomes in insects. They break apart and vesiculate (Fig. **257**). Most of the DNA is lost during this process. In *Stylonychia lemnae*, only about 2% of the micronuclear genes remain in the macronuclear anlage. This DNA is reduced to small, gene-sized fragments that are dispersed inside the macronucleus after disintegration of the vesicles.

DNA rearrangements in *Paramecium* do not appear to be a simple size reduction of the micronuclear chromosomes, as they generate a variety of different, but sequence related, macronuclear chromosomes from a unique set of micronuclear chromosomes. There is evidence that chromosomal breakage in *Paramecium*, as in other ciliates, is associated with sequence elimination;

however, the amount of eliminated sequences appears to be highly variable from one copy to another.

Sexual Processes During the Life Cycle of Protozoa

Sexual reproduction essentially involves the fusion of two haploid gametic nuclei (originating from meiosis) to form a diploid zygote. It has not been detected in certain groups of protozoa (e. g., euglenoid flagellates, choanoflagellates, trichomonads, schizopyrenids, and most testate amoeba). It is unknown if sexuality in these groups has not evolved or if it is secondarily lost, or if sexual stages have just been overlooked.

In many protozoa, the gametic nuclei occur in special, sexually differentiated cells, the gametes, that may be amoeboid or flagellated. Usually two types of gametes (+/– or male/female, respectively) are formed. If sexual differentiation is already expressed in the parental cells, these are called gamonts. The gametes may be morphologically identical (= isogametes) or different (= anisogametes). The fertilization of an immotile "egg" cell by a motile "male" gamete is called oogamy. Fusion of gametes (gametogamy) is distinguished from fusion of gamonts (gamontogamy). Fusion of gametic nuclei produced by the same parental cell is called autogamy (comp. Fig. **259**).

The life cycle of many protozoa includes a regular alternation between vegetative and sexual reproduction. The alternation of generations is homophasic if the number of chromosomes remains constant when the type of reproduction changes. For example, this is the case in the Apicomplexa (see Fig. **81**). Alternation of generations in many Apicomplexa is associated with host alternation. Foraminifers have a heterophasic alternation of generations, as meiosis occurs in the middle of the life cycle.

Vegetative and generative cells may exist together in some colonial protozoa. This kind of cellular differentiation can be observed, for example, in the volvocids Eudorina and Volvox (see Fig. **143**), and in some ciliates (e. g., Zoothamnium alternans; comp. Fig. **269 a**).

The gametes of haploid protozoa (e. g., volvocid flagellates, chrysomonads, most dinoflagellates, some hypermastigids, oxymonads, and Apicomplexa) may arise directly (e. g., some species of Chlamydomonas, Gonium) or by mitotic divisions from the vegetative cell. In Chlamydomonas, biflagellated isogametes are formed

which carry species- and mating-type specific glycoproteins at the surface of their flagella. These glycoproteins cause the flagella of compatible (+/–) cells to agglutinate. The cells aggregate pair–wise and contact each other by a small papilla. After cell fusion, the zygote may develop into a thick–walled resting stage. Meiosis takes place in the zygote resulting in haploid vegetative cells.

The larger colonial volvocids (Eudorina, Volvox) are oogamous. Sexual reproduction in Volvox carteri is initiated by a so-called sexual inducer protein (a 30 kDa glycoprotein) that is expressed in response to a rise in temperature in the environment. It is effective in concentrations as low as 6×10^{-17} M. Therefore the amount of inducer protein secreted by one individual colony may be sufficient to initiate sexual development in all other colonies in the pond. In the presence of the sexual inducer, the gonidia undergo a modified embryogenesis. A haploid vegetative cell may give rise to multiple ("male") microgametes or to a single, immobile macrogamete. In some species of Volvox, both types of gametes are produced in the same colony (e. g., in Volvox globator); in others, different colonies of the same clone (e. g., Volvox aureus) or different clones (e. g., Volvox perglobator) are involved. The first divisions of the zygote are accompanied by meiosis. Further mitotic divisions lead to the formation of colonies of haploid vegetative cells.

Vegetative reproduction in the Apicomplexa starts with the development of haploid, infective cells called sporozoites (Fig. **258 a**). These arise from the zygote. The sporozoites invade the host cells (except in some gregarines that are extracellular parasites). Here they become feeding stages (= trophozoites). In most gregarines, the trophozoites transform directly into gamonts that give rise to gametes. In other Apicomplexa, the trophozoites grow into large cells (Fig. **258 b**). These are called schizonts, as they undergo a special type of multiple cell division (schizogony) which may be preceded by a number of nuclear divisions. The host cell bursts, and a large number of small infective cells, called merozoites (Fig. **258 c**), are released. These may repeat the cycle of growth and multiple division, leading to an enormous increase in the number of infective stages. Some cells become gamonts (Fig. **258 d, e**) instead of schizonts. Next the phase of sexual reproduction starts. Usually one gamont (the so-called microgamont) by schizogony produces many small microgametes (Fig. **258 f**), while the other, the macrogamont, directly transforms into a macrogamete. Micro-

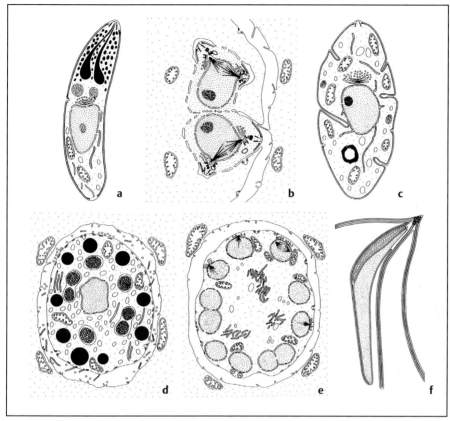

Fig. **258** Changes of shape and structure during the life cycle of Apicomplexa: **a** sporozoite; **b** part of a schizont; **c** merozoite; **d** macrogamont; **e** microgamont; **f** microgamete (after several authors).

and macrogamete fuse, forming the diploid zygote. This usually develops into a resistant spore and may later infect a different host. The transition from one stage to the next is accompanied by characteristic morphological and ultrastructural changes.

Meiosis in diploid organisms (ciliates, certain dinoflagellates, opalinids, a number of hypermastigid and oxymonad flagellates, actinophryids, several amoeboid protozoa, many green algae, diatoms, probably myxozoa, and labyrinthulids) occurs before the development of gametes or gametic nuclei, respectively.

Opalina ranarum (see Fig. **71**), an endobiont in the digestive tract of frogs, has many similar, diploid nuclei. The life cycle (Fig. **43b**) includes the formation of multinucleate cysts which are passed out with the feces. The cells hatch in the intestine of a tadpole. They may divide meiotically to produce uninucleate gametes of unequal

size. The fusion of a microgamete with a macrogamete results in the formation of a zygote. The zygote encysts and may infect another tadpole.

The diploid actinophryid heliozoa usually perform autogamy. *Actinophrys* may encyst in response to starvation (Fig. **259**). The encysted cell divides into two diploid cells. Their nuclei undergo meiosis, leading to four haploid nuclei in each cell. Only one nucleus in each cell survives. Fusion of the cells leads to the formation of a diploid zygote. This may give rise to the vegetative cell under appropriate environmental conditions. Autogamy has been found also among other protozoa (e.g., some ciliates, some foraminifers, the testate amoeba *Arcella*).

The life cycle of many foraminifers includes alternating stages of diploid, multinucleate agamonts and haploid, uninucleate gamonts. These two phases may look very similar. The modes of sexual reproduction in foraminifera are quite

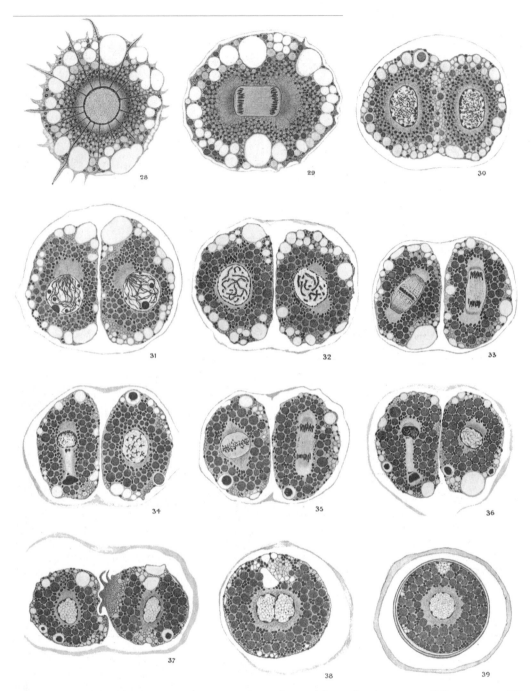

Fig. **259** Sexual reproduction (= autogamy = pedogamy) in *Actinophrys sol* (28). After encystation the nucleus divides meiotically (29–36). The gametes (37) fuse and form after karyogamy (38) the zygote (39) (from Bělař: Arch. Protistenkd. 46 [1923] 1). Magn.: 750×.

diverse, even among closely related species. They involve gametogamy *(Elphidium crispum)*, gamontogamy *(Glabratella sulcata)*, and autogamy (some species of *Rotaliella*).

In shelled foraminifers, gamonts and agamonts often differ in size and shape of the test, and in the size of the first chamber. The first chamber of the gamont is larger (= megalospheric generation) than that of the agamont (= microspheric generation) in most dimorphic species (Fig. **260**). A third generation, a megalospheric schizont, has been reported from many species of foraminifers. In *Heterostegina depressa*, the schizont is formed by a microspheric agamont. Thus a "trimorphic" life cycle of foraminifera is suggested that includes the succession of three distinct forms (microspheric agamont, megalospheric schizont, megalospheric gamont).

Other organisms that carry out an alternation of haploid and diploid phases are some haptomonads (e. g., *Hymenomonas carterae*) and some filamentous green algae (e. g., *Cladophora*). The reasons behind this pattern are not known, but as a strategy, it combines the benefits of sexuality with asexuality.

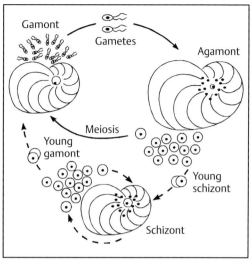

Fig. **260** Foraminiferean life cycle. The dimorphic life cycle (complete arrows) includes an alternation between a haploid gamont and a diploid, multinuclear agamont. A hypothetical trimorphic life cycle (broken arrows) additionally includes a schizont (after Röttger).

Ciliate Conjugation

Ciliate conjugation can be regarded as a special type of gamontogamy. Two cells of complementary mating types fuse, usually for a short time (Fig. **261**) and the cells exchange haploid gametic nuclei.

Not all members within several morphologically defined species can crossbreed with each other. The term syngen was adopted to refer to all members of a morphologically defined species which may or do interbreed. They therefore represent genetically isolated species. However, the identification of these species is very difficult using phenotypic characteristics.

Syngens include different numbers of mating types. For example, the morphological species *Paramecium caudatum* includes sixteen syngens with two mating types each. The syngens in *Paramecium bursaria* consist of four to eight mating types. A cell of one mating type can conjugate with members of all other mating types in each syngen. Hence, the mating system in *P. caudatum* is called bipolar, whereas in *P. bursaria* it is called multipolar.

Molecular analytical methods (e. g., of isoenzyme patterns or gene sequences) have become available that distinguish between syn-

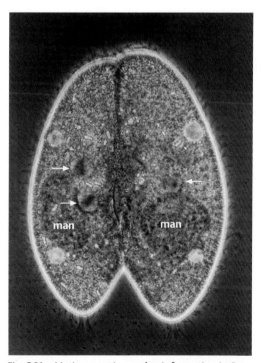

Fig. **261** Mating reaction and pair formation in *Paramecium caudatum*. The macronuclei (man) of both partners are still visible, while the micronuclei (arrows) of living individuals are rather inconspicuous (courtesy of H.-D. Görtz, Stuttgart). Magn.: 350×.

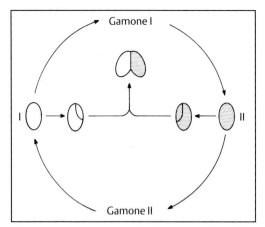

Gamone I

I

II

Gamone II

Fig. **262** Function of soluble gamones in *Blepharisma:* mating type I secretes gamone I, which induces in the complementary mating type II the conjugation reaction and the secretion of gamone II. On the other hand, gamone II evokes the conjugation reaction in mating type I (after Miyake).

gens and allow relationship to be determined. Therefore, other types of classification may be expected. The syngens of the *Paramecium aurelia* group are now recognized as 14 real genetic species, each with its own name. The situation in *Paramecium caudatum* is more complicated, as genetic exchange may occur between some syngens.

The mating types of some ciliates are not stable, and they may change spontaneously (e. g., in *Blepharisma japonicum*). In a strain of *Paramecium micronucleatum* ("cycler"), the activity of the gene (mt^+) determining the mating type is controlled by a circadian clock. The mating type in *Paramecium aurelia* is controlled by the macronucleus. It is determined during development of the new macronucleus in the postconjugant cells, where it may be influenced by factors in the cytoplasm.

Members of most ciliate species must pass through a number of cell division cycles before they are mature for conjugation. Additionally, a number of exogenous factors, such as light, temperature, and composition of the medium, influence the readiness to conjugate.

Pairing is induced by substances that ciliates may secrete into the medium (gamones), or that may be bound to the cell surface. These enable the cells to identify potential partners and to unite pair-wise. Pairing can also be experimentally induced by chemical agents (e. g., K^+, Mg^{2+}, Cs^+, in *Paramecium caudatum*).

In *Blepharisma intermedium*, mating type I secretes the glycoproteid gamone I. This is effective at concentrations as low as 60 ng/l. Gamone I binds to cell surface receptors of cells belonging to mating type II. It changes the properties of the cell surface and swimming behavior, thus providing the basis for membrane fusion. The cells of mating type II react by secreting gamone II (blepharismon). This changes the surface of cells of mating type I (Fig. **262**) and acts as an attractant for mating type I cells. Upon contact, the partners adhere by their cilia. After some hours the cells fuse.

In *Euplotes octocarinatus*, the gamones are released from cortical ampullae. The ten mating types of this species are determined by four codominant alleles $(mt^1–mt^4)$, one for each of the four gamones $G_1–G_4$. Medium that contains the gamone of one mating type can induce conjugation in cells of certain other mating types. It can even induce pairing of cells of the same mating type to create homotypic pairs. The cells of each mating type are suggested to express receptors for all other gamones — except for the one they secrete themselves.

Pairing in many ciliates (e. g., *Paramecium aurelia, Euplotes crassus, Tetrahymena pyriformis*) must be preceded by direct contact between cells. The inducing substances in such cases have not been isolated or biochemically characterized. They are located in the membrane of certain cilia. When cells of complementary mating types touch each other, their cilia agglutinate (agglutination stage). The cells then align lengthwise in pairs. Stable connections between the partners develop, especially in the mouth area. The cilia in the connecting region degenerate. Finally, membrane fusion results in a cytoplasmic bridge between the paired cells. *Paramecium caudatum* reaches this stage about three hours after contact between the mating types (see Fig. **261**).

The pairing of cells directly influences the behavior of their nuclei. Meiotic and a number of mitotic divisions (the number depending on the species) of the micronuclei give rise to haploid nuclei (Fig. **263a**). Only two of these, the stationary and the migratory nucleus remain; the other nuclei degenerate. The migratory nuclei move into the partner cell through the cytoplasmic bridge. There each fuses with the resident stationary nucleus to form the diploid synkaryon. Thereafter the cells separate. The macronuclei degenerate in the course of conjugation. The original set of macronuclei and micronuclei is reconstituted from the synkaryon in the postconju-

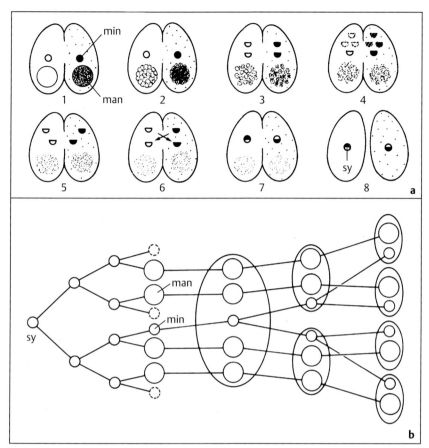

Fig. **263**   General nuclear events in conjugation of ciliates (**a**) and metagame nuclear divisions (**b**) in *Paramecium caudatum*. The micronucleus (min) divides meiotically (2–4). Three of the resulting 4 haploid nuclei degenerate (5); the surviving micronucleus divides by mitosis into two gamete nuclei (one stationary and one migratory nucleus) (5). The migratory nuclei are exchanged (6) and fuse with the stationary nuclei to form a synkaryon in both partners (7). Because the macronuclei (man) undergo degeneration, each of both exconjugants (8) contains only one diploid nucleus (sy). b After multiple metagame nuclear and cell divisions the normal nuclear features are restored (after Grell).

gant cells. Depending on the species, the synkaryon differentiates into the micronuclear and the macronuclear anlagen directly, or after a number of nuclear and cellular divisions (Fig. **263 b**).

The development of the postconjugant cell into the vegetative cell in many species is a complex morphogenetic process that may involve subsequent rounds of reorganization of the so-matic and oral infraciliature (e. g., in hypotrich ciliates). Mitotic divisions of the synkaryon may lead to the formation of supernumerary nuclei which become degraded. Experiments on *Paramecium* indicate that the decision if a nucleus becomes degraded or develops into a micronuclear anlage or a macronuclear anlage depends on its position in the postconjugant cell.

The conjugation partners in most species (e. g., *Euplotes octocarinatus, Paramecium caudatum*) are morphologically indistinguishable. Cells of different sizes may pair in sessile ciliates (e. g., peritrichs and chonotrichs). Some of these undergo a "total conjugation" where the whole nuclear and cytoplasmic content of a small, motile microgamont becomes incorporated into the larger, sessile macrogamont.

Morphogenesis and Reproduction

(by Maria Mulisch, Köln)

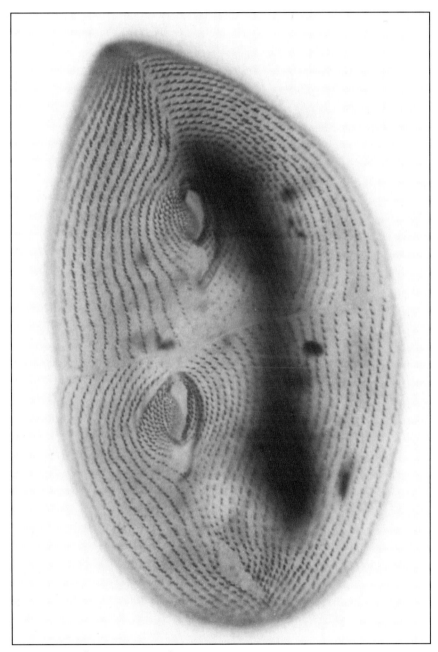

Fig. **264** Morphogenetic stage of *Paramecium*
(from Iftode et al.: Development 105 [1989] 191). Magn.: 1000×.

Many protozoa can change their structure in response to internal or external factors. If these changes are accompanied by resorption and/or construction of parts of organisms or organelles, the process is generally called morphogenesis (Fig. **264**).

Facultative Changes in Cellular Morphology

Changes in cell morphology may be caused primarily by external factors (e. g., salinity, temperature, pH of the medium, food supply). These facultative changes may include the formation of cysts, changes in cellular dimensions, or transformations of one cell type (e. g., amoeboid cell) into another (e. g., flagellated cell).

Encystment and Excystment

Many protozoa resist unfavorable conditions (e. g., desiccation or starvation) by forming cysts. These cysts are usually dormant stages with thick, highly resistant walls. Resistant cysts are common among soil protozoa. They are called "spores" in parasitic protozoa (e. g., Apicomplexa) where they often represent infective stages. Slime molds (acrasids, dictyostelids, Myxogastra; see Figs. **63, 64, 66**) and some ciliates (Sorogena; see Fig. **185**) will, under certain circumstances (e. g., starvation), produce fruiting bodies consisting of many encysted cells. The cysts may be dispersed by air in a dried state when the fruiting body breaks apart.

Other protozoa produce cysts with thin, relatively permeable walls. These include reproductive cysts (in which cells divide), digestive cysts (in which cells digest), and mating cysts (in which fertilization takes place). Some protozoa, such as the vampyrellid filose amoeba *Lateromyxa gallica*, may form digestive cysts, reproductive cysts, and resting cysts during the life cycle.

The factors causing encystment differ between species. The schizopyrenid *Naegleria gruberi* encysts on drying. *Acanthamoeba* encysts in response to an excess of oxygen, absence of food, or transfer to a mineral medium containing calcium and magnesium salts. It is supposed that all of these factors block the S phase by interfering with DNA synthesis. At the onset of encystment, the cell rounds up. Many of the cell components (mitochondria, storage granules) are degraded, and increased pulsation activity of the contractile vacuole leads to dehydration of the cytoplasm. This is followed by the formation of a layered cyst wall and the synthesis of stage-specific RNA and enzymes. Finally, almost all metabolic activity ceases, and in the dormant state the cyst can survive for several years.

The cyst wall may be formed by exocytosis of cyst wall precursors and/or synthesis of fibrils of chitin or cellulose from the cell surface. Golgi-derived precursors appear to be involved in cyst wall formation, for example, in the parasitic flagellate *Giardia lamblia*, in actinophryid helio-zoans (e. g., *Actinophrys sol*), in some heterotrich (e. g., *Stentor, Fabrea, Blepharisma*), and peritrich ciliates *(Epistylis)*. They apparently are not involved in cyst wall formation of the ciliates *Hyalophysa chattoni* and *Euplotes muscicola*.

Encystment in ciliates includes a more or less pronounced dedifferentiation of cortical structures, depending on the species. The cortex remains unchanged in e. g., *Frontonia* and *Euplotes*. In *Pseudomicrothorax*, the paroral ciliature is disorganized. The buccal ciliature disappears in *Bursaria*. In *Colpoda*, the buccal structures and the somatic cilia are resorbed. Dedifferentiation of the cortical organization during encystment is most pronounced in members of the Stichotrichia: mature cysts completely lack cilia, kinetosomes, and the infraciliature.

The signalling pathway leading to excystment is not clear. Experimental studies on the ciliate *Colpoda* suggest that soluble substances in the medium (organic acids, sugars) may penetrate the cyst wall and stimulate the cell to excyst. The excystment percentage in young cysts of *Acanthamoeba castellani* and the ciliate *Histriculus cavicola* has been found to be significantly lower than in old cysts. Therefore, the presence of endogenous excystment-suppressing factors has been discussed. These may accumulate during encystment and may be degraded during aging of the cyst. Activation of excystment in *Acanthamoeba* depends upon RNA synthesis and protein synthesis. Increased intracellular activity is followed by an osmotic uptake of water. The cell may emerge through a pore in the cyst wall, or it may perforate the cyst layers using specific enzymes. Ciliates which in the encysted state have dedifferentiated cortical structures, undergo excystment morphogenesis. The oxytrichid *Coniculostomum monilata* needs at least three morphogenetic cycles to acquire the species-specific ciliature characteristic of the vegetative cell.

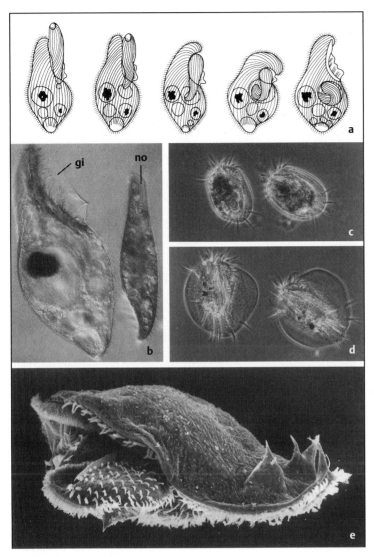

Fig. **265**   Cannibalism in *Blepharisma* (**a**) and *Onychodromus quadricornutus* (**e**). **b**comparison between gigantic (gi) and normal individuals (no) of *Blepharisma*. **c** and **d** normal *Euplotes octocarinatus* cell (c) and influenced by *Lembadion* factors (d) (a after Nilsson; b courtesy of D. J. Patterson, Sydney; c and d from Kuhlmann and Heckmann: Science 227 [1985] 1347; e from Wicklow; J. Protozool. 35 [1988] 137). Magn.: a 60×, b 140×, c and d 200×, e 350×.

Polymorphism

Individuals of the same species or of genetically identical cells (clones) that can adopt different morphologies (morphotypes) are said to be polymorphic.

Ciliates of many genera (e. g., *Blepharisma, Stylonychia, Dileptus*) are polymorphic. When starved, they may change into dwarfs that are up to one hundred times smaller than the normal forms. *Blepharisma* may change into greatly enlarged giants (Fig. **265 a, b**) if one food type (bacteria) is replaced by larger food organisms (for example, ciliates). The dimensions of the compound organelles (ciliary rows, oral structures) in both dwarfs and giants change in proportion to cell size, whereas the sizes of individual organelles (e. g., kinetosomes) remain constant. If the oral structures are enlarged to the size of individuals of the same species the cells may become cannibals. The giant cells revert to their normal size as soon as smaller food becomes available. The reduction in size is rapidly achieved by shortening the time between successive cell divisions.

If fed on bacteria, *Tetrahymena vorax* develops a disproportionately small oral opening (micro-

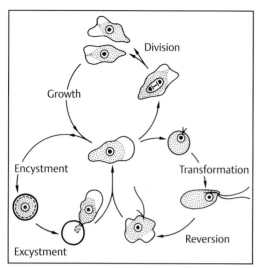

Growth

Division

Encystment

Transformation

Excystment

Reversion

Fig. **266**   Life cycle of *Naegleria gruberi* (after Fulton).

stome form). In contrast, the oral apparatus enlarges if fed on ciliates (macrostome form). The growth phase of the culture is apparently important for determining the size of the oral apparatus. Macrostome cells usually appear in the stationary phase of a culture and may become cannibals after prolonged starvation. The change in morphology thus enables some individuals to survive at the expense of others.

Species of the ciliate *Euplotes* change their morphology in the presence of certain predators (for example *Euplotes octocarinatus* in the presence of the ciliate *Lembadion*). The cell cortex of *Euplotes octocarinatus* forms wing-shaped lateral extensions and dorsal ribs (Figs. **265 c, d**), that prevent its ingestion by *Lembadion*. The reorganization of the cytoskeleton of *Euplotes* is induced by substances secreted by the predators. The inducing substance secreted by *Lembadion* is probably a polypeptide.

The schizopyrenid genera *Naegleria* and *Tetramitus* are capable of radical changes in morphology (Fig. **266**). Under favorable conditions, they feed and reproduce as amoeboid cells. These may encyst in response to starvation or desiccation. A decrease in the ionic strength of the medium causes the cells to transform into flagellates. The differentiation process includes the expression of certain genes (e. g., for tubulin, calmodulin) and cytoskeletal changes, like the synthesis, assembly, and positioning of basal bodies and the outgrowth of flagella. The flagellated stage of *Naegleria* is not able to reproduce. It is a temporary and facultative stage of the life

cycle which may retransform into the amoeboid stage under appropriate conditions. In *Tetramitus*, however, both stages can feed and reproduce.

Cyclical Changes in Morphology

Different body forms may be obligatory and regular stages of the life cycle in many protozoa. They are usually associated with a change of habits. Parasitic organisms develop stage-dependent structures which, for example, enable them to survive outside of the host, to invade the host cell, or which protect them against proteolytic enzymes, or against the immune response of the host. The stages are usually described by specific terms. Feeding stages (= trophonts) may alternate with nonfeeding stages (e. g., cysts). Sessile protozoa may produce motile, nonfeeding stages (= swarmers) for distribution.

The heliozoon *Clathrulina elegans* produces flagellated swarmers which transform into amoeboid forms. After settling, each secretes a stalk and a capsule, and adopts the morphology of a heliozoon (see Fig. **165**).

Adult Suctoria (see Figs. **127, 128**) are also usually attached by a stalk to the substrate. Their tentacles are used for prey capture and food uptake. The division products include a motile swarmer, which in contrast to the sessile cell, is ciliated. The swarmer settles on the substrate (often by secreting a stalk), resorbs its cilia, and forms tentacles. This morphogenetic process is called metamorphosis.

The folliculinid heterotrichs (see Fig. **113**) are ciliates that live inside a chitinous lorica. The oral apparatus is shaped like two wings edged by a long adoral zone of membranelles. The sessile cell (trophont) divides into one mouthless swarmer with only a short membranellar spiral, and one trophont (Fig. **267**). The swarmer leaves the old lorica and secretes a new one at a different place (see Fig. **179**). Thereafter, it resorbs the membranellar spiral and transforms into the trophont (Fig. **267**).

Many parasitic protozoa change their morphology when they invade a new host or host tissue (see Figs. **57, 58**). For example, the kinetoplastid flagellate *Trypanosoma* may occur as a leishmanial form, a leptomonad form, a crithidial form, or a trypanosomal form (see Fig. **57**). The leptomonad form and the crithidial form occur predominantly in invertebrates, the trypanosomal form in vertebrates. The leishmanial form occurs intracellularly in vertebrates but extracellularly in invertebrates. If the crithidial form of *Trypanosoma mega* is transferred into the serum of vertebrates, it transforms into the trypanosomal form that is the characteristic form in vertebrates. Experiments such as this indicate that the transformation is induced by the changed environment.

Apostome ciliates, parasites of diverse invertebrates, develop a number of polymorphic forms during complex life cycle (Fig. **268**). The trophont feeds at the cost of the host. It grows but does not reproduce. It then transforms into a so-called tomont which by subsequent divisions, usually inside a cyst, produces small cells called tomites. The tomites are released and become fixed to transporting hosts, where they encyst. They are now called phoronts and give rise to trophonts. The phoronts of *Spirophya subparasitica*, for example, are attached to copepods. Trophonts develop and grow inside Hydrozoa after these have ingested the copepods.

Alternating phases between solitary and multicellular stages can be observed in some protozoa. The colonies of the ciliates *Zoothamnium alternans* (Fig. **269 a**) and *Z. arbuscula* (see Fig. **136**) are initiated by a swarmer called a ciliospore. This settles on the substrate by secreting a stalk and it may encyst during this process. A branched, tree-shaped colony develops by a specific series of divisions which are not accompanied by complete separation of the division products. All cells of the colony are connected by a common stalk (Fig. **269 b**). The cell divisions may produce unequal daughter cells: small individuals (microzooids) develop at the lateral branches of the colony while larger cells develop only at the bases of the lateral branches. Only terminal cells can divide, but the macrozooids may transform into free-swimming swarmers by developing a marginal ciliary band (telotroch). These are the ciliospores that may form new colonies. Microzooids may transform into motile gamonts that conjugate with certain macrozooids. This process also gives rise to ciliospores.

Some cellular slime molds (Dictyostela) feed and reproduce as small amoebae. Upon starvation, the cells initiate a well-defined program of multicellular development (Fig. **270**). Some hours after depletion of the food source, a few cells start emitting pulses of a signal molecule. The signal molecule in *Dictyostelium discoideum* (see Fig. **64**) is cyclic adenosine monophosphate (cAMP). Surrounding cells respond to the cAMP signal by moving towards the source of the signal (by chemotaxis) and they too begin to secrete cAMP. After this, the cell is unresponsive to further cAMP pulses for several minutes. The

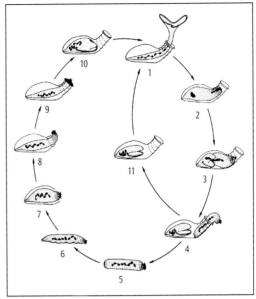

Fig. **267**   Life cycle of *Eufolliculina uhligi*: 1 trophont; 2–4 fission; 5 swarmer; 6–10 lorica formation and metamorphosis; 11 reorganization (from Mulisch and Patterson: Protistologica 19 [1983] 235).

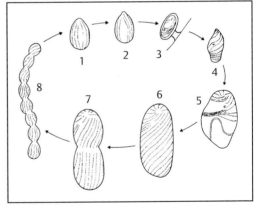

Fig. **268**   Life cycle of the apostome ciliate *Polyspira delagei*: 1 protomite; 2 tomite; 3 encysted phoront on gills of a crayfish; 4 young trophont; 5 adult trophont; 6 tomont; 7–8 successive divisions (palintomy) (after Lwoff).

cells join with each other to form streams; finally, all streams merge at an aggregation center (see Fig. **65**). An aggregate consists of tens of thousands of amoeba. These form a multicellular mold, which may produce a migrating pseudoplasmodium (slug). Eventually, the slug trans-

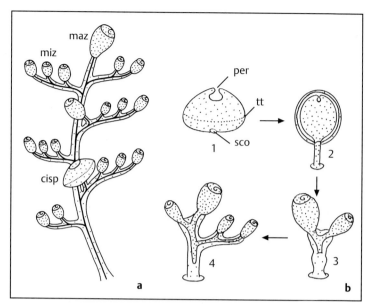

Fig. **269** Life cycle of *Zoothamnium*: **a** schematic drawing of a colony with microzooids (miz), macrozooids (maz) and ciliospore (cisp) of *Z. alternans*. **b** colony formation in *Z. arbuscula*: 1 ciliospore with telotroch (tt), peristome (per) and scopula (sco); 2 settlement and encystation; 3 products of the first division; 4 micro- and macrozooids after the second division. (a after Summers, b after Furssenko).

forms into a fruiting body (sorocarp). It consists of a stalk formed by dead, vacuolated stalk cells, and of an anterior mass of spores. After excystment, each spore may give rise to a single amoeboid cell.

Dictyostelium discoideum cells adopt their fate as a stalk or spore cell very early in the developmental program. There is evidence that the cells are predisposed to prestalk or prespore differentiation at the onset of starvation. Those cells that become starved while in an early phase of the cell cycle tend to become prespore cells, while cells starved late in the cell cycle tend to become prestalk cells. Both cell types differ in cytoplasmic calcium content, adhesiveness, and sensitivity to cAMP. Prespore and prestalk cells are randomly distributed in early aggregates, but they separate from each other during development of the mold. Prestalk cells migrate to the tip, and prespore cells occur throughout the base. Molecular biological methods have resulted in the identification of anterior-like cells (ALC) and of serveral types of prestalk cells (Fig. **270**). Cells appear to become stalk cells by the sequential expression of certain genes (pstA and pstB). The most advanced cells (pstAB) in stalk differentiation are situated in the center of the tip of the slug. When migration of the slug ceases, they synthesize a hollow extracellular tube. More and more prestalk cells migrate through this tube towards the bottom, and as they exit, they become differentiated to stalk

cells that enlarge and die. This process (somehow resembling gastrulation in animal embryogenesis) lifts the posterior spore cells above the substrate. If the prestalk region is experimentally separated from the slug, both parts develop into properly proportioned fruiting bodies. Thus, although the prestalk and prespore cells are molecularly, biochemically, and histologically distinct from each other and from the vegetative cells, they can still change their fate by dedifferentiation and redifferentiation.

Differentiation can be induced in cultured cells by combinations of extracellular cAMP and a lipophilic, low molecular weight protein family called DIF (differentiation inducing factor). DIF-1 stimulates amoeba to become prestalk cells, while high concentrations of extracellular cAMP are needed to express prespore-specific genes. Additionally, ammonium and adenosine, which both are present in the slug, act as morphogenetic substances. There is evidence suggesting that these signals act in the intact aggregate to regulate the proportion and position of stalk and spore cells.

Discoideum murocoides exhibits sexual or asexual development, depending on the culture conditions. During the sexual cycle, macrocysts are formed. Macrocyst formation and subsequent division appear to include sexual fusion and meiosis. The process is induced by a potent plant hormone, ethylene, which is produced in various species of cellular slime molds.

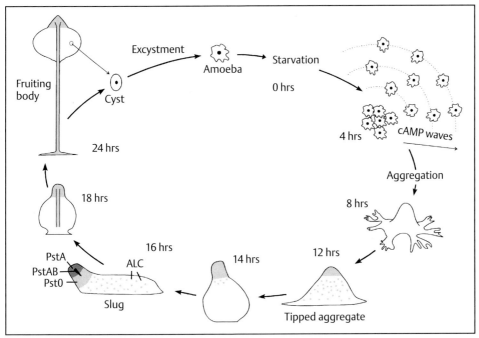

Fig. 270   Asexual life cycle of *Dictyostelium discoideum*. Upon starvation, individual amoebae aggregate in response to waves of cAMP and progress through a series of multicellular stages to form a fruiting body. The shaded regions represent prestalk or stalk cells. ALC = anterior-like cells, PstA = prestalk A cells, PstAB = prestalk AB cells; PstO = prestalk O cells (after Powell-Coffman and Firtel).

The life cycle of Myxogastra (e. g., *Physarum*) is more complex. The spores give rise to haploid amoebae, which may transform into flagellated forms. Amoeboid forms, as well as flagellated forms, may fuse pair-wise forming amoeboid zygotes. The zygote grows to a multinucleated plasmodium and finally to an excessively branched plasmodial network (see Fig. **66**). Under some circumstances, some strands of the network may give rise to stalked fruiting bodies that project above the substrate and contain many spores.

Cell Division

Most protozoa reproduce by division into two or more equal daughter cells. In contrast, some protozoa (e. g., sessile organisms such as the ciliate *Eufolliculina*; see Fig. **267**) produce daughter cells that differ in size and/or morphology. The daughter cells may also differ structurally from the parental cell (e. g., in organisms with a complex life cycle such as Apicomplexa). A diversity of species undergo division inside a reproductive cyst. An incomplete separation of the daughter cells may lead to the development of colonies (e. g., in many sessile peritrich ciliates) or of chains of cells (see Fig. **138**).

Binary Fission

Division into two similar daughter cells is the most common form of reproduction among the protozoa. Repeated binary fissions without intervening growth phases (palintomy) is found in the life cycle of some parasitic protozoa (e. g., *Toxoplasma*). This results in a large number of very small daughter cells.

In order to produce two identical daughter cells, all organelles must be duplicated before or during division, and they must be equally sorted between the division products. It is often necessary for the sorting to follow a species-specific pattern. Contractile elements (especially actin and myosin) arranged in a ring-shaped area beneath the fission furrow are involved in the separation of the cells (cytokinesis). Karyokinesis occurs in the course of cytokinesis.

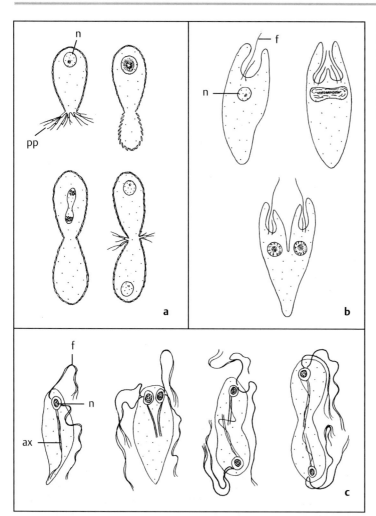

Fig. **271**   Binary fission of a testate amoeba (*Euglypha*, **a**), of a flagellate (*Euglena*, **b**), and of a hypermastigid flagellate (*Devescovina*, **c**). ax = axostyle; f = flagellum; n = nucleus; pp = pseudopodia (after Grell).

Cells without any recognizable polarity, such as heliozoans, usually separate in the plane of the nuclear division. In *Actinophrys sol*, some of the axopodial microtubules disappear during cell division. The axopodia in *Dimorpha mutans* arise from a cluster of microtubule organizing centers (MTOCs) called an axoplast. During mitosis, the axoplast separates into two parts and then dissociates. Most of the axopodia disappear. Axoplast and axopodial microtubules reappear at the end of telophase.

The cell division plane in testate amoebae is defined by the opening (= pseudostome) for the pseudopods in the test. Through this opening, a cytoplasmic extrusion emerges and produces a new shell as a mirror image of the old one (Fig. **271 a**).

In *Arcella vulgaris*, the test is produced from preformed elements, the thecagenous granules derived from the Golgi apparatus. Prior to division, the thecagenous granules and the two nuclei aggregate in the cytoplasm near the pseudostome. This area protrudes through the pseudostome and grows to form a bud. The nuclei divide mitotically during bud formation. The thecageneous granules become aligned beneath the plasma membrane and are then extruded simultaneously. The extruded elements swell, fuse in a well defined pattern, and assemble into a continuous layer made of a keratin-like protein. The test is molded by the bud and by pseudopodia which form a temporary dome around the new test. After completion of the new test, the two tests remain attached to each other with their

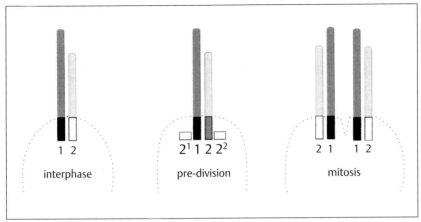

Fig. **272**   Scheme of the flagellar development cycle in *Nephroselmis olivacea*. The interphase cell has a long (1) and a short flagellum (2). Two new basal bodies appear before cell division. They give rise to short flagella during mitosis. The short flagellum of the parental cell elongates and becomes the long flagellum of a daughter cell (after Melkonian).

pseudostomes. The cytoplasm shuttles between the tests until cell division occurs.

In flagellated cells, the insertion site of the flagella defines the anterior pole of the cell; the division furrow develops along the longitudinal axis of the cell. The number of basal bodies duplicates before cell division. The new basal bodies give rise to associated cytoskeletal elements and to the new flagella. Biflagellated species such as *Euglena* now develop two pairs of flagella. In the course of nuclear division, which is perpendicular to the longitudinal axis of the cell, the flagellar pairs move apart and are finally separated by the longitudinal division furrow (Fig. **271 b**).

Most phototrophic flagellates bear flagella which are structurally and/or functionally different. These differences are accompanied by a heterogeneity in the root structures attached to the flagellar basal bodies. There is evidence that the heterogeneous flagellar apparatus is the result of the development of the flagellar structure over more than one cell cycle. *Nephroselmis olivacea* (Fig. **272**), for example, has one short and one long flagellum. Two new basal bodies are produced before cell division. They give rise to flagella, which become the short flagella of the daughter cells. The parental short flagellum elongates and thus transforms into the long flagellum of one daughter cell. Consequently, the

basal bodies of *Nephroselmis olivacea* give rise to short flagella in the first generation, while they give rise to long flagella in the second generation. Light microscopical and ultrastructural observations on *Nephroselmis* and other phototrophic flagellates, suggest that their basal bodies require more than one cell generation to mature. The organization of the flagellar apparatus appears to depend on the developmental state of the basal body.

The multiflagellated Hypermastigida have a few "privileged flagella" or basal bodies at certain sites (e. g., three to four in *Joenia*; see Fig. **49**). These alone duplicate before cytokinesis; all others are resorbed. The flagella and their associated root structures, including the replicated axostyle, draw themselves towards opposite poles of the cell. The daughter cells thus separate at their posterior ends (Fig. **271 c**).

Asymmetric cell division is thought to be the origin for cellular differentiation in *Volvox*. The colonies of *Volvox carteri* consist of many small flagellated cells and few immobile larger cells. Only the large cells (reproductive cells, gonidia) can divide and create new colonial organisms, while the small cells (somatic cells) finally die. Both cell types arise by asymmetric cell divisions during embryogenesis; they represent distinct cell lines. If (by mutations or by microsurgical ex-

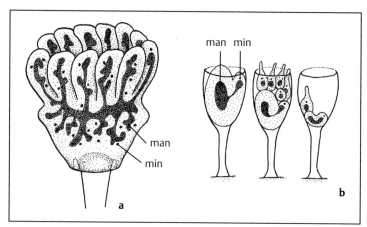

Fig. **273** Gemmation (= budding) in *Ephelota* (**a**) and *Tachyblaston* (**b**). man = macronucleus; min = micronucleus (after Grell).

periments) large cells (more than 8 μm in diameter) are created at an area of the embryo which normally produces small somatic cells, these cells always become reproductive cells. It is therefore suggested that the fate of a *Volvox carteri* cell is determined only by its size.

In ciliates, cell division is associated with complex processes of pattern formation. Ciliates usually divide perpendicular to their longitudinal axis (see Fig. **110 a**). Whereas the daughter cell originating from the anterior cell part (proter) normally maintains the old organellar complexes (usually the oral apparatus, sometimes also contractile vacuole and cytopyge), new complexes must be produced and assembled for the posterior daughter cell (opisthe). The daughter cells of most ciliates have the same orientation. In contrast, division in some ciliates involves stages with the axis of proter and opisthe strongly shifted (e. g., *Halteria, Tintinnidium*). In *Opercularia* and *Tintinnopsis subactuta*, division appears to be longitudinal.

Budding

The separation of one or several smaller daughter cells from a large mother cell is called budding or gemmation. This process occurs in many sessile protozoa where the buds often differentiate into motile swarmers. Exogenous budding in many Suctoria, peritrichs, and chonotrich ciliates involves an external protrusion of the parental cell, which is then detached as the progeny. *Ephelota gemmipara* reproduces by multiple budding. The anterior area of the cell gives rise to a crown of buds. Each receives a part

of the highly segmented macronucleus, and a micronucleus (Fig. **273 a**); the latter having been produced by numerous mitotic divisions. The buds detach simultaneously as motile ciliated swarmers. Multiple buds are produced sequentially in *Tachyblaston ephelotensis* (Fig. **273 b**).

Endogenous budding occurs in some members of the Suctoria. In *Acineta tuberosa*, one or several small daughter cells develop inside an internal "brood pouch" of the mother cell. The plasma membrane at the apical pole invaginates to separate an internal cytoplasmic region, which includes a section of the macronucleus (see Fig. **126**). Basal bodies in a field located in the separated area proliferate and give rise to the ciliature of the motile swarmer.

Multiple Fission

Multiple fission can result in the rapid production of many cells. It often occurs in parasitic forms that overwhelm the immune response by flooding the host with large numbers of infective stages.

In coccidians (e. g., *Eimeria*, see Fig. **87**) and other members of the Apicomplexa, vegetative reproduction and the formation of gametes are achieved by schizogony. After several mitotic nuclear divisions, the cell divides into a corresponding number of daughter cells. The mode of development of the daughter cells varies between genera. In *Eimeria*, the nuclei migrate to the periphery of the cell (see Figs. **258 b, e**); the sporozoites (or merozoites) separate all around the cell by a process that resembles budding (see Fig. **88**). A central residual body is left behind. In other genera (e. g., *Sarcocystis*), the sporozoites develop inside the cytoplasm of the parental cell.

Large multinucleated protozoa such as *Pelomyxa* may separate without associated nuclear divisions into smaller individuals. The number of nuclei later increases as the cell grows. In foraminifers, the gamonts form by schizogony (associated with meiosis) of the agamont (see Fig. **156**).

The life cycles of the astome and apostome ciliates (see Fig. **268**) include a mode of reproduction known as palintomy. They undergo a series of repeated (cross) divisions without intervening growth phases. Chains of cells may be formed by incomplete separation of the daughter cells. Some genera (e. g., *Radiophrya*) distally pinch off small cell parts (see Fig. **138**); others separate into subunits of identical size (e. g., *Haptophrya*). Palintomy in *Foettingeria* (see Fig. **139**) occurs inside a reproductive cyst.

Control of the Cell Division Cycle

The essential processes of the eukaryotic cell cycle (DNA replication, mitosis, cytokinesis) have been found to be triggered by a central cell cycle control system (Fig. **274**). This is regulated at certain checkpoints of the cycle by feedback from the processes that are being performed. For example, the cell is not permitted to enter mitosis until the nuclear DNA has been replicated.

The control system is based on two families of proteins. The first is the family of the cyclins (named cyclins because they undergo a cycle of synthesis and degradation during the cell cycle). The second is the family of cyclin-dependent protein kinases (Cdk). Cyclin forms a complex with Cdk. This activates Cdk to phosphorylate certain target proteins. At least two complexes regulate the normal cell cycle: One at a late G_1 checkpoint, just before S phase, and the other late in G_2 just before M phase (Fig. **274**). The general principle is as follows. Mitotic cyclin is synthesized and accumulates gradually during G_2. It associates with Cdk to form a so-called M phase promoting factor (MPF), which becomes activated and triggers a number of processes to drive the cell into mitosis. The degradation of mitotic cyclin inactivates MPF, enabling the cell to exit from mitosis. The assembly of a related complex of a CdK protein and G_1 cyclin is thought to trigger the events that lead to DNA replication.

The cell-cycle control system has been especially well explored in yeast cells. Recent observations on animal and plant cells indicate

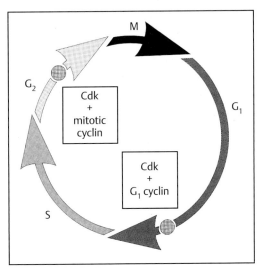

Fig. **274** The standard eukaryotic cell cycle. Two major checkpoints (indicated by shaded circles) regulate the entry into the DNA-synthesis phase (S) or into the mitotic phase (M), respectively. Complexes of cyclin-dependent protein kinases (Cdk) and cyclin trigger the downstream processes of the cycle, G_1, G_2 = gap phases (no DNA synthesis).

that the basic mechanism is universal. It therefore should apply also for protozoa.

Cdks have been identified in some protozoa, e. g., the ciliates *Tetrahymena thermophila* and *Paramecium tetraurelia*. Their role in the cell division cycle, however, is yet unclear. There is also evidence for the existence of cyclin in protozoa (e. g., in *Trypanosoma brucei*, *Physarum polycephalum*, and in the ciliates *Stylonychia lemnae* and *Tetrahymena*). Division in *Tetrahymena* can be synchronized by treatment with heat shocks. These have been shown to destroy a heat labile protein that accumulates before the M phase and appears to be necessary to allow the cell to enter mitosis. The protein disappears after mitosis. It is therefore suggested to be a cyclin.

Cytoplasmic extracts of the ciliates *Tetrahymena pyriformis*, *T. thermophila*, and *Paramecium caudatum* can reinitiate meiosis when injected into amphibian oocytes. The meiosis-reinitiating factor (MRIF) differs from the M phase promoting factor (MPF). The latter has not been detected in *Tetrahymena* or *Paramecium*. MRIF of *Tetrahymena* is a heat-labile, Ca^{2+}-sensitive, soluble protein. It is present in meiotic and mitotic stages, but also in cells at stationary growth phase.

Pattern Formation in Ciliates

Ciliates have a complex pattern of cortical subunits with precise arrangements of the oral apparatus, contractile vacuole(s) and cytopyge. Each cell has an anterior-posterior polarity and a left-right asymmetry. Both asymmetries can be seen not only in the general organization of the cell, but also in the structure of each cortical unit. Each cortical unit involves a kinetid – one or two basal bodies with associated root structures (see Fig. **174**). As kinetids or single cytoskeletal elements can be specifically stained by silver impregnation or immunofluorescence techniques, their development and arrangement can be easily observed by light microscopy (see Fig. **264**). Ciliates (especially *Stentor, Paramecium, Tetrahymena*, and some hypotrichs) are therefore used as model organisms to study the principles and mechanisms of pattern formation. Additionally, certain developmental processes in ciliates, such as oral development (stomatogenesis), provide useful criteria for the determination of phylogenetic relationships. As the subsequent stages in stomatogenesis have responded to adaptive selection, comparative studies on pattern formation may be useful for erecting a natural system of the ciliates. Only a few examples of pattern formation in ciliates will be presented here.

Regeneration, Reorganization, and Division

Ciliates can usually replace and reconstruct cell parts which have been experimentally removed. However, only fragments containing at least a portion of the macronucleus can develop into complete cells that grow and reproduce.

Like metazoa, ciliates differ greatly in their capacity to regulate their pattern. For example, *Stentor* recreates its normal (though smaller) morphology even from small fragments during one generation, but the total reconstruction of structures removed from *Paramecium* may take several cycles of cell division. Development in *Stentor* (and, e. g., in *Tetrahymena*) appears to be regulated throughout the entire cell, whereas in *Paramecium*, pattern is determined on a more local scale.

Oral structures may need to be regenerated without experimental intervention. They become degraded and are later reassembled by stomatogenesis. This process is termed physiological regeneration or reorganization. It especially occurs in heterotrich ciliates, such as *Stentor*. Here it may be associated with the nuclear phenomena which are typical of cell division (condensation and renodulation of the macronuclear chain, mitosis of micronuclei). The purpose of oral reorganization is possibly to replace disturbed mouthparts, or to adapt the size of the oral apparatus to a changed cell size (for example, to a smaller cell size after starvation).

A number of morphogenetic phenomena are associated with cell division. These include degradative processes (e. g., resorption of mouthparts) as well as assembly and positioning of new organelles and organellar complexes. New and old parental structures, such as the oral apparatus, may undergo dramatic morphogenetic changes. The somatic kinetids must be duplicated, and the cytoskeleton has to be reorganized.

Detailed observations of certain cytoskeletal elements (e. g., kinetodesmal fibrils) by immunofluorescence microscopy have produced a "fate map" for the cortex of *Paramecium*. It is now known from which areas of the parental cell certain cortical units of the daughter cells originate (Fig. **275**). Morphogenetic processes (e. g., proliferation of basal bodies) in *Paramecium* start from the parental oral apparatus and from the future fission line and spread around the whole cortex. This developmental pattern is believed to be regulated by waves of morphogenetic signals in the cortex. A transcellular wave of phosphorylation of the ciliary rootlets preceding their disassembly has been shown in dividing *Paramecium* cells. It appears to be coordinated with the putative morphogenetic signals. The signals are interpreted differently in different cortical fields because the fate and organization of cortical units vary depending on their position in the cell.

In hypotrich ciliates, the infraciliature is renewed at each morphogenesis (division, regeneration, reorganization, excystment). Nevertheless, the cortical pattern of the parental cell is transmitted to the next stage even when the parental infraciliature is totally resorbed, before the new one is assembled. *Oxytricha* maintains its morphotype, for example a doublet configuration, even if passed through encystment and excystment. It is still unclear how the pattern in a cell establishes itself without the participation of any structural landmarks. Recent immunocytochemical studies on *Paraurostyla weissei* indicate that material associated with the basal bodies during interphase aligns to form tracks during morphogenesis. These tracks are suggested to act in guiding cellular components towards assembly and to provide a scaffold for organellar patterning.

Regeneration, reorganization, and cell division share some features of morphogenesis, and the boundaries between these processes are ill defined. The regeneration of an oral apparatus in *Stentor* (and other heterotrichs) may lead to cell division; early dividers may end up in reorganization. In most morphogenetic processes in ciliates, the proliferation of basal bodies is accompanied by nuclear changes (mitosis of micronuclei, condensation and decondensation of macronuclear chains). This suggests that common signals might act upon regeneration, reorganization, and cell division. One of the mechanisms discussed today is that the activation of a cyclin – dependent protein kinase triggers the cytoskeletal and nuclear changes in cilate morphogenesis.

Stomatogenesis

Ciliates with a complex oral ciliature do not modify an existing oral apparatus by intercalation of newly formed structures. Cell division and damage to the oral structures both generally lead to the development of a new oral apparatus by a process called stomatogenesis.

The first visible step in stomatogenesis is the production of new kinetosomes in a specific region of the cortex. These cluster to form the oral primordium. All oral components are subsequently completed and then arranged in the precise pattern found in this ciliate. The origin of the kinetosomes, as well as the course of stomatogenesis, differ between different ciliate groups.

Ciliates with an oral ciliature resembling the somatic ciliature are regarded as the most primitive. In these, the oral primordium arises from fragments of certain somatic rows of cilia (kineties). This mode of formation of the oral primordium is called telokinetal and is especially obvious in the ciliate groups that were united by this feature in the former taxon the Kinetofragminophorea. Fragments of the kineties may transform directly into the oral ciliature of the posterior daughter cell *(Colpoda)*, or the transformation may follow a complicated rotation and migration *(Trithigmostoma*; Fig. **276 a**). The transformation involves changes in structure and position of the kinetids. For example, in *Trithigmostoma* the single somatic units (monokinetids) become pairs (dikinetids) (Fig. **277**). Whereas in *Colpoda* the whole oral ciliature arises from fragments of kineties, in *Furgasonia* only the left (adoral) ciliature has this origin (Fig. **278 a**). The right (adoral) ciliature in *Furgasonia* arises from kinetosomes of the original

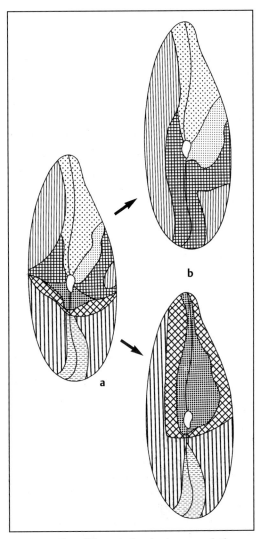

Fig. **275** The differently-batched areas of the somatic cortex of *Paramecium* (**a**) change during division and are differently distributed between the daughter cells (**b**): the areas anterior of the cell meridian reappear in the proter, the posteriorly located areas in the opisthe. In anterior (dotted) and posterior regions (horizontally dashed), the cortex exhibits no changes during divisions (after Iftode).

paroral ciliature that splits during divisional stomatogenesis.

In *Paramecium*, the whole oral ciliature arises from paroral kinetosomes in the parental cell. Hence, the elements for the new oral apparatus develop inside the old one. This is termed buccokinetal. Separation of the two oral fields and their positioning in the daughter cells requires complicated shifts and rotations.

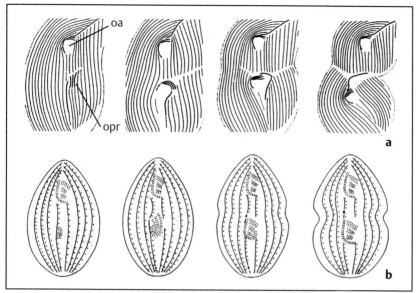

Fig. 276   Modes of stomatogenesis during cell division: **a** telokinetal formation of anlagen in *Trithigmostoma steini;* **b** parakinetal formation of anlagen in *Tetrahymena.* oa = oral apparatus, opr = oral primordium (a from Hofmann and Bardele: Europ. J. Protistol. 23 [1987] 2).

In *Tetrahymena,* the basal bodies of one somatic kinety (k1) proliferate, forming a field of basal bodies. At first there is no evident pattern (= anarchic field), but later they become oriented and aligned to form membranelles (Figs. **276b, 278b**). A similar pattern of development is observed in hypotrichs (e. g., *Paraurostyla;* Fig. **278c**) and heterotrichs (e. g., *Stentor*). In these ciliates, however, several (stomatogenic) kineties (in heterotrichs) or cirri (in hypotrichs) give rise to the anarchic field. The paroral, as well as the adoral ciliature, emerge from the anarchic field. Thus, both are of somatic origin. This type of formation of the oral primordium is called parakinetal.

In most cases, new kinetosomes are formed at the base of preexisting ones. The oral kinetosomes in some ciliates arise without visible association with somatic or oral ciliature (apokinetal formation of the oral primordium). The encysted cells in *Oxytricha fallax* contain no basal bodies, but the excysted cells are ciliated, indicating that kinetosomes form *de novo.*

New basal bodies first appear as short probasal bodies with incomplete triplets oriented at a right angle to the "parental" kinetosome. After completion, they become attached to the plasma membrane where they might give rise to cilia (see Fig. **277**). The filamentous and microtubular elements associated with the basal body are

then formed. The oral ciliary units assemble into monokinetids, dikinetids, or polykinetids (see Figs. **277, 278**) depending on the species and on the area of the cell. During the assembly of the adoral membranelles of *Tetrahymena* and *Paraurostyla,* for example, double-rowed "promembranelles" appear first. Subsequent sequences of basal body proliferation (one in heterotrichs, two in hypotrichs) give rise to further ciliary rows — until the final number of ciliary rows in each membranelle is achieved (Fig. **278c**). The cytostome where the food vacuoles are formed develops at the conclusion of stomatogenesis when the oral apparatus has reached its final position in the cell.

Ciliate stomatogenesis occurs continuously in a temporally and spatially coordinated series of steps. Several different categories of factors are involved controlling pattern formation. These include genetic mechanisms, cell-wide morphogenetic fields, and interactions with neighboring structures. A number of models based on genetic and experimental studies help to explain the underlying mechanisms. However, a model that explains all phenomena of pattern formation observed in ciliates has not yet been developed.

In *Stentor,* for example, morphogenetic gradients appear to be involved in the control of pattern formation. The longitudinal ciliary rows in *Stentor* are arranged in increasing distances

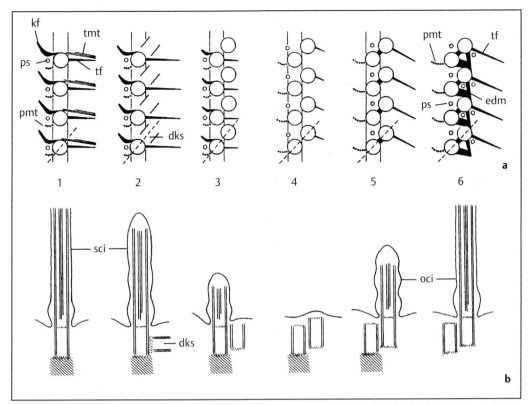

Fig. **277** Differentiation of oral from somatic kinetids in *Trithigmostoma steini* (**a** cross sections, **b** longitudinal sections of kinetosomes): 1 somatic monokinetid; 2–4 differentiation of daughter kinetosomes (dks) and dedifferentiation of somatic cilia (sci); 5 growth of oral cilia (oci); 6 oral dikinetid. The structures associated with the kinetosomes change during each step. edm = electron-dense material; kf = kinetodesmal fiber; pmt = postciliary microtubules; ps = parasomal sac; tf = transverse fibril; tmt = transverse microtubules (from Hofmann and Bardele: Europ. J. Protistol. 23 [1987] 2).

around the cell. The rows are close to each other on the ventral side, becoming more widely spaced along a clockwise gradient. In the ventral area, the most widely spaced ciliary rows meet the most narrowly spaced rows (Fig. **279**). The site where they meet, the "contrast zone," is where the oral primordium appears − even if the zone has been created microsurgically or has been grafted into another area of the cell. Experimental reversal of the contrast zone gives rise to a mirror image reversed oral apparatus (Fig. **279 c**). In addition to the circular gradient in *Stentor*, there is a basal-apical gradient. It is suggested that the (basal) foot of *Stentor* promotes the formation of the ingestion organelles (buccal cavity and cytostome); its influence decreases gradually towards the anterior of the cell. The (apical) oral apparatus is supposed to inhibit stomatogenesis. The inhibitory effect decreases when the cell grows larger and the oral area be-

comes smaller in relation to the somatic area. Stomatogenesis can take place when a critical ratio of the somatic area to the oral area is reached (leading to reorganization or division) or when the oral apparatus has been removed (leading to regeneration).

Analyses of pattern formation in a number of different ciliate species and genera have led to the proposal of two types of hypothesis. The first one is based on the concept of positional information; it assumes that the ciliate cell surface includes a cylindrical coordinate system, set up by graded signals. These signals lead to the formation of certain structures at certain locations, which correspond to specific positional values. This type of model explains many observations on pattern formation, especially in heterotrichs (e. g., *Stentor*), hypotrichs, and in *Tetrahymena*. A second type of hypothesis is derived mainly from analyses of cell division in *Paramecium*. It

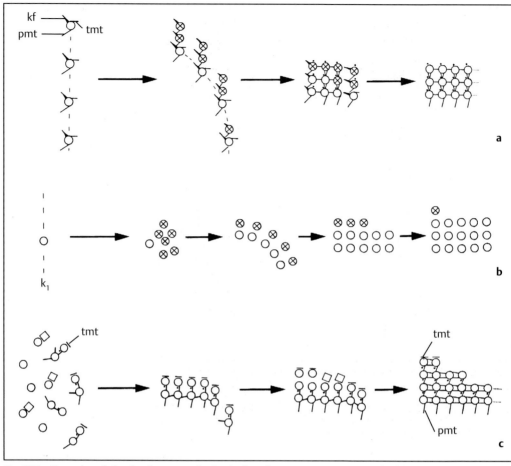

Fig. **278**   Examples of the development of adoral ciliature: **a** *Furgasonia blochmanni*; **b** *Tetrahymena thermophila*; **c** *Paraurostyla weissei*. The crossed circles and squares represent nonciliated and differentiating kinetosomes, respectively. kf = kinetodesmal fiber; pmt = postciliarly microtubules; tmt = transverse microtubules (from Eisler: Europ. J. Protistol. 24 [1989] 181).

assumes that the cell surface is composed from distinct territories that differ in terms of molecular and/or structural characteristics. Thus, the territories provide a prepattern that is used by basal bodies for both their replication and their global arrangement. Further analysis may show which model will finally explain all phenomena of pattern formation observed in ciliates.

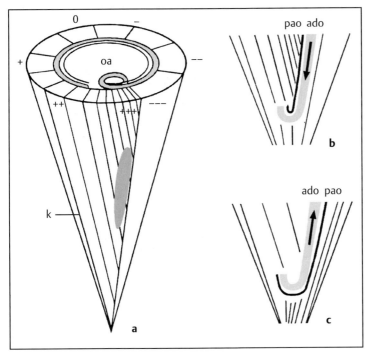

Fig. **279**   Determination of stomatogen areas (hatched) in *Stentor:* **a** schematic representation of *Stentor* with apically projected kineties (k) and circular gradients (oa = oral apparatus). The oral anlage develops at the place with maximal strip contrast (---/+++). **b** and **c** posterior part of the oral primordium during formation of the ingestion area. b arrangement of paroral (pao) and adoral organelles (ado) and beat direction of cilia (arrow) at normal strip contrast, c at opposite contrast (after several authors).

Molecular Biology

(by Günther Steinbrück, Tübingen)

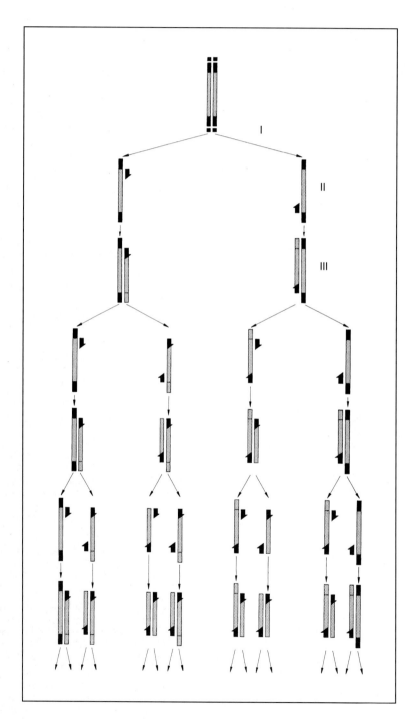

Fig. **280**   Scheme of the *polymerase chain reaction* (PCR), one of the most important basic methods for actual molecular biological investigations. PCR produces an amount of DNA that doubles in each cycle of DNA synthesis and includes a uniquely sized DNA species. Three steps [separation of DNA strands by heating (I), annealing of primer (II), DNA synthesis (III)] constitute each cycle (adapted from Alberts et al.).

For some time, technical difficulties have prevented many desirable molecular biological studies on protozoa, although methods have been considerably improved, even for difficult parasitic protozoa, e.g., trypanosomes or *Plasmodium* species. The polymerase chain reaction (PCR) (Fig. **280**) now provides the formerly remote possibility of making many molecular investigations with nothing more than a single protozoan cell. Thus, many molecular problems can now be attacked, promising considerable progress in our understanding of molecular mechanisms in eukaryotes. It is likely that molecular studies on protozoa will increasingly contribute to the solution of molecular biological problems and the detection of new molecular phenomena.

In the last few decades it has been shown that, compared to higher organisms, protozoa show a striking diversity not only in morphology and ecological adaptations, but also in biochemical pathways and molecular mechanisms. Research on protozoa has profited in two different ways from the development of new techniques in certain areas of molecular biology:

– A few protozoan species became model organisms for the study of basic aspects of the molecular biology of eukaryotes.
– The deciphering of unusual and unique molecular peculiarities in some protozoan groups contributed fundamental new aspects to our understanding of molecular evolution.

In this chapter some of the most striking findings in molecular biological research with respect to protozoa will be presented.

Variant Surface Glycoproteins (VSGs) of Trypanosomes

African trypanosomes cause important diseases, such as sleeping sickness in humans, and Nagana disease in cattle. The life cycle of the best-investigated species *Trypanosoma brucei* consists of several successive steps (see Fig. **57**). In mammals the parasites live mainly in the bloodstream. When a tsetse fly ingests the flagellates with a blood meal, the parasites differentiate into the so-called procyclic, or insect form. The cell surface of this form is covered with a single type of invariant glycoprotein, the procyclin. From the hindgut of the fly, the parasites move to the salivary glands of the fly and transform there into the nondividing metacyclic form.

The tsetse fly infects a mammal with the metacyclic form of trypanosomes in the saliva.

In this form variant surface glycoproteins (VSGs) cover the bodies of the parasite cells. One VSG out of a limited reservoir of about 14 VSGs is expressed in each parasite. These VSGs show a very high switch rate. The metacyclic VSG coat preadapts the trypanosome for life in the mammalian host. In the bloodstream of the mammalian host, the parasites express at any given time one VSG out of a repertoire of 1000 or more different VSG genes.

When the parasites have invaded the mammal, the immune system of the host begins to attack the parasites. The strategy of the trypanosomes to evade the attack of the host's immune system is to change the antigenic surface coat. With a frequency of 10^{-2}–10^{-6} per division, new VSG coats are expressed. The immune system is confronted with new antigenic epitopes, and the infection shows characteristic parasitemic waves. The majority of the parasites are attacked successfully by the immune system, but those cells which have changed the VSG coat can escape and give rise to the next parasitemic burst.

The basic molecular features of the VSG variation have been studied intensively and are relatively well known. Antigenic switching can be produced by several different mechanisms. A simplified scheme showing the most frequently found mechanism is given in Fig. **281**. The genome of *T. brucei* consists of about 20 larger chromosomes and about 100 minichromosomes. More than 1000 different transcriptionally silent basic VSG gene copies (BC-VSGs) are found in tandem arrays at internal positions within the chromosomes. Several VSG gene expression sites are located close to telomeres, the structure of which will be explained later. Only one of the expression linked copies of the VSGs (ELC-VSGs) is transcribed at any time. An interesting aspect concerning this point was the detection of an unusual DNA modification in VSG genes. β-glucosyl-hydroxymethyluracil was found in *T. brucei* DNA. This base has not been found in any other organism, so far. It is speculated that this unique kind of modification might be involved in the regulation of the activation of VSG genes in expression linked sites.

Switching between the expression of different VSG genes can be based on either differential transcriptional control of the telomeric expression sites, or on different types of DNA rearrangement events. One possible mechanisms is shown in Fig. **281**. Alternatively, a less frequently found mechanism of replacement of the active VSG genes is reciprocal exchange or telomere conversion.

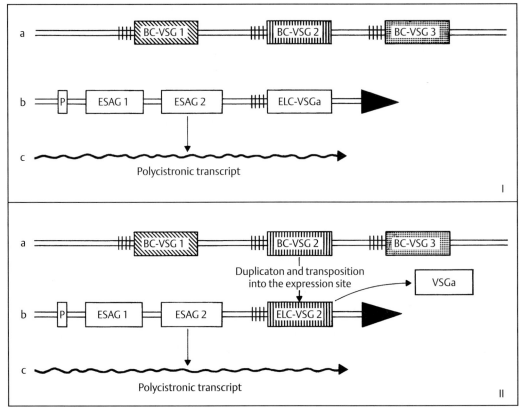

Fig. **281** Simplified scheme of a mechanism responsible for antigenic variation in trypanosomes. Several different mechanisms can produce antigenic switches. This figure gives a schematic view of the most frequently found case of duplicative transposition. I) Situation before the antigenic switch. **a)** Tandem arrays of transcriptionally inactive basic copy VSG genes (BC-VSGs, boxes) are located at internal positions within chromosomes. The short vertical lines represent 70 bp repeats that flank the VSG genes. **b)** An expression linked copy of a VSG gene (ELC-VSGa) is located at a transcriptionally active VSG gene expression site close to a telomere. The large black arrowhead indicates the telomere. **c)** ELC-VSGa is transcribed into a polycistronic transcript together with the expression site associated genes ESAG1 and ESAG2 under the control of a single promoter P. II) Antigenic switching occurs by duplicative transposition. **a)** The formerly silent basic copy BC-VSG2 is duplicated and transposed to the telomeric gene expression site. **b)** The previously expressed copy of VSGa is lost and replaced by VSG2. **c)** The activated copy of VSG2 (ELC-VSG2) is now transcribed into a polycistronic transcript (after several authors).

In contrast to most other eukaryotic genes, the ELC-VSGs are transcribed together with other *expression site associated genes* (ESAGs) into a large polycistronic transcript. The promoter of this unit is located about 50 kbp upstream of the VSG gene.

Another unique peculiarity of trypanosomes is that VSG and procyclin genes are transcribed by a RNA polymerase I (pol I). This type of RNA polymerase transcribes only ribosomal genes in all other eukaryotes. As a consequence of this pol I transcription, the primary transcripts of these genes do not receive a cap structure at their 5'ends. Cap structures of eukaryotic mRNAs play important roles in the initiation of translation, and in transport and stability of mRNA molecules.

In trypanosomes, mRNAs receive a cap structure by a unique maturation process called trans-splicing. Short *mini-exon donor* RNAs (MED RNAs) are transcribed from repetitive sites found elsewhere in the genome and receive a cap. These MED RNAs are spliced together with the coding exons resulting from the polycistronic transcripts (Fig. **282**).

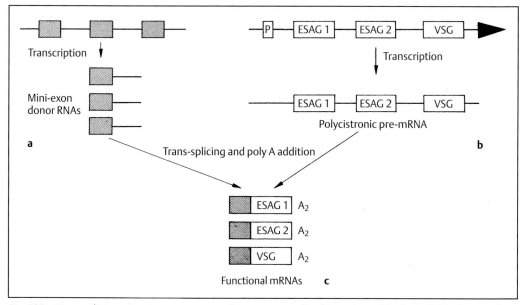

Fig. **282** Trans-splicing in trypanosomes. Genes coding for mini-exon donor RNAs are transcribed into the corresponding RNAs (**a**). VSG genes and expression site associated genes (ESAGs) are transcribed from a common promoter (P) into a large polycistronic pre-mRNA (**b**). Then, the miniexon donor RNAs are spliced together with the individual genes from the large precursor (**c**). Thereby functional 5'ends are added to the exons of the ESAGs and VSG transcript, and poly A tails are formed at their 3'ends (after several authors).

Considerable progress has also been made in the analysis of the VSG protein structure and the mechanism of VSG attachment to the cell membrane. Complete amino acid sequence data are available for several VSGs. The VSG chains are arranged on the surface of the trypanosome cells like stalks in a cornfield. Different oligosaccharides are found, depending on the type of trypanosome, on the external exposed surface of the VSGs and not the exposed ones within the antigenic coat. Surprisingly, the exposed oligosaccharides do not serve as immunogenetic epitopes; only the exposed protein parts of the VSGs are immunodeterminants. Most interestingly, in the surface coat of trypanosomes it has been shown definitively for the first time that the surface proteins, the VSGs, are covalently linked via ethanolamine to glycosylphosphatidylinositol, serving as a membrane anchor (GPI-anchor) (Fig. **283**). Many membrane proteins in other cells have now been shown to be anchored to the membrane by covalent linkage to a glycolipid anchor.

Kinetoplast DNA Networks and RNA Editing

In addition to the remarkable survival strategy offered by antigenic variation, trypanosomes show another molecular peculiarity that was first detected in trypanosome mitochondria and later in other organisms also — the phenomenon of RNA editing.

Trypanosome cells possess a single giant mitochodrion which has a special part called the kinetoplast (see Figs. **53–56**). Kinetoplasts contain an unusually large amount of DNA. This mitochondrial DNA is organized in about 10 000 minicircles and about 50 maxicircles concatenated into a large network (see Fig. **56e**). The maxicircles carry the information for mitochondrial genes. The function of the minicircle DNA remained unknown for a long time. Many of the mitochondrial genes found in the maxicircle DNA seemed to be nonfunctional because they lacked conventional translation initiation codons or contained frameshifts. Sequence analysis of the corresponding mRNAs showed functional coding regions. This meant that a nonsense message in the DNA was converted into

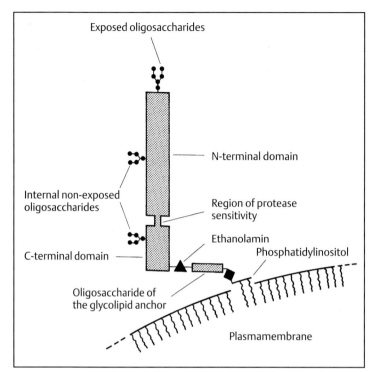

Exposed oligosaccharides

N-terminal domain

Internal non-exposed oligosaccharides

Region of protease sensitivity

C-terminal domain

Ethanolamin

Phosphatidylinositol

Oligosaccharide of the glycolipid anchor

Plasmamembrane

Fig. **283** Schematic structure of a VSG of *Trypanosoma congolense* (adapted from Risse).

sense in the mRNA. This process is called RNA editing. In trypanosomes, Uridinemonophosphate (UMP) residues are post-transcriptionally inserted into the pre-mRNA. In some genes more than 50% of the coding regions are not found in the genes but are produced by editing. In some cases U residues are also deleted. The amount of editing differs for the same gene from species to species. Whereas in one species a certain mRNA is edited dramatically, the same gene in another species may not be edited at all. Recently, it was shown that the information necessary for the correct insertion or deletion of U residues is encoded in the minicircles, whose function remained a mystery for a long time. Short, so-called guide RNAs are transcribed from minicircle DNA and provide the information for the editing process. Base pairing between pre-mRNA and guide RNA is necessary close to the site where editing starts. A putative editosome seems to stabilize this complex and is involved in UMP addition and removal (Fig. **284**).

Editing has since been found to occur in other organisms. It occurs in the mitochondria of *Physarum*, of plants, in viruses and even in a mammalian nucleus. In these other organisms, however, mechanisms and modifications are different from those found in trypanosomes.

Whether the process of RNA editing is just a relic of the prebiotic RNA world (which seems rather unlikely), or whether it provides additional evolutionary flexibility or has another unknown functional significance, remains to be answered.

Molecular Peculiarities of Ciliates

Having been intensively studied with respect to their molecular biology, the ciliates have turned out to be of special interest in several respects.

Genome Structure and Reorganization

A striking characteristic of ciliates is the presence of nuclear dualism. Each ciliate cell contains one or several small micronuclei, which have the germ line function, and one or several large and DNA-rich macronuclei, which are responsible for the somatic or vegetative functions. Micronuclei contain normal eukaryotic chromosomes and can divide by mitosis or meiosis. On the other hand, macronuclei contain no normal visible chromosomes, and they divide by a process called amitosis. The presence of germ line and somatic nuclei within one cell and the

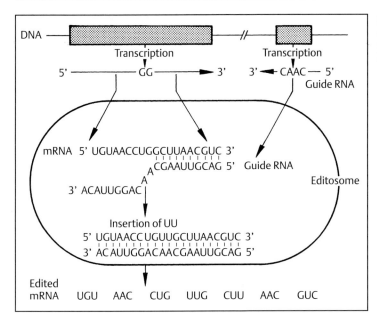

Fig. **284**   RNA editing in mitochondria of trypanosomes (adapted from Schuster and Brennicke).

suspicious dissimilarity of the two kinds of nuclei stimulated relatively early molecular investigations on genome organization and gene structure in ciliates.

Of the thousands of ciliate species, only a few in the order Hymenostomatida, mainly *Tetrahymena* and *Paramecium* species, and some species of the subclasses Stichotrichia and Hypotrichia, mainly *Stylonychia, Oxytricha* and *Euplotes* species, have been studied in detail. Molecular comparison of micro- and macronuclear genomes in these species has led to the detection of surprising and unexpected molecular differences in the organization of germ line and somatic genomes.

Both micro- and macronuclear genomes are closely related developmentally, since, following the sexual processes of conjugation or autogamy, a new macronucleus develops from a micronucleus. The developing macronucleus is called the macronuclear anlage. During this developmental process, dramatic changes occur in the genome organization. In the developing macronuclear anlage of *Tetrahymena*, 10% to 20% of the micronuclear sequences are eliminated in a complex process. The presumably chromosome-sized micronuclear DNA is fragmented into about 200 different DNA molecules ranging in size from about 20 kilobase pairs (kbp) up to more than 1500 kbp (Fig. **285**).

The smallest DNA molecules are dimers of the rRNA genes. The single rRNA gene of the micronuclear genome is excised and duplicated into a dimeric palindrome, which carries the two gene copies in a head-to-head arrangement. During the fragmentation process, telomeres are added to the ends of all DNA molecules and the molecules are amplified to different copy numbers.

Even more dramatic is the reorganization process during macronuclear development in hypotrich species. In the macronuclear anlage of these species polytene chromosomes are formed by several rounds of replication of the micronuclear chromosomes (see Fig. **257**). Differential replication can lead to unequal gene copy numbers. Up to 98% of the micronuclear DNA sequences are eliminated in *Stylonychia*, smaller amounts in other species. Repetitive as well as single copy sequences are eliminated. Transposon-like elements are eliminated in a first step. The DNAs is finally fragmented into small gene-sized DNA molecules ranging in size from about 100 base pairs up to about 15 kbp (Fig. **286**). Each fragment receives, as in *Tetrahymena*, specific telomeres, and is amplified to a gene-specific copy number which can range from a few hundred to more than a million copies per macronucleus. The general organization of macronuclear gene-sized DNA molecules as they are found in stichotrich and hypotrich ciliates is shown in Fig. **287**.

More than 50 genes of ciliates have been cloned and sequenced in their full length so far. Sequence elements needed for the regulation of

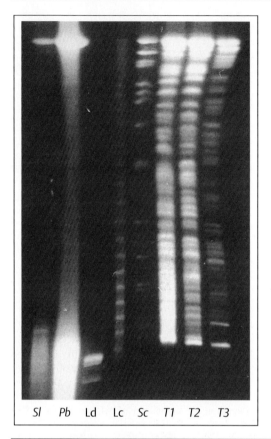

Fig. **285** Fragmented macronuclear DNA of *Tetrahymena* can be separated by the technique of pulsed field electrophoresis. Macronuclear DNA of two strains of *T. thermophila* (T_1, T_2), and the closely related *T. pyriformis* (T_3) is separated by this technique in an agarose gel. For comparison, macronuclear DNA of *Stylonychia lemnae* (*Sl*) and *Paramecium bursaria* (*Pb*) as well as yeast chromosomes (*Saccharomyces cerevisiae*, *Sc*), concatemeres of phage lambda (*Ls*) and digested lambda DNA (*Ld*) are included. The lambda concatemeres are multiples of 49 kbp. Macronuclear fragments of the *Tetrahymena* range in size about 20 kbp up to more than 1000 kbp. The much smaller DNA of the other ciliate species runs out of the gel under these conditions (after Horstmann and Steinbrück).

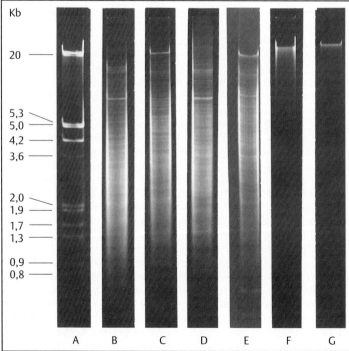

Fig. **286** Macronuclear and micronuclear DNA of different hypotrich ciliates is separated by conventional agarose gel electrophoresis and the gel then stained with ethidium-bromide. Lane **A** is marker. Lanes **B–E** show macronuclear DNAs of hypotrich ciliates. Lane **F** shows macronuclear DNA of the penialid ciliate *Paramecium* and lane **G** contains micronuclear DNA of a hypotrich species. The macronuclear DNAs of the hypotrich species show a size range from less than 0.5 kb to more than 10 kb. The species differ by differences in the faint banding patterns. Macronuclear DNA of the non-hypotrich species and micronuclear DNA is of much higher molecular weight and, therefore, cannot be separated into discrete size classes with this technique.

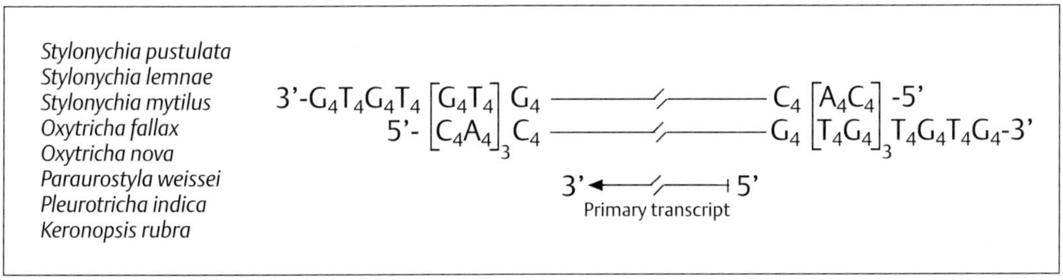

Stylonychia pustulata
Stylonychia lemnae
Stylonychia mytilus
Oxytricha fallax
Oxytricha nova
Paraurostyla weissei
Pleurotricha indica
Keronopsis rubra

$$3'\text{-}G_4T_4G_4T_4 \left[G_4T_4\right] G_4 \underline{\quad/\quad} C_4 \left[A_4C_4\right]\text{-}5'$$
$$5'\text{-} \left[C_4A_4\right]_3 C_4 \underline{\quad/\quad} G_4 \left[T_4G_4\right]_3 T_4G_4T_4G_4\text{-}3'$$

$$3' \longleftarrow \underline{\quad/\quad} 5'$$
Primary transcript

Fig. **287** General structure of gene-sized macronuclear DNA molecules of hypotrich ciliates. Each gene-sized molecule carries a single coding region which is flanked by short non-coding sequences containing cis-regulatory elements. At the ends of all molecules identical telomeres are found with protruding 3'single-stranded extensions.

transcription and putative origins of replication could be characterized in the non-coding parts of the gene-sized pieces of hypotrichous genes. In some cases, corresponding micronuclear and macronuclear versions of the same gene could be analyzed in detail. These comparisons presented another surprising feature. Some micronuclear versions of hypotrichous genes are interrupted by so-called internal eliminated sequences (IESs) that are eliminated during the formation of the macronuclear copy of that gene. Even more surprising, the macronuclear destined sequences (MDSs) that form the final macronuclear version of the gene are found in a different order and orientation in the micronuclear genome. This was first detected in the actin genes of an *Oxytricha* species. Sequences necessary for the elimination of the IES elements and for the unscrambling of the MDSs have been characterized in detail. Specific short repeat sequences are found at the MDS-IES junctions which are responsible for the correct splicing and elimination of IESs.

After the detection of fragmented macronuclear genomes in hymenostomatid, stichotrich, and hypotrich ciliates, other groups of ciliates were checked for the occurrence of fragmented macronuclear genomes. In several other groups of ciliates, some of them not closely related to hypotrichs, a very similar macronuclear genome organisation with gene-sized DNA molecules was found. In all ciliate species investigated using molecular methods, the macronuclear genome has been found to contain DNA fragments of sub-chromosomal size. Thus, DNA fragmentation during macronuclear development seems to be a general characteristics of most if not all ciliates.

The functional significance of these strange reorganization processes is not understood. In a few cases alternative processing could be found. This means that different gene-sized macronuclear DNA molecules are derived from the same micronuclear precursor. Whether these reorganization processes lead to greater genetic variation in the species and therefore a better adaptation to environmental change remains to be shown.

Telomeres and Telomerases

The unique genome organization of ciliate macronuclei, consisting of thousands of independent linear DNA molecules, made them the favorite material for studying the structure and formation of telomeres. Telomeres are protein-DNA structures that stabilize the ends of eukaryotic chromosomes and allow complete chromosomal replication. In addition, they may have an essential function in the association of chromosome ends with the nuclear envelope during meiotic prophase. Chromosomes with broken ends lacking telomeres are unstable and fuse end to end, leading to unstable chromosome forms.

The second function of telomeres — allowing the ends of chromosomes to be replicated completely — solves a special problem of DNA replication in eukaryotes. DNA polymerases proceed only in the 5'- to 3'-direction and require a primer for the start of the synthesis. From this mechanism one can predict that

Table **7**   Telomere repeats found in protozoa

Group	Organism	telomeric repeat
Ciliophora	Tetrahymena	TTGGGG
	Glaucoma	TTGGGG
	Paramecium	TT(T/G)GGG
	Stylonychia	TTTTGGGG
	Oxytricha	TTTTGGGG
	Paraurostyla	TTTTGGGG
	Onychodromus	TTTTGGGG
	Urostyla	TTTTGGGG
	Keronopsis	TTTTGGGG
	Pleurotricha	TTTTGGGG
	Euplotes	TTTTGGGG
Kinetoplasta	Trypanosoma	TTGGGA
	Crithidia	TTGGGA
Apicomplexa	Plasmodium	(C/T)TTGGGA
Myxogastra	Physarum	TTGGGA
	Didymium	TTGGGA
Dictyostela	Dictyostelium	AG_{1-8}
Chlorophyta	Chlamydomonas	TTTTGGGA

during each round of replication the 5'-ends of the lagging strands of the replication forks are shortened. In succesive rounds of replication, large amounts of DNA would be lost from the chromosome ends. Telomere structures prevent this loss.

In higher eukaryotes, studies of the synthesis and function of telomeres were found to be extremely difficult due to their low abundance in the genomes. The fragmented and highly amplified macronuclear genomes of ciliates contain from tens of thousands up to millions of telomeres per macronucleus, depending on the species investigated. Therefore, structure and synthesis mechanisms were unravelled first in ciliates. Since telomere structures are very similar in all organisms, studies on telomeres in ciliates have been shown to be relevant to most eukaryotes.

The first telomeres to be sequenced and studied at the molecular level were the telomeres at the ends of the rDNA molecules of *Tetrahymena thermophila*. Subsequently, telomeres of macronuclear genes of several hypotrich ciliate species were analyzed in detail. All these telomeres consist of short tandem repeats $(3'-G_4T_4-5')_n$ in hypotrichs and $(3'-G_2T_4-5')_{50-70}$ in *Tetrahymena*. In ciliates and some other protozoa the telomere sequences end with a single-stranded 3'-overhang (Fig. **287**). The G-rich strand is synthesized by the enzyme telomerase (telomeric terminal transferase), a ribonucleoprotein which exhibits a reverse transcriptase activity.

Telomerase activities were first characterized in *Tetrahymena* cell extracts and were later also identified in hypotrich ciliates. This enzyme activity is now known for several other organisms, including man. In ciliates, telomerase activity in necessary for the formation of telomeres at the ends of the DNA fragments, which are formed during macronuclear development as described above, and its activity counteracts the shortening of chromosome ends during replication of the numerous macronuclear DNA fragments.

Telomerase contains an essential RNA molecule which serves as the template for the synthesis of the telomere repeats (Table **7**). The telomerase RNA components have been cloned from *Tetrahymena* and *Euplotes*. The *Tetrahymena* telomerase RNA contains the sequence CAACCCCAA which is complementary to the TTGGGG repeat of the telomeres of *Tetrahymena*. Similarly, the *Euplotes* telomerase RNA contains the sequence CAAAACCCCAAAACC, which is complementary to the TTTTGGGG repeat of *Euplotes* telomeres.

The proof that telomerase RNA provides the template for telomere repeat elongation came from elegant mutation experiments. Cloned telomerase RNA genes were mutated by single base exchanges. *Tetrahymena* cells were transformed with these mutated genes lying on high copy number vectors. DNA sequencing showed that in the transformed *Tetrahymena* cells, telomeres were synthesized at the ends of macronuclear DNA molecules which contained the altered sequence corresponding to the mutation in the telomerase RNA gene. These experiments demonstrate convincingly that the telomerase is involved in normal telomere synthesis and that the telomerase RNA component serves as the template for this synthesis. *Tetrahymena* cells containing the mutated telomerase RNAs exhibit striking changes in morphological phenotype and in senescence.

The single-stranded overhangs of the 3'ends of ciliate telomeres are an unusual and suspicious feature. Since similar single-stranded tails have been found in slime mould *Didymium* and in yeast, these tails may be a general character of telomeres. Such a G-rich single-stranded tail can form a hairpin structure, and two hairpins can interact with each other forming an unusual four-stranded structure, which is held together by nonclassic base-pairing of quartets of guanine bases. This can result in vitro in intra- or inter-molecular end-to-end aggregation of the DNA molecules. It is debated whether such G quartet structures also play an essential role in vivo in chromatin organization. It has been shown that *telomere binding proteins* (TBPs) are involved in telomere aggregation in vivo. Such TBPs could be characterized in some hypotrichous ciliate species.

Whereas ciliates and other lower eukaryotes exhibit permanent telomerase activity, such activity has been found in humans only in germ line cells and is apparently absent from somatic cells. As a consequence, the length of human telomeres decreases with cellular age. But recently it was shown that in some human ovarian carcinoma cells, telomerase is re-activated. These cells maintain short stable telomeres and become immortal cell lines. It has been suggested that telomerase inhibitors might be effective antitumor drugs. Ciliate cells, in which telomerase is activated permanently, might be useful test organisms for finding suitable inhibitors. Similarly, it has been suggested that telomerase be explored as a target for drugs against eukaryotic pathogenic or parasitic protozoa. Since telomerase activity is absent from somatic cells of the mammalian host, a drug that binds to telomerase selectively should affect only the eukaryotic protozoan parasite. Thus, studies on the structure of telomeres and function of telomerase, which were started on ciliates, may have unexpectedly far-reaching consequences.

Self-Splicing Introns

The macronuclear genome organization of ciliates favors studies of gene expression. In an investigation on the processing of ribosomal RNA precursor molecules in *Tetrahymena* a quite unexpected and exciting detection was made.

The rRNA genes of *Tetrahymena* exist as linear palindromic dimers and are amplified to about 10 000 copies per macronucleus. Each half of the dimer contains the coding sequences for 17 S, 5.8 S, and 26 S rRNA and is transcribed as a 35 S primary transcript (35 S pre-rRNA). In *T. thermophila* the 26 S rRNA coding region is interrupted by an intervening sequence of 414 basepairs in length. Many eukaryotic and a few prokaryotic genes are interrupted by stretches of noncoding sequences called intervening sequences or introns. Transcripts of such genes must undergo RNA splicing, a cleavage–ligation reaction to produce the mature functional form. By investigating the splicing mechanism which removes the 414 bp intron from the pre-rRNA of *Tetrahymena* it was shown that this intron was excised from the precursor without any assistance from protein enzymes. Careful control ex-

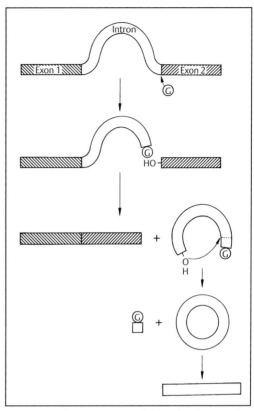

Fig. **288**   Splicing reaction of the self-splicing intron found in the ribosomal DNA of *Tetrahymena thermophila*. The primary transcript consists of two exons which are separated by an intron. A guanosin or guanosin-nucleotide molecule (G) attacks one exon/intron boundary and cleaves off the exon 2. In a second transesterification the two exons are connected with each other and the intron is spliced out. Then the intron forms a cycle and cleaves off a short piece of RNA which carries the G residue. Finally, the circle is opened to a linear molecule. All these complex reactions occur without any participating protein enzyme (adapted from Cech).

periments were carried out to prove that tightly bound protein contamination did not produce the astonishing result. All experiments lead to the same result: the intron must have autocatalytic properties. These introns are called self-splicing introns or ribozymes.

The reaction requires guanosine or guanosine nucleotides as cofactors and consists of two transesterifications (Fig. **288**). The intron is spliced out as a linear molecule and is later converted to a circle which is finally linearized again. Introns of this type, now called group I in-

Table **8**   Coding properties of the codons TGA, TAG, and TAA in ciliates. In nuclear genes of all other organisms these codons are used for translation stop.

Organism	TGA	TAG	TAA
Tetrahymena	stop	Gln	Gln
Paramecium	stop	Gln	Gln
Stylonychia	stop	Gln	Gln
Oxytricha	stop	Gln	Gln
Paraurostyla	stop	Gln	Gln
Euplotes	Cys	stop	stop
Blepharisma	?	?	stop

trons, were found later in fungal mitochondrial genes, in rRNA genes of the slime mould *Physarum* and in chloroplast genes of plants. It was shown that in addition to these self-processing reactions the intron RNA can act as a true enzyme with other polynucleotides as substrates. Thus, the dogma that only proteins can act as enzymes was broken. RNA can also have catalytic properties either in autocatalytic reactions or acting as a true enzyme. This finding throws a totally new light on the early steps of prebiotic evolution. It solves the question of who came first — the protein or the nucleic acid.

Codon Usage

Ciliates show another molecular peculiarity in their use of the genetic code. It is textbook knowledge that the genetic code is universal, i.e., all organisms use the 64 possible codons in an identical manner for specifying amino acids in protein synthesis. But sequence analysis of genes in different ciliate species revealed that ciliates make an exception to this rule. Ten years ago, several research groups found almost simultaneously, that in *Tetrahymena, Paramecium, Stylonychia,* and *Oxytricha* species, the two codons TAA and TAG, which are universal stop codons in nuclear genes of all other organisms, code for glutamine (Table **8**). The only stop codon found in genes of these species is TGA. It was shown that *Tetrahymena thermophila* has three tRNAs for glutamine, two of them recognizing the codons TAA and TGA. Sequence comparisons of the tRNAs lead to the conclusion that the unusual tRNAs evolved from a normal glutamine tRNA by anticodon mutation.

But the story of difference in codon usage became even more complicated. *Euplotes* and *Blepharisma* species use the codon TAA, which codes for glutamine in other ciliates, as a stop codon as in all other eukaryotes and prokary-

otes. And the codon TGA, the only stop codon in the other ciliate species codes for cystein in different genes of *Euplotes* species. Since *Euplotes* species belong to the Stichotrichia and are phylogenetically related to *Stylonychia* and *Oxytricha* species, a mutational switch back to the original meaning of the TAA codon must have occurred, as well as a second switch, in the case of the meaning of the TGA codon. The causes that might have lead to these changes in the genetic code of ciliates are unknown. It has been speculated that this special codon usage might be a barrier to the adaptation of non-ciliate viruses to ciliates, since so far, no true viruses have been found in ciliates.

Molecular Immune Escape of Malaria Parasites

Malaria is one of the most frequent causes of serious illness in many tropical and subtropical regions of the world. Four species of apicomplexan protozoa cause malaria in humans: *Plasmodium falciparum, P. vivax, P. ovale,* and *P. malariae.* Several aspects of the molecular biology of *Plasmodium* and related species are currently being intensively investigated, e.g., molecular structure and cellular localization of putative target antigens for vaccine development. Recombinant DNA and monoclonal antibody techniques have enhanced the knowledge of suitable vaccine target antigens and the structure and variability of the underlying genes.

Strategies for control and even eradication of malaria were quite successful for many years in some countries. These strategies involved the use of antimalaria drugs and potent pesticides directed against the insect host. But in most tropical regions, malaria could not be eradicated, and the number of cases is currently increasing worldwide. Drug-resistant strains of the parasites arose in many places. In addition, the appearance of insecticide-resistant mosquito strains, as well as the severe health and environmental problems caused by large scale use of pesticides, has prevented a breakthrough against malaria. Therefore, the development of a potent vaccine is of outmost importance for millions of people worldwide. This urgent need has stimulated interest and research on putative potent antigens, which might be used to develop a molecular vaccine.

Malaria parasites have a complex life cycle comprising sexual and an asexual phase (see Fig. **92**). The parasites live in humans during most of their life cycle, intracellularly enlosed by a parasitophorous vacuole in red blood cells (see Figs. **93** and **94**). Therefore, promising antigens are likely to occur mainly on the surface of the stages of parasites that live extracellularly. In recent years, more than 20 surface antigens (S-antigens) have been characterized by molecular methods in *Plasmodium* species. In most cases the gene of the relevant protein has been cloned, and the cloned gene is expressed in a suitable translation system. An interesting molecular feature of many of these antigens is the finding that very often peptide as well as gene regions have a repetitive structure (Fig. **289**).

The S-antigen of *P. falciparum* was the first malaria antigen found to have tandem repeats of oligopeptide sequences. These S-antigens have been shown to exhibit considerable variation between different strains of *P. falciparum* populations. The S-antigen consists of a single exon with a central block of tandem repeats. Different alleles differ with respect to sequence, size, number, and reading frame of these repeats. This variation results in a high serological diversity.

Repetitive amino acid sequences are known from some other eukaryotic proteins e.g., silk-

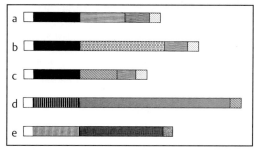

Fig. **289** Sequence variation in surface antigen genes (S-antigen genes) of *Plasmodium falciparum*. Different shadings indicate different sequences (adapted from Kemp).

protein, keratin, or collagen. But the high number of repetitive structures in these parasitic surface proteins may provide a hint towards a defence strategy of these parasites directed against the immune system of the host. This system is reminescent of the VSG-switch system of trypanosomes mentioned above, but uses different molecular mechanisms.

Behavior of Protozoa

(by Hans Machemer, Bochum)

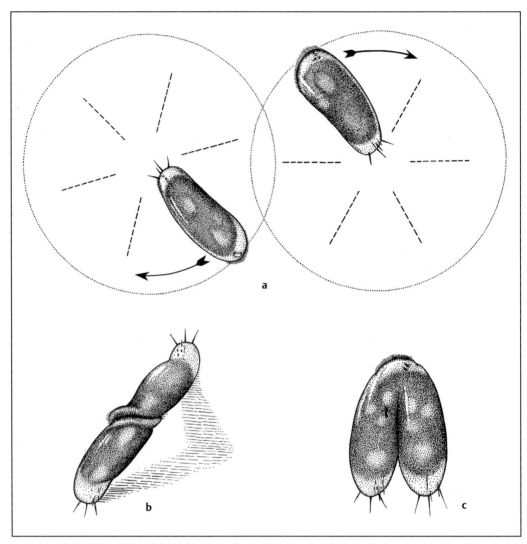

Fig. 290 *Stylonychia mytilus*. Mating play preceding conjugation. **a** Jerky rotation in circles by the two partners. **b** Touching of the peristomes. **c** Final fusion (from Grell: Z. Naturforsch. 6b [1951] 45). Magn.: 200×.

An organism which can respond to an intrinsic and/or extraneous signal generates behavior (Fig. 290). In the protozoa, as well as in the higher organisms, there occur spontaneous, internally regulated types of behavior, and those activities which are induced by stimuli from the outside world. Reports on behavior most often deal with motor responses because the motility of cells can be easily observed and recorded.

A typical example of an internally regulated behavior would be the directional changes in stichotrich and hypotrich ciliates, which occur even after the cells have been equilibrated in a stimulus-free environment. The rapid back and forth movements of *Stylonychia* or *Euplotes* are under the control of the membrane potential, which fluctuates greatly in this group of ciliates (an invisible behavior at the membrane level). Because of the fluctuation, the potential often passes the threshold for excitation, each time eliciting a Ca^{2+} action potential, which then causes a synchronized reversal response of the cilia.

Movements of cells in the absence of external stimuli are spatially randomized, that is, they have no preferential direction. Swimming protozoa (as well as bacteria) commonly take a helical course about a straight axis, thereby compensating local asymmetries in force generation. A population of predominantly forward swimming cells tends to distribute evenly in a given body of water. On the other hand, frequent directional changes (reversals) tend to cause aggregation of the cells in a population. Time proportions of forward swimming over reversals in the unstimulated state, provide a baseline of behavior in a particular protozoan organism.

A directional change of a protozoon may be elicited equally well by an external stimulus, such as a head-on bump into a mechanical obstacle, or a chemical (thermic, photic) stimulus. In these cases, the reversal is termed a phobic or shock response (Fig. **291 d**). Because a shock response is nondirectional with respect to the stimulus source, a successful removal of a cell from a repellent stimulus can only result from sequences of shock responses and swimming, that is, by means of trial and error.

The behavioral output of a cell is often the product of both internal and external factors. The integration of various stimuli tends to generate complex forms of behavior. Separation and indentification of the signals modulating protozoan behavior is a task for experimental research.

Stimuli from the outside world may be one-dimensional, that is, they are characterized by intensity only, such as a gradual increase in overall temperature or illumination (Fig. **291 a, b, d**). Often, an intensity is conveyed together with directional information. For instance gravity, or radiant light of a particular (range of) wavelength, impinges on the organism in the manner of a stimulus vector, giving important clues to environmental properties (Fig. **291 c**). Protozoa can generate adequate behaviors in response to those "simple" and "vector-type" stimuli which may be crucial for survival. The stimulus-dependent motor responses are subdivided into kineses (velocity regulation and randomization of direc-

tions by reversals; Fig. **291 a, b**) and taxes (orientational movements; Fig. **291 c**). A third class of responses to stimuli, tropisms, applies to sessile organisms only; tropisms are commonly growth responses with respect to stimulus direction.

Kinesis

A kinesis is conventionally observed under gradient-type and/or diffuse conditions of, for instance, light, temperature, and chemical stimulation. The relative intensity of the stimulus is the signal regulating the motor output of the cell. A raised velocity response is termed direct kinesis (or positive kinesis), a reduced velocity is termed inverse kinesis (or negative kinesis). Velocity regulation (orthokinesis; Fig. **291 a**) differs from regulation of the frequency of induced directional changes (shock responses, klinokinesis; Fig. **291 b**). Previously, a novel type of nonorientational and stimulus-vector sensitive kinesis was identified in the graviresponses of ciliates.

A kinesis results from repeated comparison of stimulus intensities in the time domain. This enables a cell to respond to spatio-temporal stimulus gradients (chemokinesis, thermokinesis etc.). Protozoa, which happen to be oriented "uphill" of a chemical attractant, may increase the proportion of forward swimming periods and decrease the rate of shock responses analogous to the biased random walk of gradient-detecting bacteria. In the case of "downhill" swimming with respect to an attractant, the shock response rate may increase, and forward swimming periods may decrease. In addition, the cell can (unlike prokaryotes) regulate the speed of locomotion, which goes up with an uphill-attractant orientation and down with downhill-attractant orientation. Inverse behavioral responses occur with gradients of a chemical repellent. An attractive or repulsive property of a stimulus can vary with concentration or intensity. For instance, very low and very high stimulus intensities may repel a population of cells, which eventually accumulates within a range of intermediate (optimal) intensity. It can be concluded that the effect of kinetic behavior is indirectly orientational, and that orientation by kinesis and accumulation within a spatio-temporal gradient, are probabilistic in nature.

Taxis

A taxis is a directly oriented movement with respect to a vector-type stimulus source (toward

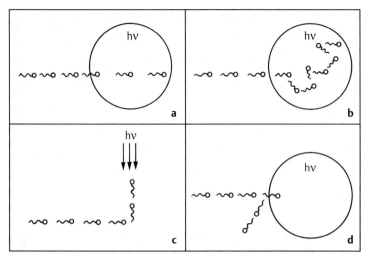

Fig. 291 Types of stimulus-induced motor responses of a protozoon as exemplified by photic stimulation (hv). **a** (Ortho-) Kinesis: Swimming velocity modified inside illuminated area (time intervals indicated by distance between successive positions of cell). **b** Klinokinesis: Directional changes increased in-side illuminated area. **c** Taxis: The cell turns directly toward (or away from) the light source. **d** Shock response: Upon passing the border between dark and illuminated area, the cell generates directional changes unrelated to the light source. After Song and Poff.

the source: positive taxis; away from the source: negative taxis). In protozoa, taxes are restricted to a few stimulus modalities (gravity, light, electric field) because a vector can be identified only by means of spatial discrimination. While a multicelluar organism may be supplied with devices for directional photoreception, or with statocysts for identification of what is "up," the small volume of a cell offers little room for spatially organized receptors. It is quite interesting to note that some structural and functional properties of the protozoan organization facilitate the performance of taxes (see galvanotaxis, gravitaxis, phototaxis). Conventionally, a taxis is deduced from behavioral data irrespective of whether a pathway of signalling was identified. Often, morphological and physiological cues give supportive evidence, but there are only few cases where cellular signalling steps, serving to establish a tactic response, are understood in some detail (see galvanotaxis, phototaxis).

Behavioral Changes of Time

In addition to immediate causation of a motor response, some stimulus-induced behaviors may change with time. Following a rapidly modified and then stationary stimulus level, behavior readapts, after the passage of seconds, minutes, or hours (Fig. 292). Without adaptation, a change in ionic composition of a small water well (due to evaporation, or dilution by rain) would disrupt the fine-tuned control systems of the cell, prohibiting perception of additional stimuli and walks along stimulus gradients. The behavior of the cell depends upon bioelectric regulation of ciliary motion (Fig. 293) and on cytoplasmic contractions. Such regulation implies a reference (resting) value of the membrane potential. The resting potential controls membrane ion channels and enzyme cascades under homoiostasis, and is a starting condition for stimulus-induced shifts of the transmembrane voltage, signalling the intensity of a particular stimulus modality.

In some protozoa, time-dependent behavioral changes have been documented that faintly resemble learning in higher organisms. For instance, strong mechanical agitation sensitizes a population of *Stylonychia* to shy away from walking on rough surfaces, but this ability to discriminate between rough and smooth surfaces relaxes in the course of one hour (desensitization). Moreover, repetition of a mechanical shock applied at regular time intervals to contractile ciliates such as *Spirostomum* or *Stentor*, leads to

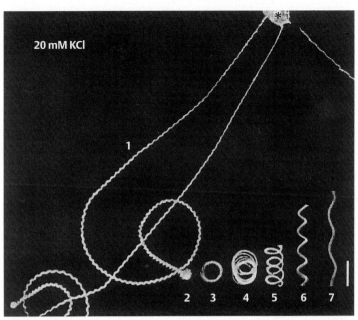

Fig. **292** Swimming traces of *Paramecium* recorded under dark-field illumination, illustrate adaptation to chemical stimulus. Cells released (*) by pipette into a homogeneous solution of 20 mM KCl were initially depolarized so that they swam backward rapidly (1). Backward swimming slowed down ending, after 40 s, with stationary spinning (2). In the following minutes, spinning transformed to slow circling (3) with a gradual increase of a forward swimming component (4–7). Stage 7 represents the normal pattern of an unstimulated, forward swimming cell (after Parducz). Compare with Fig. **282** to identify the gradual transition from strong depolarization to the resting state of the *Paramecium* membrane (after Parducz).

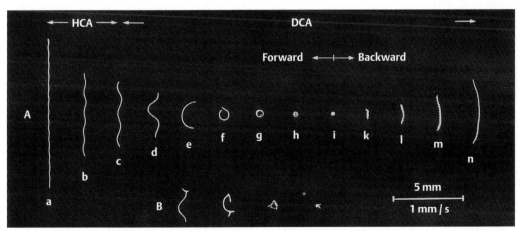

Fig. **293** Voltage dependence of swimming patterns in *Paramecium* (HCA: hyperpolarization-induced ciliary activation; DCA: depolarization-induced ciliary activation). All traces were recorded during 5 s dark-field illumination so that the length of a trace indicates the swimming velocity. Traces are arranged to illustrate the transformation of locomotion with a gradual shift of the membrane potential from hyperpolarization (at left) to strong depolarization (at right). A Continuous locomotion (a–h: forward; i: stationary spinning of cell; k–n: backward swimming). B In the weakly depolarized state (d–i) forward swimming may be interrupted by directional changes (see kinks in traces). Note that an increase in the rate of directional changes coincides with decrease in the forward swimming rate (from Machemer and Sugino: Comp. Biochem. Physiol. 94A [1989] 365).

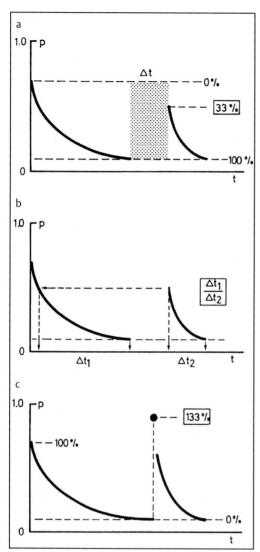

Fig. **294** Habituation in *Spirostomum*, as represented by the probability (p) of contraction following periodically applied mechanical shocks (t; time spanning about 12 min). a The contraction probability decreased with time, but during a shock-free period (Δt), retention wanes to a fraction of the previous habituated state. b During a second series of shocks, cells habituate faster (potentiation). c Habituation breaks down upon application of a very strong shock (dishabituation); upon resumption of the shock paradigm, habituation starts again (from Machemer: BIUZ 18 [1988] 122).

gradual abolition, in the course of several minutes, of the rapid contraction response (Fig. **294**). This behavior includes a number of characteristics of habituation as studied in

higher organisms. On the other hand, the association of a conditioned stimulus with an unconditioned stimulus (associative learning), has never been proven beyond doubt in protozoa.

Behavioral Responses

In the following, some examples are given to illustrate behavioral responses to stimuli. The examples are given by stimulus modality (voltage, mechanical force, heat, chemical, and light).

Responses to Electric Field (Galvanotaxis)

An electric DC field established across an electrolyte can specifically excite cells immersed in this solution: Due to ohmic voltage division along transcellular series resistances, the side of a cell facing the cathode is depolarized, and the side facing the anode is hyperpolarized. Such extracellular electric stimulation is commonplace in excitable tissues such as nerve and muscle, but only in motile ciliated or flagellated cells, can it generate a peculiar behavioral response called galvanotaxis (Fig. **295 a**).

In *Paramecium*, a voltage gradient of, for instance, 2V/cm causes arbitrarily oriented cells to reorient their anterior ends and directly swim toward the cathode (Fig. **295 b**). Galvanotaxis is a key in understanding electric (= voltage-dependent) regulation of ciliary activity. At the same time, the voltage-dependent modulation of ciliary responses (see Fig. **293**) explains why ciliates can generate a precise taxis under DC-field conditions.

In a galvanotactic *Paramecium*, the number of normally beating cilia (near resting potential or hyperpolarized) exceeds the number of reversed cilia at the cathodal side. It follows that the cell swims forward. Cells oriented at right angles to the isopotential lines of the electric field and heading for the cathode (orthodromic orientation), are symmetrically stimulated and produce an inherently stable ciliary activity (Fig. **295 b, I**). On the other hand, ciliary propulsion is laterally out of balance in those cells which meet the isopotential lines at oblique orientation. This turns the cell around until the orthodromic orientation is established in a negative feedback manner (Fig. **295 b, II–IV**). Depending on voltage sensitivity and the proportion of cilia (or flagella) beating in reverse, galvanotactic swimming may also occur in the direction of the anode *(Opalina, Chilomonas)*. Some groups of ciliates employing specialized cilia of diverging sensitivity show an oblique galvanotaxis (*Spiro-stomum,* Climacostomum) or even a galvano-taxis oriented parallel to the isopotential lines (*Loxodes*).

Mechanoresponses

There are two basic subdivisions in mechano-responses: those which occur upon touch (a phasic stimulus) and those which occur in conjunction with the persistent pull of gravity (a tonic stimulus). Touching by microneedle, or even directing the water current from a capillary toward the anterior end of *Stentor*, induces a depolarization and rapid contraction or, after repeated stimulation, even a backward swimming response of this commonly sessile ciliate. In various other ciliates, depolarizing mechanore-ceptor channels are primarily located near the anterior cell end. Thus, the avoiding reaction of *Paramecium*, a shock response following, for instance, anterior collision with a mechanical obstacle, is triggered by a mechanically induced depolarization (comp. Fig. **213**). Analysis of the local distribution of mechanosensitivity in ciliates (*Paramecium, Stylonychia*) has established an additional hyperpolarizing sensitivity, which is predominantly located near the posterior cell end (Fig. **296**). In agreement with the principles of electric sensitivity of cilia (see galvanotaxis), touching *Paramecium* posteriorly induces the cilia to augment their normal beating mode ("escape response") so that the cell swims away in the forward direction.

Metazoan sense organs often include a cilium or ciliary rudiment. By inference, it has been suggested that cilia in protozoa sense mechanical disturbances. Electrophysiological analysis has shown that this is not the case. Removal of all cilia in live *Paramecium* did not interfere with the production of depolarizing or hyperpolarizing mechanoreceptor potentials (Fig. **296**). Moreover, local stimulation of the elongated and immobile cilia at the posterior end of *Paramecium caudatum* established that these cilia do not generate receptor responses per se, but passively transmit mechanical disturbances to the highly sensitive membrane of the posterior cell soma. There are ciliary organelles, as in hypotrich ciliates (the dorsal bristle complex), where a direct or indirect role in mechanosensory transduction has been suggested (Fig. **297**).

A well-known graviresponse is negative gravitaxis (formerly: negative geotaxis), which may be easily observed in *Paramecium* and some other ciliates (Fig. **298a**). Numerous investigators claim that this response results from sheer mechanics, such as due to separation of the "center of propulsion" from the "center of gravity" in a cell (gravity-buoyancy principle). This interpretation would be difficult to apply to surface-gliding *Loxodes*. In this ciliate, precise vertical upward gliding alternates with precise downward gliding, so that a population of *Loxodes* maintains its position in space (neutral gravitaxis). Neutral gravitaxis was also previously documented in equilibrated *Paramecium* and *Didinium* after integration of the orientational and velocity responses. Mechanical and physiological effects of gravity are based on the difference in density between cytoplasm and the surrounding water, or between components of specialized organelles (Müller's vesicles of loxodid ciliates; see Fig. **112**). Experiments accounting for the rate of gravity-induced (passive) sedimentation in *Paramecium* have shown that active propulsion rates in upward swimming cells are increased, and active propulsion rates in downward swimming cells are reduced as compared to horizontally swimming cells (gravikinesis). This behavior is exhaustively explained, accounting in *Paramecium* for gravitransduction in the actual "lower" soma membrane (Fig. **298b**), the distribution of mechanoreceptor conductances (see Fig. **296**), and voltage dependency of ciliary responses (see Fig. **293**). Gravikinesis, as based on topographical gradients of sensors, has been confirmed in other ciliates. Gravikinesis differs from the conventional definition of kinesis in that it accounts for the vector property of the stimulus, and tends to neutralize its sedimenting effect.

Two more responses, rheotaxis and thigmotaxis, are closely associated with mechanical input. Many ciliates can swim against the direction of a water current (positive rheotaxis). A parsimonious explanation of rheotactic behavior assumes that the geometric center of propulsive effort is anterior to the center of gravity of a cell (see gravitaxis). Thigmotaxis describes a tendency of an organism to settle in an area of maximal and persistent contact to solid surfaces. This behavior operates by means of trial and error rather than orientation. In ciliates, thigmotactic behavior appears to distinguish between "preferred" and "ignored" solid objects, so that chemoreception by contact, rather than by mechanical deformation, may be a crucial step in thigmotaxis.

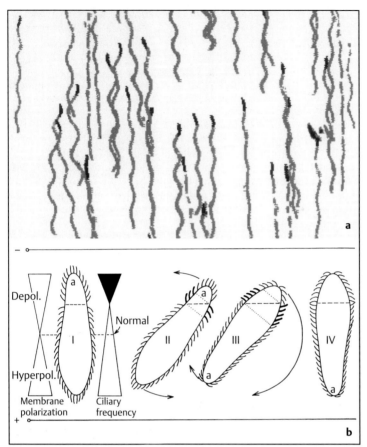

Fig. 295 Galvanotaxis in *Paramecium*. **a** Cells in linear DC voltage gradient (cathode: up) swim forward toward cathode (end of traces enhanced). **b** Electromechanical basis of galvanotaxis. I A cell exposed to the DC voltage gradient faces the cathode (–) with its anterior end (a). This orthodromic orientation results from bipolar gradients of membrane polarization (depol. = depolarized; hyperpol. = hyperpolarized) and ciliary frequency responses to membrane polarization (white gradient: cilia beat posteriorly; black gradient: cilia beat anteriorly). The cell swims forward because the posteriorly beating cilia outnumber the anteriorly beating cilia. II, III Since cilia along isopotentials (dashed) beat in the same mode, cells at oblique orientation generate laterally diverging forces (see enhanced cilia), which turn the cell around toward an orthodromic orientation. IV Theoretically, an antidromic orientation could generate swimming toward the anode (+), but a minor imbalance of this labile state quickly reorients *Paramecium* from IV to III and eventually to the stable orthodromic state I (from Machemer: Galvanotaxis: Grundlagen der elektromechanischen Kopplung und Orientierung bei *Paramecium*. In: G. H. K. Zupanc (Hrsg.): Praktische Verhaltensbiologie, Parey, Berlin 1988).

Thermoresponses

It has been repeatedly demonstrated that a population of *Paramecium* accumulates in an area of optimum temperature that is close to the culturing temperature. In agreement with the concept of electrophysiological control of biased random walk along stimulus gradients, cells augment swimming velocity and reduce reversals when swimming toward the optimal temperature, whereas they show frequent reversals when they swim away from the optimum.

Overall application of heat generates a complex bipolar potential response of the membrane. However, local application of a heated wire to the anterior cell end results in a depolarization, and local application of the same wire to the posterior end results in a hyperpolarization, if the experimental temperature is more negative than the culturing temperature. Analysis of

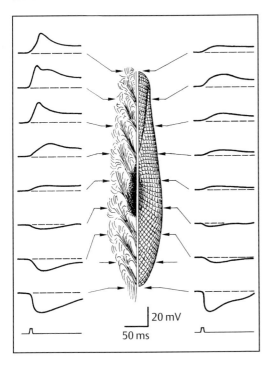

Fig. **296** Topology of mechanical sensitivity in *Paramecium* (left: ciliated cell; right: deciliated cell, anterior end up). Shown are the sites of pulse deformation (bottom traces) of the membrane by microneedle (arrows) and the potential response following a single pulse. It is seen that anterior stimulation depolarizes, and posterior stimulation hyperpolarizes the cell membrane. Due to graded distributions of mechanoreceptor channels, depolarizing (Ca^{2+}-dependent) sensitivity decreases in the posterior direction, and hyperpolarizing (K^+-dependent) sensitivity decreases in the anterior direction. Comparison of left and right column of traces shows that the soma membrane is mechanically sensitive, whereas graded action potentials can only be generated in the presence of ciliary membranes (see depolarizing traces at left) (after Ogura and Machemer).

20 mV

50 ms

the electrophysiological basis of thermoreception in *Paramecium* has shown, in agreement with findings in vertebrates and invertebrates, a close similarity to mechanoreception: depolarizing, Ca^{2+}-dependent thermoreceptor channels accumulating near the anterior cell end, and hyperpolarizing, K^+-dependent thermoreceptor channels accumulating near the posterior cell end. These data might indicate the physical identity of mechano- and thermosensitive conductances. On the other hand, the same conductances gave an inverse current response to rising temperature: thermoreceptor currents decreased, but mechanoreceptor currents increased, suggesting separate thermo- and mechanotransduction channels, which may nevertheless share some important molecular properties.

Chemoresponses

Cells can meet a chemical substance dissolved in water during active locomotion, and due to diffusion of the signalling source. Because chemical gradients are difficult to detect by instantaneous spatial discrimination, chemoresponses of cells are limited to kineses (biased random walk). In *Paramecium*, experiments have shown that some organic compounds such as acetate, lactate, folate, inosine monophosphate, and cyclic

AMP can act as attractants indicating, presumably, that bacterial food is nearby. Repellents such as quinidine, acetic acid, $BaCl_2$, and KOH are less clearly identified by molecular properties since their repulsive action depends greatly on the reference solution and degree of equilibration to the solution.

Chemoresponses in protozoa can interfere with cellular signalling at different levels: (1) changes in inorganic ion concentrations may directly modify electric conditions of the membrane potential; (2) organic signalling molecules and O_2 may bind to receptor proteins in the body (= soma) membrane causing intracellular messengers to change channel conductances or to activate an ion-transporting (electrogenic) pump; (3) walking along a concentration gradient requires adaptation to the previously experienced signal intensity. In *Paramecium*, time-dependent adaptation to rising (or falling), concentrations of KCl is based on readjustment of the depolarized or hyperpolarized membrane potential by regulation of the number of open K channels. It is to be expected that mechanisms of electric adaptation apply to organic as well as inorganic attractants.

Chemical signals guide, for instance, the sporozoites of *Plasmodium*, after release from the ookinete and passive transport in the body

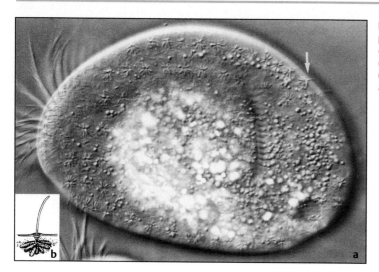

Fig. **297** Dorsal bristle complex of *Euplotes vannus*. Elongated ampullae encircle a cilium (**a**, arrow, and **b**) (b after von Gelei). Magn.: a 900×.

hemolymph, allowing it to accumulate in the salivary gland of the mosquito. Observations have established that the sporozoites use chemokinetic random-walk modulation for detection of the optimal concentration of a signalling substance.

While chemical signalling may be crucial for the approach of cells to food sources, small particulate food appears to be discriminated and selected by ciliates according to size alone. *Paramecium* ingests nondigestible materials such as carmine, carbon, or iron, as well as bacteria. *Didinium*, which feeds preferentially on *Paramecium* and ignores most other ciliates, identifies its prey by contact: the predator scans a given volume of water by erratic swimming to meet a potential prey by chance. It is thought that after collision, extrusive organelles, the pexicysts, check for "good" or "bad" taste, as represented by the molecular composition of the cell surface. Compliance leads to extrusion and penetration of the deadly toxicysts, which immediately disable the electric control of ciliary movement of the prey, *Paramecium*.

Mutual recognition of sexually differentiated cells may be organized or even guided by signalling molecules (see Fig. **279**), which are either incorporated in ciliary membranes *(Paramecium)*, secreted into the medium to induce chemokinesis of the target partner *(Blepharisma, Euplotes)*, or work in both ways *(Chlamydomonas)*. Secreted attractants (here: gamones) are members of the diverse family of sexual pheromones, which at extreme dilution ($\geq 10^{-13}$ M), act with high specificity, and have long been known to exist in insects. In protozoa, the response to chemical identification and sexual compatibility is often an agglutination and inactivation of flagella or cilia, followed by cell fusion.

Photoresponses

Many protozoa, including colorless ones, can respond to visible light. *Amoeba proteus* increases the rate of locomotion with a step-up in illumination intensity (direct photokinesis), relaxing to the normal rate in the course of about 10 min. (adaptation). The *Chlorella*-containing and the *Chlorella*-free *Paramecium bursaria* accumulate in an illuminated region, reducing the swimming velocity or even coming to a standstill (inverse photokinesis). A circadian change in photoaccumulation coupled to the resting membrane potential has been demonstrated. Very high intensities of white light induce *Paramecium bursaria*, as well as the colorless *Paramecium multimicronucleatum*, to accumulate in the shaded area due to an increase in velocity and decrease in reversals in the illuminated area (direct photokinesis). In *Paramecium*, spectral sensitivities have been identified for depolarizing and hyperpolarizing phototransduction, suggesting that photokinesis follows the common pattern of an electrically modulated random walk. Depolarizing photosensitivity of *Paramecium bursaria* was maximal in the soma membrane of the oral groove. Hyperpolarizing responses were seen only in normal, ciliated cells, indicating the possible involvement of ciliary membranes in this type of phototransduction.

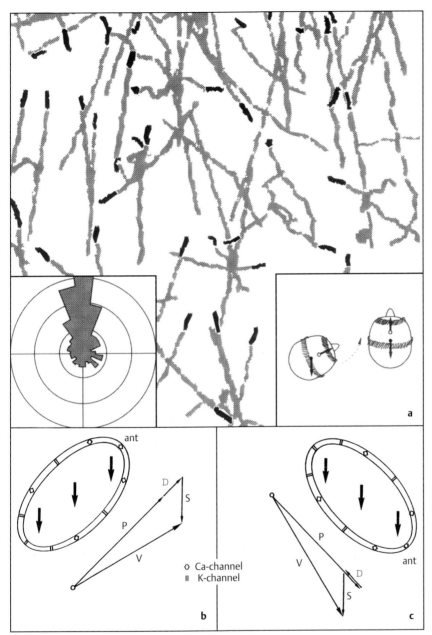

Fig. **298**   Graviresponses in ciliates. **a** Negative gravitaxis in *Didinium*. The emphasized ends of traces show that the majority of cells take an upward swimming orientation (see histogram of orientation at left). This behavior may be explained (inset at right) assuming that the center of gravity in *Didinium* (filled circle) is behind the geometric center of locomotory effort (open circle). **b, c** Gravireception and kinesis in an upward swimming (b) and downward swimming ciliate (c; ant.: anterior cell end) is based on the raised density of the cytoplasm as compared to that of the environment. Two opposing gradients of depolarizing (Ca²⁺-dependent) and hyperpolarizing (K⁺-dependent) mechanically sensitive channels occur in the membrane (see Fig. **296**). b During upward swimming, pressure exerted by the cytoplasm on the lower membrane activates predominantly hyperpolarizing gravireceptors. This induces the cell to raise the rate of active propulsion (P) by Δ (see Fig. **293**). Then, the observed swimming velocity (V) is the vectorial sum of P, Δ and the sedimentation rate (S). b In downward swimming cells, activation of depolarizing gravireceptors leads to a reduction in forward propulsion (P – Δ). The value of Δ is calculated from measurements of V, P and S. Behavioral data in *Paramecium*, *Didinium* and *Loxodes* suggest that the gravikinetic vector, Δ, always opposes sedimentation (negative gravikinesis) (from Machemer and Bräucker: Acta Protozool. 31 [1992] 185).

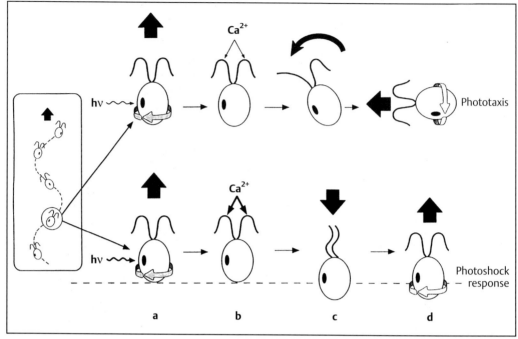

Fig. **299** Phototaxis as compared to a photoshock response in *Chlamydomonas*. The biflagellate cell swims along a helix rotating about its translational axis (inset). Thereby, the eyespot near the equator of the cell continuously scans the environment. Phototaxis: A light beam illuminates the unilaterally shaded eyespot when it comes to face the light (**a**). Alternation of the light signal in the rhodopsin-type photoreceptor generates a periodically rising (depolarizing) receptor potential followed by an action potential, which causes transient influx of Ca^{2+} into the flagella (**b**). A difference in motor response of the flagella to the messenger, Ca^{2+}, translates into a turning motion of the cell (**c**), which persists until the photoreceptor receives the same illumination during helical swimming of *Chamydomonas* parallel to the direction of the light beam (**d**). Photoshock response: With a very strong change in light intensity (a), the Ca^{2+} action potential is sufficiently large to cause a saturating increase of Ca^{2+} in the flagella (b). This induces both flagella to switch from cilia-type breaststroke activity to flagella-type undulatory motion inducing backward swimming (c). Stimulus termination, or adaptation to the persistent light stimulus leads to removal of excess Ca^{2+} from the flagella which resume normal activity (d; compare with Fig. **281**) (from Witman: TICB 3 [1993] 403).

Antibody labelling suggests that rhodopsin-like photoreceptor molecules occur in the ciliary and somatic membranes.

Sharp transition between shadow and illumination under natural circumstances can change a light intensity by several orders of magnitude. Photosensitive cells passing "step-down" or "step-up" generate a shock response. The mechanism of photoshock was analyzed in the biflagellate *Chlamydomonas* (Fig. **299**). Photoreceptor transduction depolarizes the membrane to generate a Ca^{2+} action potential. Rapid increase of Ca^{2+} to a high level ($\geq 10^{-6}$ M) in the flagella induces *Chlamydomonas* to switch from forward propelling "ciliary type" beating to transient backward propelling "flagellar-type" beating. The normal beating activity and forward swimming in a random direction is resumed after intraflagellar Ca^{2+} is renormalized toward the resting level (10^{-8} M). Photoshock responses are also seen in colorless ciliates. The mechanism of sensory-motor coupling resembles the avoiding reaction of *Paramecium* (see Fig. **213**) and may apply, in principle, to all ciliated or flagellated protozoa.

Localized accumulations of receptor pigments are not required for the generation of photokinesis or photoshock but may be a prerequisite for phototaxis. Movements with respect to light direction have been well studied in autotrophic protozoa. The eyespot-bearing *Chlamydomonas* rolls about its longitudinal axis, while it heads for the light source (Fig. **299**). The eyespot (stigma) is located near the cell equator and contains a rhodopsin-like protein which is unilaterally lined with light-absorbing granules. With laterally impinging light, the photoreceptor is periodically shaded and illuminated during axial rotation of the cell. A change in light intensity is thought to rapidly shift the membrane potential via modulation of a photoreceptor conductance. The flagellar responses to voltage-dependent Ca^{2+} influx are slightly different in the two flagella, causing reorientation of the cell toward the light. A fully symmetric beating of the flagella (and hence absence of a turning motion) occurs, while the cell swims directly towards the light source (positive phototaxis) or away from it (negative phototaxis).

Phototaxis has been documented in a few ciliates such as the pigmented *Stentor, Fabrea* and *Chlamydodon*, and the colorless *Ophryoglena*. The latter histophagous ciliate shows quite spectacular responses to incident light in that the sign of phototaxis may change during the reproduction cycle (Fig. **300**). A positive phototaxis is seen in the free-swimming theronts. Theronts, after taking up food, transform to become large trophonts until, several hours after feeding, they settle down as negatively phototactic tomonts. An intermediate, migratory stage of *Ophryoglena* is barely, if at all, photosensitive. It is speculated that the strongly refractive, reniform Lieberkühn's organelle (watchglass organelle) in the oral cavity of *Ophryoglena* has a function in perceiving the direction of light.

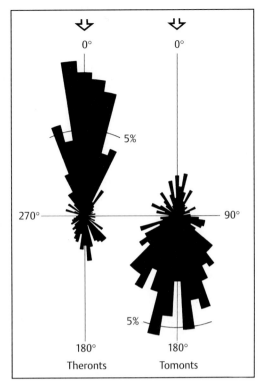

Fig. **300** Life-cycle dependent change from positive phototaxis to negative phototaxis in the ciliate *Ophryoglena*. The circular histograms show proportions of swimming orientations to incident white light (arrows at top) in theronts and tomonts of *Ophryoglena* (from Kuhlmann: Europ. J. Protistol. 29 [1993] 344).

Ecology of Protozoa

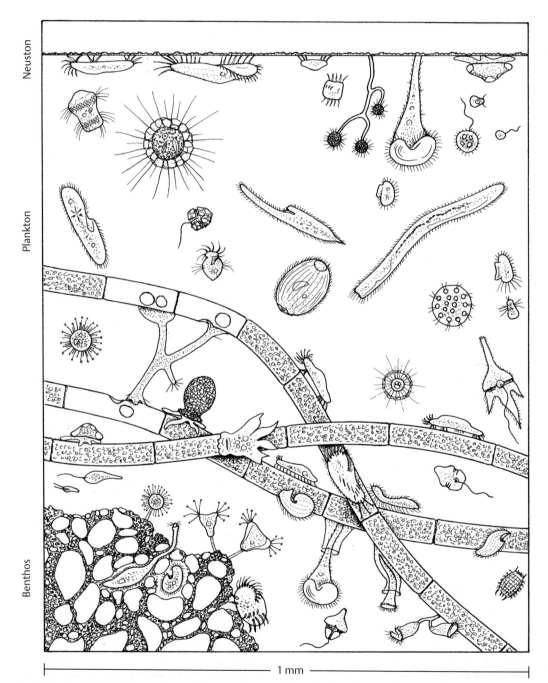

Neuston

Plankton

Benthos

|— 1 mm —|

Fig. **301** Spatial distribution of unicellular organisms in aquatic habitats. Representatives of all groups of free-living protozoans live on different substrates (phase border air-water, benthos, detritus, aquatic organisms) or in the planctic community (after several authors).

It is not simple to study the ecology of protozoa, and the task is especially complicated because the organisms are of microscopical dimensions. Moreover, it is often difficult to exactly identify the organisms in their natural habitat. Free-living benthic communities live preferentially in microhabitats (Fig. **301**), and they change rapidly in abundance as well as species composition. However, one of the most important obstacles is the correct identification of the protozoa in the field, a problem which can be surmounted only with much practical experience in the field of taxonomy. This could be one explanation for the fact that knowledge of the ecology of protozoa, compared to other organisms, increases relatively slowly. Nevertheless, it is increasingly evident that protozoan communities, especially those of the plankton, play an important role in the natural environment.

The recent free-living protozoa live in an environment which is continuously influenced by their pro- and eukaryotic relatives, as well as by themselves. As a consequence, the environment in which they grow and multiply is the product not only of abiogenic, but also of biogenic forces. Examples of the contributions of protozoan activity to the shaping of the surface of our earth are not only the silicate and calcium carbonate deposits (as skeletons or loricae) forming, for example, the limestone Alps, but also the photosynthetic activity of unicells which is responsible for the O_2- and O_3-containing atmosphere of our planet.

It would be unjustified to attribute the influence of protozoa on the biosphere only to bygone epochs. It is becoming increasingly evident that the dynamics of the carbon balance of the biosphere depend greatly on the structure and function of heterotrophic and photosynthetically active planktonic microorganisms.

Besides the importance of global environmental aspects, we should bear in mind that the study of the ecology of protozoa is increasingly estimated as being a historical scientific discipline which is also intimately associated with considerations of the evolution of organisms.

Factors Influencing Distribution

The appearance and distribution of protozoa are determined by abiotic and biotic factors. Abiotic factors are of physical and chemical character; biotic factors are, for instance, food resources, competition, and predator-prey relationships. The sum of the abiotic factors is crucial in determining whether or not specific protozoa can live in a given habitat, whereas the biotic factors are of primary importance in determining the relative abundance of specific populations.

Abiotic Factors

Considered as a whole, protozoa are tolerant to a wide range of physical and chemical environmental factors. Consequently, they are found in a wide variety of biotopes and habitats (Fig. **302**). Most information about the influence of abiotic factors is the result of laboratory investigations, combined with measurements obtained from the natural environment. The abiotic data, taken from the natural habitat, refer mainly to large water bodies and do not necessarily reflect what is of relevance to the microhabitats in which protozoa live.

It is known from laboratory experiments that the cysts of the ciliate *Colpoda* can persist at very low temperatures. However, this does not mean that species of this genus are typical for arctic regions. On the other hand, the absence of certain species in a given habitat tells us nothing about their demands. For instance, it has been shown that numerous typical fresh water ciliates can also survive in sea water if they adapt slowly, and to gradually higher salinities, as demonstrated in petri dish cultures.

Abiotic paramenters include chemical factors such as humidity, concentration of ions, pH, concentration of dissolved gases, and physical factors such as light, temperature, and the movements of water. As the influence of individual factors cannot be isolated from each other, but rather act as a whole (often in conjunction with biotic factors), it is difficult to evaluate their specific importance or effect. For instance, an increase in light intensity is normally connected with an increase in temperature, which causes a lower solubility of gases in water, as well a higher metabolic rate and a greater demand by protozoa for dissolved oxygen. Even if certain organisms can tolerate these conditions, nothing can be predicted about their chances of survival. The tolerance of *Paramecium multimicronucleatum* to increasing temperatures increases only with a simultaneous decrease in food (bacteria), or an increase in the ambient pH. Nevertheless, basic knowledge concerning the most important limiting factors regulating the occurence of protozoa is a necessary prerequisite for understanding the ecology of protozoa.

Water

The occurrence of protozoa depends on at least some free water. it is of secondary importance, whether this surrounding fluid is fresh water, salt water or the tissue/organ fluid of other organisms; and it is of no general relevance whether, for instance, the fluid covers grains of soil as a thin film, or the surface of leaves as drops of dew. Completely dry biotopes cannot be colonized by protozoa. However, numerous unicells are able to produce cysts in which they can survive desiccation for long periods. Some even form sclerotia, in which they survive for years in a dormant state known as cryptobiosis. By this, it becomes understandable that even deserts serve as habitats for the persistence of protozoa. They become active trophonts only for a very limited period, after some rain has fallen. During these few hours, the protozoa excyst, ingest food, grow, multiply, and reencyst. This can be observed not only in deserts, but also in temperate climates. Here, such protozoa can be observed, e. g., on the bark of trees, leaves, mosses, and lichens which are only occasionally covered by water. Slime molds and certain ciliates such as *Colpoda* are examples of very advanced adaptation to these extreme conditions. The normally dry habitat of "air" is also used by protozoa, for example for the distribution of cysts or spores. In these cases, sorocarps or sori are formed, which emerge a few millimeters from the substratum, allowing the dispersal stages to be transported by the wind (slime molds [see Figs. **63, 64, 66**], the ciliate *Sorogena* [see Fig. **185**]).

Temperatures below freezing can also be regarded as water-free conditions, as water at these temperatures is solid and cannot be used as a solvent for metabolic activity. Moreover, solid surroundings do not allow movement of protozoa. As a consequence, it is not surprising that dry-resistant cysts are also often frost resistant. Frost has at least two dangerous effects on cells. Firstly, intracellular ice crystals can grow which damage the cellular membranes, causing great disturbance to the metabolism or even death of the cell; secondly, changes in the ionic milieu can lead to denaturation of, e. g., enzymes. The protozoa concerned are able to excrete water actively before periods of dryness or frost. Their cysts normally have a complex wall. The process of sclerotization of the acellular slime molds and, as suspected, of certain terrestrial amoebae (e. g., *Thecamoeba terricola*), is interpreted as a reaction to acute water stress.

However, some protozoa are able to escape the hazards of ice crystal formation without entering a resistant cyst stage by adapting to very high salt concentrations of the surrounding medium. This is the case in unicells living inside blocks of ice. During freezing, water crystallizes as pure H_2O. This means that inside a forming ice block, small channels of 100 μm to several millimeters in diameter remain and are filled with a concentrated brine (Fig. **303**). This causes a lowering of the freezing point. The concentration of NaCl in this fluid might increase to 144‰ (at – 10 °C), compared to 35‰ of normal sea water. This channel system is inhabited by certain protozoa which not only survive, but even find optimal conditions there. Besides several diatoms, dinoflagellates mainly of the genera *Amphidinium, Gymnodinium*, and *Glenodinium* are found, as well as some chrysomonads, foraminifers, and ciliates.

To survive in these conditions, protozoa modify the composition of their cytoplasm by the accumulation of sugars (trehalose, mannitol, glucose), glycerine, amino acids and their derivates (proline, β-alanine, taurine), methylamines, or β-dimethylsulfonioproprionate. These organic compounds are not necessarily synthesized by the protozoon itself, but may be sequestered from its food. At least under experimental conditions, heterotrophic protozoa fed with the extremely salt tolerant flagellates of the genus *Dunaliella* (Chloromonadina), survived deep freezing and cryopreservation better than control organisms fed on normal food.

In habitats of very low osmotic pressure, such as rain water, condensation, or spring water, the intracellular milieu is hypertonic and water enters the cell continuously. This would eventually cause the cell to burst, but protozoa are able to reduce the intracellular concentration of osmotically active substances by converting them into osmotically inactive substances. Some can also stiffen the cell wall or intracellular cortical structures, and most are capable of the active expulsion of water via contractile vacuoles.

Temperature

Most protozoa live at temperatures ranging from the freezing point of water to a maximum of about 40 °C. Nevertheless, certain protozoa tolerate extreme temperatures: in thermal springs, the ciliate *Oxytricha fallax* lives at temperatures from 41 °C to 56 °C; and in sea water at –2 °C, species of the ciliate genus *Euplotes* have been found (Fig. **304**).

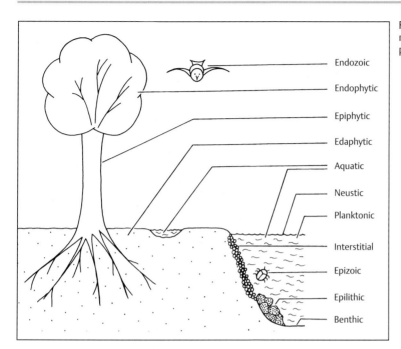

Fig. **302** Summary of the most important habitats of protozoa.

Endozoic

Endophytic

Epiphytic

Edaphytic

Aquatic

Neustic

Planktonic

Interstitial

Epizoic

Epilithic

Benthic

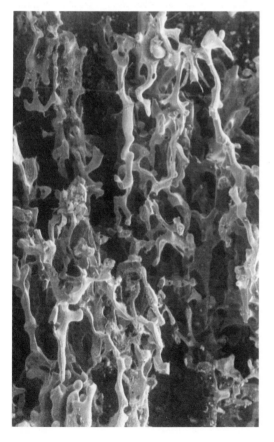

Short-term gradual warming to 35 °C is tolerated by numerous species. In nature, these temperature shifts occur due to solar irradiation in habitats such as shallow bog ponds, stands of *Sphagnum*, or tidal ponds. However, due to the nocturnal drop in temperature or tidal flows, the average temperature is in the tolerable range. On the other hand, certain species have their growth optima at relatively high temperatures as, for instance, some of the free-living species of the genus *Naegleria* (Schizopyrenida). In these species, the maximum division rate takes place at 40 °C *(Naegleria fowleri, N. australensis)*. Consequently, under optimal conditions they can live as facultative endoparasites, even in warm blooded animals. This preadaptation to elevated temperatures is one of the prerequisites for a protozoon to survive and develop inside birds and mammals (Table **9**).

During extended, calm periods in open water, massive aggregations of certain unicells may occur, e. g., in the North Sea. Besides heterotrophic dinoflagellates such as *Noctiluca scintillans* (Fig. **305**), mainly autotrophic haptomonads are the causative agents of these blooms, as, in

Fig. **303** Resin-filled brine channels inside an Arctic ice lump (from Weissenberger et al.: Limnol. Oceanogr. 37 [1992] 179). Magn.: 15×.

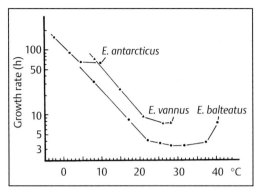

Fig. **304**   Growth curves of various species of *Euplotes: E. antarcticus* (from Antarctica), *E. vannus* (from Denmark), and in *E. balteatus* (from Florida). The curves demonstrate that rather large temperature ranges are tolerated, but that the temperature for optimum growth is related to the temperature characteristics of the habitant from which cells were originally isolated (after Lee and Fenchel).

Table **9** *Naegleria* spp. and their optimal growth temperatures

Naegleria species	Temperature (°C)	Pathogenicity (in mice)
N. gruberi	≤ 37	–
N. fowleri	45	++
N. lovaniensis	45	–
N. austr. australiensis	42	+
N. austr. italica	42	++
N. jadini	≤ 30	–

the early spring of 1988, the "killer-alga" *Chrysochromulina polylepis*. Similar, but not necessarily toxic phenomena occur occasionally during the spring along the North Sea shore and Channel shore. In these cases, species of the haptomonad genus *Phaeocystis* in particular occur in tremendous numbers, producing carbohydrate foams reaching a height of one meter (Fig. **306**).

Oxygen

Most protozoa depend on dissolved oxygen in the surrounding medium. However, protozoa do not need such high concentrations of oxygen as metazoa, due to their microscopic size, and the short diffusion distances which are consequently involved. As a result, many species can survive in habitats with extremely low concentrations of oxygen. Some are even able to exist in oxygen-free surroundings under reducing conditions, such as in sludge. Protozoa found in these habitats are, for instance, ciliates of the genera *Metopus* and *Caenomorpha* or free-living diplomonads of the genera *Trepomonas* and *Hexamita*. Many of the protozoa living in these habitats harbor methanogenic bacteria in their cytoplasm. These are able to metabolize the H_2 produced as a waste product by the protozoa, and to convert it into methane (Fig. **307**). At the same time, many protozoa, especially flagellates which do not contain mitochondria or any other redox organelles (e. g., the diplomads) are found in such habitats.

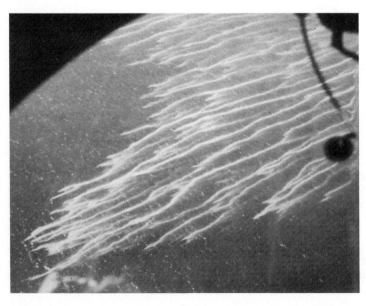

Fig. **305**   Aerial photograph of extensive blooms of *Noctiluca scintillans* in the North Sea (courtesy of J. Voß, Landesamt für Wasserhaushalt und Küsten, Schleswig Holstein, Kiel).

Fig. **306** Foam formations on the beach of the island of Norderney (Germany) caused by the abundance of *Phaeocystis* during May 1981 (courtesy of H. Michaelis, Norderney).

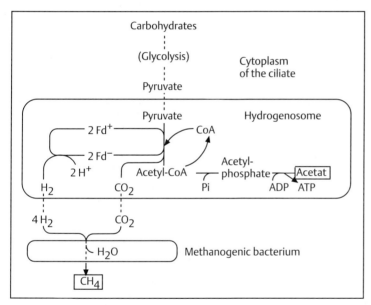

Fig. **307** Interactions between hydrogenosomes and endobiotic methanogenic bacteria in anaerobic protozoans (after Wagener).

Conspicuously, numerous ciliates living in habitats with extremely low oxygen concentrations, or in stagnant waters, contain cytoplasmic unicellular green algae of the genus *Chlorella*, called zoochlorellae. These can occur obligatorily, as in *Climacostomum virens* or *Paramecium bursaria*, or facultatively, as in certain *Vorticella* or *Frontonia* species. For the ciliates, the advantage is that they are provided with photosynthetically produced sugars and probably also with oxygen derived from photosynthesis.

Other ciliates, which are especially adapted to low concentrations of oxygen, follow the migrating oxic-anoxic boundary in mono- and dimictic lakes. These organisms, especially *Loxodes* and *Spirostomum* species, spend most of their time in the sediment, but following stratification of the water column and the development of deep water anoxia, they swim up into the water column, to the border between the oxygen-rich and oxygen-free water. Thus, the populations migrate vertically over a distance of several meters during the period of summer stratification (Fig. **308**). The efficient gravitactic orientation of *Loxodes* is probably controlled by a special organelle, the Müller's body (see Fig. **112**).

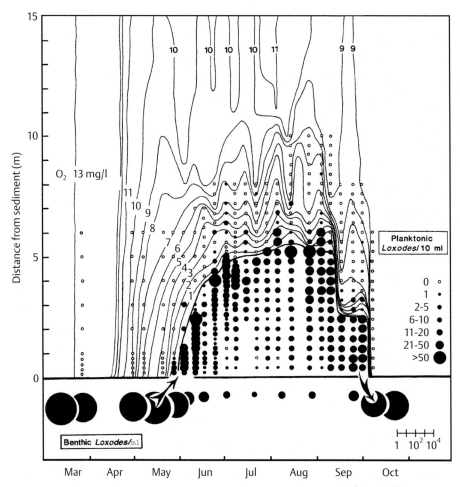

Fig. 308 Vertical distribution of *Loxodes* populations in an eutrophic lake (Esthwaite Water, Lake District, England) depending on different O₂ concentrations above the sediment. During summer stratification (June–September), the critical oxygen concentrations (0–2 mg/l) reaches to about 5–6 m above the sediment. *Loxodes* tracks this source of oxygen during the summer, although cells also spend periods living in the anoxic hypolimnion. When water column stratification collapses, all *Loxodes* return to the sediment (beginning of October) (from Finlay and Fenchel: Freshwater Biological Association Annual Report 54 [1986] 73).

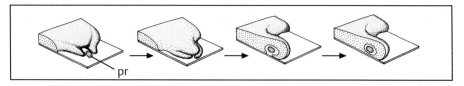

Fig. 309 Phagocytosis of a prey organism (pr) by *Amoeba proteus* (after Haberey and Stockem).

Substrates

The existence of substrates and their physical and chemical nature is of importance for sessile organisms, as well as for those which need a sur-face for locomotion. From the physical point of view, a low stream velocity is important and can be observed close to surfaces in running waters. The peripheral regions and "dead zones" in

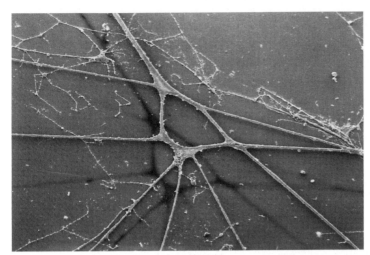

Fig. **310** Partial view of the pseudopodial network of *Reticulomyxa filosa* which can be spread out — as demonstrated by the electron shadow — also above the substrate. Magn.: 100×.

streams and rivers are important habitats for protozoa.

In nearly all biotopes, one part of the community consists of sessile protozoa. In the vicinity of sediment particles and aggregates, especially high numbers of species can be observed due to the high abundance of microbioal food particles and potential nutrients.

Sessile filter feeders use mechanical fixation to ensure that the ciliary motion results in the collection of food particles instead of locomotion. Simultaneously, the cell needs to be elevated from its substrate to allow the creation of adequate water flow. This is ensured by stalks or elongated body forms, e. g., in choanoflagellates, and ciliates such as *Vorticella* and *Stentor*. The effectiveness of the filtering activity can be heightened by the formation of colonies, which can become suspended in the water, e. g., spherical colonies of certain chaonoflagellates and small colonies of ciliates such as *Ophrydium versatile*.

Crawling amoebae overwhelm their prey by creating a trap formed by pseudopods pressed against the substrate (Fig. **309**). The plasmodia of *Reticulomyxa* (Granuloreticulosa) form with their reticulopodia large networks to catch large prey organisms such as *Volvox* colonies or rotifers (Fig. **310**).

Certain sessile ciliates attach, exclusively or at least preferentially, to organic or living substrates (Fig. **311**). This is particularly true for the ciliate genus *Spirochona*, members of which have been found, so far exclusively, on the gill plates of gammarids; or for certain Apostomatia, which are attached to the surfaces of crustaceans. Certain suctorians live on animals or plants in a species-specific way, e. g., *Tokophrya lemnae* living on rhizomes of species of *Lemna*. The evolutionary significance of this high degree of specialization might be seen in the avoidance of competition. The mechanisms of recognition of the substrate, however, are unknown.

Biotic Factors

As mentioned above, the most important biotic factors are food resources, competition with other organisms, and also predator-prey relationships.

Species with generation times of several weeks or months, such as foraminifers, survive only in very stable habitats. Normally, they represent one link in a rather complex biocoenose which is predetermined by the limitations of a given habitat. In contrast to these, protozoa with reproductive times of a few hours make rapid use of resources. Such populations, be they amoebae, flagellates, or ciliates, tend to collapse as rapidly as they developed.

The first step in establishing a new biocoenose is the settlement of a new habitat. Experiments with artificial substrates in fresh water ponds have shown that within three weeks, about 60 different species settled on these substrates. On the other hand, under extreme situations, a new settlement might be established very slowly, if, for instance, a new habitat is distant from a source of inoculating species. This is illustrated by the newly formed island Surtsey, which originated in 1963 south of Iceland due to volcanic activity: in water butts filled with fresh water, not more than 15 species have been found within two years.

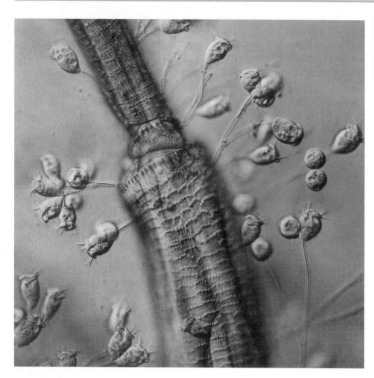

Fig. **311**   Peritrich ciliates on the antenna of a phyllopod (from Hausmann and Rambow: Mikrokosmos 74 [1985] 208). Magn.: 150×.

The "pioneer species" are normally small flagellates such as choanoflagellates and bodonids, which are followed by small amoebae and ciliates. Their food is bacteria, cyanobacteria, and unicellular algae e. g., diatoms. The composition of the community of food organisms is greatly influenced (quantitatively, as well as qualitatively), by the feeding activity of the protozoa. In the course of the establishment of these biocoenoses, the habitat is modified, forming new ecological niches, especially for predatory protozoa.

The sequence of the appearance of different species is called succession. Laboratory experiments have shown that protozoan biocoenoses are unstable, and that they can change very rapidly and dramatically, often within one day (Fig. **312**). The succession possibly reflects the exhaustion of specific food resources, as well as the dominance of new predators. In a simple food chain of three organisms, the influence of a predator on a population of bacterivorous ciliates is demonstrated (Fig. **313**). If the carnivorous ciliate *Hemiophrys* is added to a growing culture of *Vorticella*, the number of *Vorticella* will be drastically diminished, whereas the *Hemiophrys* population reaches a maximum. After exhaustion of the prey, the numbers of the predator collapse.

The *Vorticella* population grows again due to the consumption of the abundance of bacteria. The cycle starts again. Similar predator-prey successions are known for the ciliates *Didinium nasutum* – *Paramecium caudatum* and *Trachelius ovum* – *Ophrydium versatile*.

In natural biotopes, the situation is more complex due to the influence of additional factors on the population densities of various species. One of these factors is, for instance, the appearance of metazoan predators in protozoan biocoenoses. Some of the protozoa will be ingested, for instance, by larval fish, water snails, platyhelminths, or oligochaetes; others will die due to competition for food with rotifers (Fig. **314**).

Protozoan populations can live in habitats in which few if any metazoa exist. As shown for the marine interstitial, the microhabitats might be too small for metazoa to enter. Other habitats with high numbers of protozoa may have extreme physical or chemical characteristics, such as anaerobic systems where few if any metazoa can survive. Finally, very dense populations of protozoa are found in artificial habitats e. g., the "activated sludge" of biological sewage treatment plants. In these plants, the sewage is pumped into large tanks in which the organic material is degraded by bacteria that develop very dense populations. This food resource is used by protozoa, especially sessile, filter-feeding ciliates. Since the flow of water is relatively high in these plants, metazoan predators cannot develop dense populations.

In cases of food resource depletion, protozoa have developed characteristic survival strategies. One, as already mentioned, is the formation of resting cysts. Sessile species may become temporarily motile by active swimming or by passive flotation to other habitats. Examples of these strategies are the formation of the telotroch by *Vorticella* (see Fig. **133**) and the transformation into flagellated stages by schizopyrenids (see Figs. **62, 266**), or into floating stages (radiosa forms) by naked amoebae (see Fig. **26**). Others become completely immotile and switch to a minimal metabolic rate which provides only 2–4% of the energy normally needed for survival (e. g., the palmella stage of euglenoids). Another possibility is the secretion into the surrounding medium of metabolites which inhibit growth. This phenomenon is called, e. g., bacteriostasis, fungistasis, and ciliostasis. Conspecific consumption is also known through the development of macrostomes capable of cannibalism (see Fig. **265**).

Symbiosis between *Chlorella* species and protozoa is, in all likelihood, not a reaction to food depletion, but an adaptation. These endosymbionts each live inside a vacuole in the cytoplasm of the protozoa, and they are normally not attacked by the protozoan digestive system. In extreme situations, however, when food becomes limited, they may be digested by the host. Under normal conditions, the host benefits from this cohabitation by receiving products of photosynthesis for its own metabolism. In stands of *Sphagnum* of high moors (which receive their water exclusively by rainfall and which are consequently extremely lacking in organic nutrients), mainly green protozoa, i. e., *Chlorella*-bearing species are found: the testate amoeba *Hyalosphenia papilio*, the naked amoeba *Mayorella viridis*, the heliozoon *Acanthocystis turfacea*, or the ciliate, *Climacostomum virens*.

As far as is known, the abundance of one species, as well as the species richness in a protozoan habitat, depends mainly on the available food. Niche overlap between species is also minimized by specialization on particular diets. Examples include large amoebae such as *Amoeba proteus* and *Chaos chaos*, which are omnivores, and ciliates such as *Pseudomicrothorax dubius*, which are highly specialized by their ability to ingest and digest only cyanobacteria belonging to

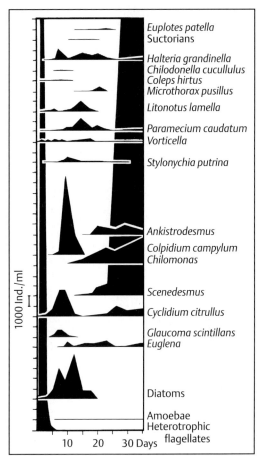

Fig. **312** Succession of protozoans from a natural sample grown in the laboratory in a medium enriched with cellulose as a source of organic material. The samples were monitored for 35 days (after Bick).

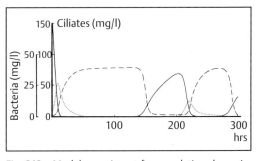

Fig. **313** Model experiment for population dynamics of bacteria (-----), *Vorticella* (——), and *Hemiophrys* (⋯⋯) (after Curds).

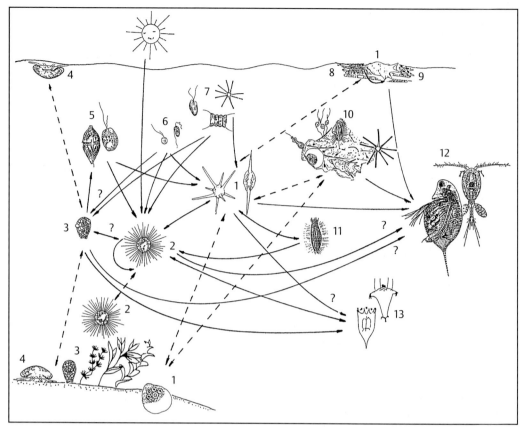

Fig. **314** Nutritional relationships between plankton organisms in a lake. Broken lines indicate changes in the habitat. 1 = naked amoebae; 2 = heliozoans; 3 and 4 = testaceans; 5 = large heterotrophic flagellates; 6 = heterotrophic nanoflagellates; 7 = phytoplankton; 8 = cyanobacterial filaments; attached bacteria; 10 = detrital floc; 11 = ciliates; 12 = mesometazooplankton; 13 = micrometazooplankton (from Arndt: Marine Microbial Food Webs 7 (1993) 3.

a single genus. The species of *Vampyrella* are probably dependent on a very few species of algae.

Heterotrophic, free-living protozoa ingest almost every type of organic matter: bacteria, cyanobacteria, other heterotrophic or autotrophic unicells, fungi and their spores, small metazoa, as well as decaying plant and animal tissues. For these protozoa, the most important limiting factor for food uptake is the architecture of the oral apparatus. Even different species of the same ciliate genus are able to filter specific size ranges of food particles (Fig. **315**), thus allowing their coexistence in the same habitat.

Aquatic Biocoenoses and Habitats

On a global scale, the aquatic biocoenoses of the oceans are the source of most biological productivity. The role that unicellular organisms play

has been explored only during the last two decades.

Firstly, it is striking that the species composition of the plankton, neuston, and benthos (see Figs. **301** and **302**) is very similar in marine and limnic habitats. Exceptions are the radiolarians and foraminifers, as well as the large, heterotrophic dinoflagellates which are found exclusively in the marine environment, especially in the open oceans. Additionally, a typical interstitial fauna, composed mainly of karyorelictid ciliates, is known only for marine habitats. Consequently, the similarities between marine and freshwater habitats concentrate mainly on autotrophic and heterotrophic flagellates, ciliates, and amoebae.

A major habitat type is the pelagic water column which is characterized by a certain uniformity of environmental parameters. Here, we do not find exclusively planktonic, but also ses-

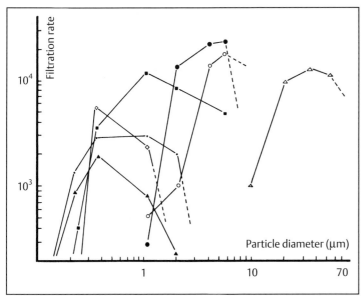

Fig. **315**   Selectivity of the filtration of latex particles of different sizes by seven ciliates. • = *Glaucoma scintillans,* ■ = *Paramecium caudatum,* ○ = *Blepharisma americanum,* ▲ = *Colpidium campylum,* ◇ = *Cyclidium glaucoma,* ● = *Euplotes moebusii,* △ = *Bursaria truncatella* (after Fenchel).

sile forms attached to suspended particles. These biotopes are inhabited by prokaryotes, eukaryotic unicells, and also by metazoa. In Table **10**, different heterotrophic organisms are listed according to their size. It becomes evident that protozoan abundance covers a range of five orders of magnitude, from the nanoplankton (2–20 μm) to the microplankton (20–200 μm). There is very little competition with metazoans. The reason for this is based on the morphology and mode of function of the filter apparatus of small crustaceans such as copepods which cannot sieve organisms the size of picoplankton. The zooplanktic groups are completed by photoautotrophic unicells of similar sizes; in the microplankton these are diatoms, dinoflagellates, and haptomonads. Much of the organic material which is provided by the phytoplankton does not function, as initially suggested, directly as food for protozoa and small crustaceans (the classical pelagic food chain), but is degraded to at least macromolecules by bacteria that support heterotrophic predators and the microbial loop. The estimates are that up to 50% of the net primary production of photoautotrophic microorganisms enters the microbial loop, at the base of which are bacterivorous nano- and microplankters. Since the heterotrophic nano- and microplankters are not faced with metazoan competitors, the biological control of bacterial populations is the responsibility of basal groups, especially of choanoflagellates. A simplified model of the planktic flow of energy is given by the mi-

Table **10**   Linear size relations between aquatic heterotrophic organisms

Size class	Dimensions	Organisms
1	0.2–2 μm	prokaryotes
2	2–20 μm	small-sized protozoa (mostly flagellates, amoebae)
3	20–200 μm	medium-sized protozoa (mostly ciliates), rotifers
4	200–2000 μm	large protozoa (mostly ciliates), rotifers, small crustaceans
5	2–20 mm	giant protozoa (e.g., foraminifers, radiolarians), cnidarians, platyhelmiths, nematodes, arthropods (mostly small crustaceans, insect larvae), vertebrate larvae
6	20–200 mm	giant protozoa (e.g., Myxogastria, foraminifers, colonial radiolarians), copepods, annelids
7	200–2000 mm	crustaceans, fish
8	2–20 m	mammals, fish

crobial food web (Fig. **316**). A more realistic but more complex scheme is demonstrated in figure **317**.

In stagnant and thermally stratified water bodies, the abundance of heterotrophic or-

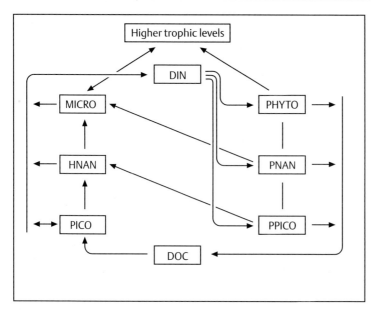

Fig. **316** Simplified scheme of the energy flow within microbial food webs. HNAN – heterotrophic nanoplankton (2–20 μm), PNAN – phototrophic nanoplanktic organisms (micro algae), PICO – heterotrophic picoplankton (0,2–2 μm, bacteria), PPICO – autotrophic picoplankton (cyanobacteria), MICRO – heterotrophic microplankton (20–200 μm), PHYTO – autotrophic plankton, DIN – dissolved inorganic nutrients, DOC – dissolved organic components (after several authors).

ganisms is structured vertically. The main elements of the microfauna in such a biotope are illustrated by a 1 hectare fresh water pond (Fig. **307**). The situation in the benthos close to the shore and in the middle of the lake is documented also in the water column at different depths. The oxygenated epilimnion extends to a depth of about 100 cm; the oxygen-free hypolimnion starts at about 180 cm and reaches down to the sediment at a depth of about 350 cm.

Compared to the pelagial, benthic habitats exhibit a versatile and heterogeneous structure. Local differences in chemical parameters, surface properties, and food resources lead to the formation of microhabitats with specific faunas. Some microhabitats exist exclusively due to their small size. An example is the marine interstitial, with grain sizes of 100 to 250 μm. Normally, food competitors and predators cannot invade the habitat due to their large body size. It is therefore mainly colonized by slender or vermiform protozoa.

These and similar benthic microhabitats might be additionally subdivided vertically. The basis for stratification horizons is mainly the gradient of oxygen, which is dictated by the degree of underlying bacterial degradation, and the flux of O_2 into the sediment. At a certain depth, the oxygen concentration is too low for aerobic microorganisms, and the redox potential becomes negative. Under these reducing, anoxic conditions, degradation is exclusively anaerobic.

Many ciliates which live in such habitats, prefer a certain redox potential and inhabit specific horizons in the sediment (Fig. **319**).

Inside the different sediment horizons, protozoa live in specific ecological niches. This is based on specialization on specific diets. For instance, some feed on anaerobic, others on aerobic bacteria, some feed on algae, protozoa and small metazoa, others on detritus. The specialization can reach such a point as detected by different species of the ciliate genus *Remanella*, which live in the same habitat but feed on differently sized food particles (Fig. **320**).

Small organisms are characterized by relatively high metabolic rates. The contribution of the protozoa to the cycle of organic material is normally very high. Experiments show that protozoa speed up the rate of degradation of sedimented plant material (Fig. **321**). The experimentally obtained result that protozoa, in cooperation with bacteria have the fastest rate of degradation, seems at first glance inconsistent because bacteria are also phagocytosed by the ciliates. However, the explanation for this apparent paradox is that due to the digestion of bacteria, the surrounding medium becomes enriched in essential nutrients which positively influence the growth and reproduction of the remaining bacteria. Furthermore, ciliates positively influence the density of bacterial populations by their ciliary activity, which causes a better circulation of dissolved nutrients and gases in the

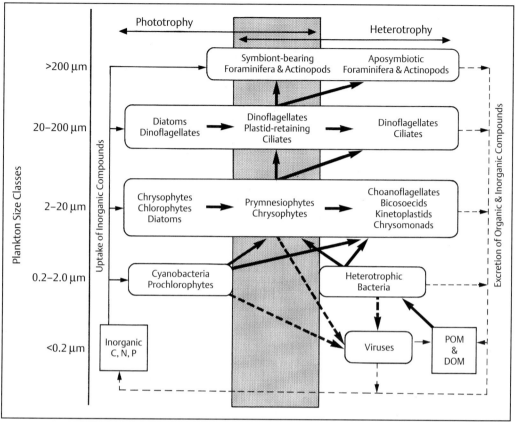

Fig. **317** Microbial food web diagram showing the size classes and generalized trophic relationships among photosynthetic and heterotrophic protists and prokaryotes in a planktonic marine ecosystem. Representative taxa are indicated in each size class and trophic mode. The solid arrows indicate the direction of energy flow as a consequence of consumption. Thin solid lines represent the uptake of inorganic compounds by primary producers; the release of organic and inorganic materials resulting from egestion and secretion is shown by thin dashed lines. Thick dashed lines represent losses of microbial biomass to lysis by viruses. The shaded area in the center indicates those organisms which conduct both photosynthesis and phagotrophy. DOM and POM represent pools of dissolved and particulate organic material respectively (from Caron & Finlay: Protozoan links in food webs, in: Hausmann und Hülsmann: Progress in Protozoology. Fischer, Stuttgart 1994).

fluid microhabitat. Additionally, larger protozoa are involved in the destruction of animal and plant material as carnivorous and herbivorous organisms. Of special importance are ciliates belonging to the genus *Ophryoglena*, as well as the macrostome forms of *Tetrahymena*. The filopodial organized stages of vampyrellids and the labyrinthulids also belong to these protozoan groups.

Terrestrial Biocoenoses and Habitats

Terrestrial habitats show some similarities to the sediments of aquatic systems. Notable differences include the availability of water and the presence of atmospheric gases. Consequently, it is not surprising that in soil (as well as in decaying leaves and on plants) the protozoan fauna has native elements with a high degree of adaptation, as well as species known from freshwater with a relatively low degree of specialization.

The upper horizon of soil is provided with water from rainfall and by ground water. The

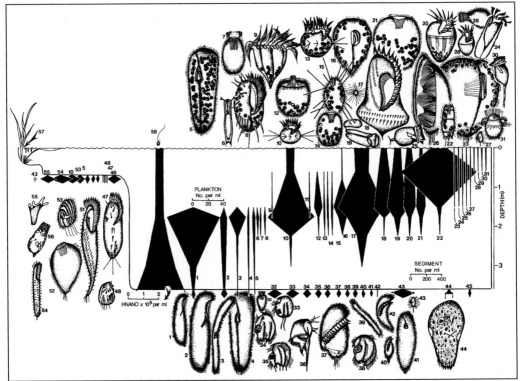

Fig. **318** Vertical distribution and abundance of ciliates and other heterotrophic protozoans in Priest Pot (Lake District, England). The information following species names includes (in μm) cell size, food type, and size of the food respectively. [Z] indicates the presence of zoochlorellae, [S] the existence of ingested but undigested plastids. With the exception of disproportionately large species (1–5 and 44), all organisms are drawn to scale (from Finlay et al.: Europ. J. Prot. 23 [1988] 205).

1. *Loxodes striatus* (200; algae; 5–50)
2. *Loxodes magnus* (400; algae; 5–50)
3. *Spirostomum teres* (300; algae; 2–10)
4. *Frontonia leucas* (300; algae; 3–50)
5. *Frontonia vernalis* (300; algae [+Z]; 3–50)
6. gastrotrichs
7. *Enchylis* sp. (60; algae; <5)
8. *Euplotes daidaleos* (80; algae [+Z], bacteria; <5)
9. *Strobilidium velox* (50; algae [+Z]; 5–10)
10. *Halteria grandiella* (50; algae [+Z]; 5)
11. unidentified testate amoeba
12. *Monodinium balbianii* (75; algae [+Z]; 5–10)
13. *Halteria grandinella* var, *chlorigella* (50; algae [+Z]; 5)
14. *Prorodon* spec. 1 (50; algae [+Z]; 2–5)
15. *Actinobolina radians* (60; algae [+Z]; 5)
17. *Acanthocystis pantopodeoides* (18; ?; <5)
18. *Hypotrichidium conicum* u, *H. geleii* (120/100; algae; 3–25/ 2–10)
19. rotifers
20. *Strombidium viride* (70; algae; 2–10)
21. *Prorodon* spec. 2 (80–100; algae [+Z]; 2–15)
22. *Coleps hirtus* (45; algae [+Z]; 2–5)
23. *Stokesia vernalis* (150; algae [+Z]; 3–30)
24. *Tintinnidium* sp. (75; algae; 2–5)
25. *Vorticella bosminae* (20; bacteria; <2)

26. *Lembadion magnum* (120; algae, HNANO; 5–25)
27. *Cyclidium* spec. (20; bacteria; <2)
28. *Strombidiumn spec.* (30; algae; 2–5)
29. *Enchelydium clepsiniforme* (60; algae; 2–5)
30. *Vorticella* spec. (40; bacteria; <2)
31. *Askenasia volvox* (40; algae; 2–5)
32. *Enchelydium amphora* (35; bacteria; <5)
34. *Epalxella* sp. (30; ?; ?)
35. *Saprodinium* sp. (40; bacteria; <2)
36. *Caenomorpha lauterborni* (70; purple bacteria; 5)
37. *Brachonella spiralis* (120; purple bacteria (-cluster); 5–20)
38. *Pelodinium reniforme* (50; ?; ?)
39. *Chaenea sapropelica* (100; ?, ?)
40. *Metopus undulans* (70; bacteria; <2)
41. *Loxocephalus luridus* (150; ?; ?)
42. *Metopus curvatus* (60; bacteria; <2)
43. various scuticociliates
44. *Pelomyxa palustris* (100–700; diverse particles; >200)
45. various naked amoebae as *Acanthamoeba*, *Vahlkampfia* etc.
46. various scuticociliates
47. *Stylonychia* (2 ssp.) (70/130; algae; 2–10)
48. *Aspidisca cicada* (30; bacteria; <2)
49. as 12
50. as 30
51. *Paruroleptus* sp. (130; algae; <20)
52. *Phagadostoma* sp. (100; agae; 5–10)
53. *Cinetochilum margaritaceum* (30; bacteria; <2)
54. *Holosticha* sp. (80; algae; 2–10)
55. naked amoebae as *Oscillosignum*, *Vahlkampfia* (25/7; algae; 2–5)
56. various testate amoebae as *Centropyxis*, *Difflugia* etc.
57. *Vorticella* as others peritrich ciliates settling on macrophytes, and choanoflagellates
58. Solitary chrysomonads (1–5 μm)

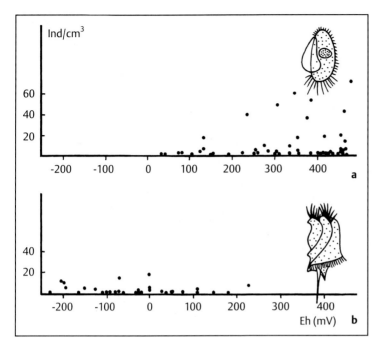

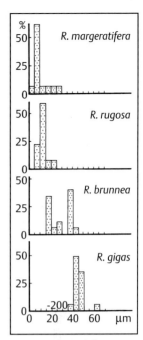

Fig. **319** Distribution of the two marine ciliates *Pleuronema coronatum* (**a**) and *Caenomorpha levanderi* (**b**) in the interstitial, according to the redox potential (after Fenchel).

Fig. **320** Size of diatoms ingested mainly by certain interstitial *Remanella* species. The ordinate shows the total individual density of each size sclass, the abscissa the absolute length of the diatoms (after Fenchel).

water accumulates in capillaries and pores with size depending on the grain size of the mineral substrates. A rich protozoan community can be observed in a 30 µm thick film of water (Fig. **322**). The minimal thickness of the water film in which small naked amoebae can survive is about 5 µm. The protozoan fauna is concentrated in the humus regions or close to the growing tips of plant roots. In both cases, the destruction of organic material is the basis of these accumulations.

The primary decomposers are — depending on the degree of aeration — anaerobic or aerobic bacteria and fungi which represent 80–90% of the total biomass. The remaining 10–20% are arthropods, other invertebrates, and protozoa. Protozoa represent 10 to 30% of the total animal biomass. However, taking the respiration rate as a measure, and by this the contribution to the overall metabolic rate, the value is in the range 35–70% (Fig. **323**). With regard to their productivity and ecological importance, the testate amoebae or the ciliates are as important for soil as earth worms.

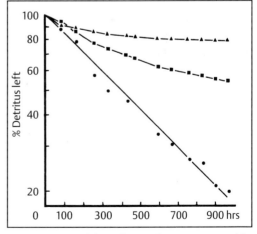

Fig. **321** Rate of degradation of organic material by bacteria (triangles), bacteria and choanoflagellates (squares), and bacteria with various protozoa (dots) (after Fenchel and Harrison).

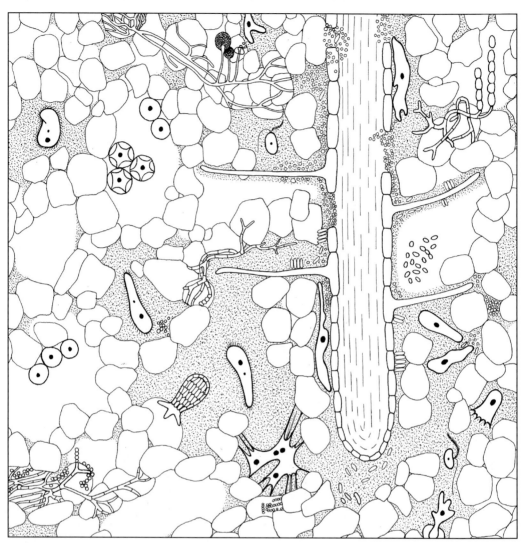

Fig. **322**   Habitats of terrestrial microfauna elements and the distribution of air- and water-filled capillary systems. The activity of unicellular organisms is found exclusively in aquatic microhabitats (dotted); chambers with air contain only encysted forms (adapted from Old).

Dominating representatives of the protozoan soil fauna are the ciliates and amoebae (testate and naked). Among the ciliates, the Colpodea, Hypotrichia, and Prostomatea are of special importance. They feed preferentially on freely swimming bacteria or suspended yeast cells, spores of fungi, etc. The amoebae ingest mainly organisms which are attached to the substrate. As is true for aquatic habitats, the soil protozoa are also highly specialized with respect to the size of their food. Additionally, predators are known which feed on organisms in the size range 2–20 μm and 20–200 μm. Extreme dimensions are observed in certain slime molds, which are able to develop under natural conditions (very flat) plasmodia with a diameter of more than 40 cm (see Fig. **28**).

Symbiosis and Parasitism

Protozoa themselves present cellular microhabitats which are suited for other very small organisms. Depending on the nature of the interactions between the organisms, the relationship is called symbiosis, if both organisms benefit from living together, or parasitism, if the host is harmed while the parasite benefits. Frequently, the type of relationship is not obvious, a situation which is called commensalism. Symbiosis is often used for all types of relationships. Other terms which reflect the spatial relationship between the particular organisms are: ectobiotic = living on the surface of another organism, endobiotic = living insider another organism, endocytobiotic = living inside a cell of another organism.

Protozoa as Commensals

Protozoa are frequently attached to the surfaces of another organisms. If the organism is motile, the situation is known as symphorism. Well studied examples are sessile peritrich ciliates and suctoria, which are transported by cnidarians, water beetles, copepods, water snails, or polychaetes (Figs. **324, 325**). The benefit to the protozoa is probably one of efficient dispersal over relatively large distances. For the phoront, the effect is neutral. Presumably, such an association could evolve into parasitism (e. g. *Ichthyophthirius multifiliis*, see Fig. **132**).

Numerous protozoa (e. g. retortamonads, proteromonads, opalinids) live inside the urogenital system of metazoa with no detectable effect upon the host. Ciliates which live in the rumen of Ruminantia are obviously involved in the decomposition of the ingested plant material (Fig. **326**). However, metazoa deprived of their protozoan rumen fauna survive without any serious problems. Consequently, entodiniomorphid ciliates (see Fig. **123**) are commensals with a very high degree of specialization. As far as is known, the have never been detected in any other habitat except for the intestine or feces of phytophagous hosts. In all likelihood, the are of terrestrial origin as are some amoebae that adapted from an anoxic life in feces and manure. In this context, marine species of the ciliate genus *Metopus* should be mentioned as an example of an analogous development because they might have a comparable history of evolution from the sludge of marine habitats into the alimentary tract of sea urchins.

Protozoa as Symbionts

The symbiotic nature of the coexistence of protozoa and metazoa is evident in the case of the flagellates, which live exclusively in the gut of termites and cockroaches. Although incompletely understood, it is apparent that the flagellates play in important role in the digestion of the cellulose material. This has been shown by

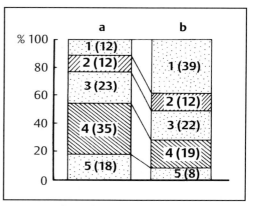

Fig. **323** Relative biomass (dry weight per square meter) (**a**) of various groups of soil organisms compared to the respective amount of organisms (**b**) 1 protozoans, 2 minute metazoa, 3 enchytraeids, 4 micro arthropods, 5 macroarthropods (after Person et al.)

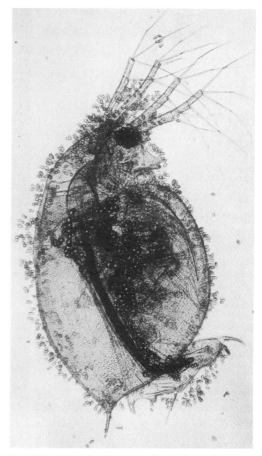

Fig. **324** Peritrich ciliates on the surface of a water flea (from Hausmann and Rambow: Mikrokosmos [1985] 208). Magn.: 50×.

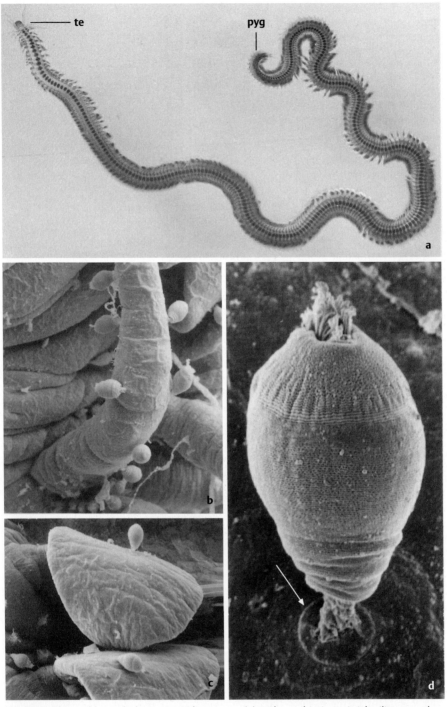

Fig. **325**   The wadden polychaete *Anaitides mucosa* (**a**) with ectobiotic peritrich ciliates on the antenna (**b**) and parapods (**c**). The ciliates adhere to the host surface by means of a stalk with adhesive disc (**d**, arrow). te = head tentacles, pyg = pygidium (from Hausmann: Mikrokosmos 69 [1980] 156). Magn.: a 2×, b 170×, c 150×, d 1700×.

Fig. **326** Lucern stalk degraded by the entodiniomorphid ciliate *Epidinium* (**a***); especially tissues with thin cell walls are phagocytosed (**b**) (from Bauchop: Appl. Environ. Microbiol. 37 [1979] 1217. Magn.: a 50×, b 100×.

heat treatment of termites, which did not harm the insects, but killed the protozoa: the termites died soon after, due to starvation. Larval termites are provided with flagellates via coprophagy by ingesting the feces of adults.

Symbiosis of great ecological importance are found in the photoautotrophic unicells living inside heterotrophic protozoa or metazoa. They live inside vacuoles, which are each separated from the cytoplasm by a membrane. The most important examples are the calcium carbonate producing protozoa (e. g., foraminifers) and metazoa (e. g., reef forming corals). In these, the endosymbionts, mainly dinoflagellates (also called zooxanthellae) upset the bicarbonate–carbonate equilibrium by consuming CO_2 for photosynthesis, thus facilitatating the precipitation of calcium carbonate (Fig. **327**). These dinoflagel-

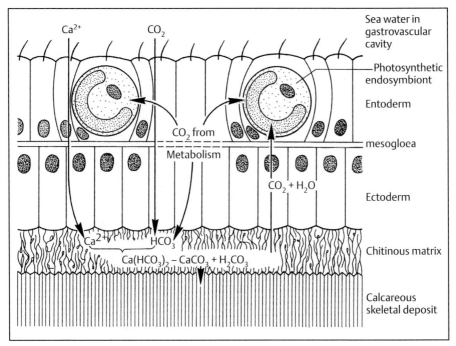

Fig. 327   Scheme of physiological interactions between corals and their endobiotic zooxan-thellae (after Remane et al.).

lates, which also occur in ancanthareans and polycystineans, are able to live autonomously as shown by the fact that the gametes of the host cells are free of symbionts and that the zygote has to internalize new algae by phagocytosis. Besides dinoflagellates, a few species of chryso-monads, chloromonads, prasimonads, hapto-monads, and euglenoids are able to live intracellularly. The benefit for the algae is the supply of CO_2 and minerals and an increased mobility.

Protozoa as Parasites

Protozoan parasites live as endobionts or endocy-tobionts in other protozoa, in metazoa, and in metaphyta. The most important diseases which are caused by protozoa, i. e., sleeping sicknes, leishmaniasis, amoebiasis, coccidiosis, and malaria, have already been discussed. According to our present knowledge, these pathogenic agents began their evolution as free-living organisms. This is indicated by so-called amphizoic species, e. g., the schizopyrenids *Naegleria*

and *Acanthamoeba*, which can transform from free-living bacterivorous amoebae to hazardous human parasites causing diseases such as meninigitis. Moreover, free-living species have been detected which are interpreted a being morphologically and ecologically transitional stages towards parasitism. An example is the protozoon *Colpodella (Spiromonas) gonderi* of uncertain systematic position (Fig. **328**). This organism shows striking similarities with the morphology of sporozoites of certain Apicomplexa. The causative agent of giardosis, *Lamblia intestinalis*, which is phylogenetically close to mitochondrion-free protozoa such as the Microspora, has to be regarded as a very ancient endobiont. An almost perfect parasitism is found in those forms which can be transferred to the next host generation via the ovary or uterus of the host (e. g., *Plasmodium*, *Toxoplasma*).

Several opportunistic protozoan parasites, which take advantage of a therapy-induced, congenital, or acquired deficiency of the immune system, have been known for decades, but are at

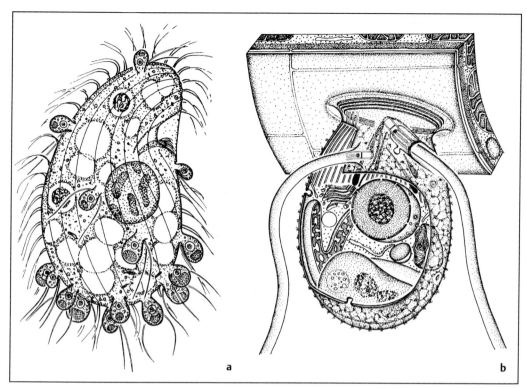

Fig. 328 **a** Several individuals of *Colpodella (Spiromonas) gonderi* during attack of a free-living ciliate *(Colpoda)*. **b** Three-dimensional reconstruction of the morphological organization which demonstrates, by the presence of alveolar pellicula, subpellicular microtubules, circularly arranged apical microtubules, conoid-like structures, micropores and micronemes, a surprising similarity with occidia (from Foissner and Foissner: Protistologica 20 [1984] 635). Magn.: a 750×, b 7000×.

present the main topic of medical discussions in connection with the AIDS problem. Especially the coccidium *Cryptosporidium parvum* causes a life-threatening persistent diarrhea in immunocompromised humans. Besides the protozoa, which have already been mentioned in this context in the systematic chapter, organisms of interest exhibit characteristics of yeast-like ascomycetes combined with morphological features of amoebae. Examples of those chimeras are species known as *Pneumocystis carinii* and *P. jiroveci*, which cause certain types of pneumonia in warm-blooded animals, where they live as amoeboid trophozoites, forming sturdy walled cysts after a meiotic division.

It might appear surprising that even in our times, no effective vaccination exists against protozoan parasites. The main problem is the complex life cycles of these protozoa, which are characterized not only by having different hosts and by changing their location in the host organisms (intra-, extracellular), but also by their ability to rapidly change their surface so that the immune system of the host can no longer recognize them as invaders. This phenomenon is known as antigen variance.

Protozoa as Hosts

As mentioned above, the presence of photosynthetically active endocytobionts is of particular benefit to the heterotrophic protozoan hosts (Figs. **329** and **331**). In almost all groups of free-living heterotrophic protozoa, examples of a facultative or obligatory association with photoautotrophic forms are known. Cyanobacteria (cyanelles) live in the testate amoeba *Paulinella chromatophora*, as well as in *Cyanophora paradoxum*, a flagellate of unknown systematic position. Green algae of the genus *Chlorella* are often present in ciliates and amoeboid protozoa (Figs. **330** and **331**). These associations are frequently relatively loose and often terminated by digestion, when the host is kept in the dark. On the other hand, associations can be observed which are very tight, insofar as at least one part-

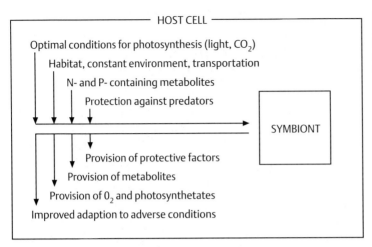

HOST CELL

Optimal conditions for photosynthesis (light, CO_2)

Habitat, constant environment, transportation

N- and P- containing metabolites

Protection against predators

SYMBIONT

Provision of protective factors

Provision of metabolites

Provision of O_2 and photosynthetates

Improved adaption to adverse conditions

Fig. **329**   Diagram of interrelations (output and input) between a heterotrophic host and a photosynthesizing symbiont.

ner cannot survive independently. For instance, the green endosymbionts of *Chlorarachnion*, a plasmodial organism of uncertain systematic position, are free of mitochondria. Certain actinopods and ciliates exist that digest photoautotrophs with the exception of their chloroplast, which for a while continue to photosynthesize for the benefit of the predator.

Of particular interest for discussions of phylogeny is the dinoflagellate *Peridinium balticum*, which harbours a relict of a former free-living euglenoid. Even more interesting are organisms such as the ciliate *Mesodinium rubrum*. The endosymbionts, which are derived from dinoflagellates, are probably unique. The fragments — one of which is the nucleus-containing portion — are located in different vacuoles of the host. As the ciliates have neither an oral apparatus nor a cytoproct, the host is absolutely dependent on its symbiont: an heterotrophic organism metamorphosed into a photoautotroph. Findings such as these are regarded as important for the understanding of eukaryotic evolution, and they have greatly influenced the development of the endosymbiont theory (see Fig. **29**).

In many protozoa, endobiotic prokaryotes are present that do not photosynthesize but are essential to their hosts. For instance, the ciliate *Euplotes aediculatus* needs, for binary fission, a particular bacterium called omikron. It is still obscure what the functional role of this endobiont might be. Bacteria, called kappa-, lambda-, sigma-, pi-, or mu-particles, live in *Paramecium*. The reproduction of the prokaryotes is regulated by the nuclear activity of the host. Vice versa, as demonstrated in the case of the lambda-particle, which provides the host with the essential folic acid, the host in not viable without the endosymbiont.

The properties of the kappa-particles have been studied in detail. These bacteria live in so called killer strains of *Paramecium aurelia*. Killers are strains of *P. aurelia* which are able, by means of their kappa particles, to kill other strains of *P. aurelia*, called "sensitives," which do not possess these prokaryotes. Kappa-particles occur in two modifications, as "brights" and "nonbrights." The brights contain a refractive body (R-body) and are no longer able to divide. The nonbrights do not have an R-body and they divide easily. The R-body is a tightly coiled ribbon which unrolls explosively. It is rather similar in morphology and mode of function to the ejectisomes.

The brights are delivered from the killer strains into the medium. If they become phagocytosed by sensitives, the R-body discharges inside the food vacuole. The vacuolar membrane is disrupted and the digestive enzymes have direct access to the cytoplasm. Simultaneously, virus particles are released during the unrolling of the ribbon. Both processes finally lead to the death of the sensitives. It is suspected that the virus particles are present in the nonbrights in the form of inactive proviruses which initiate — after their transformation into the active forms — the synthesis of the R-bodies, simultaneously preventing the division activity of the bacteria.

Other bacteria, found mainly in flagellates from anaerobic habitats *(Pelomyxa, Mastigella, Tetramitus)*, may take over some functions of the missing mitochondria. The bacteria, living in the cytoplasm of termite flagellates or rumen cili-

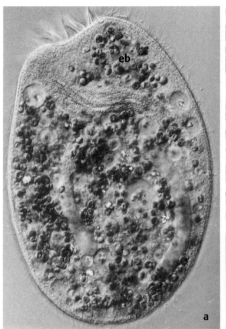
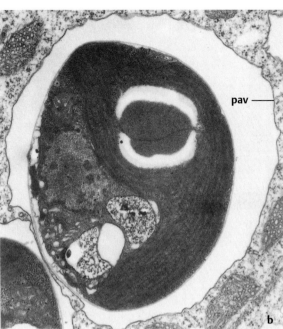

Fig. **330** The ciliate *Climacostomum virens* with endocytobiotic *Chlorella* algae in light microscopical view (**a**) and in ultrathin section (**b**). The endosymbionts (eb) are located in a perialgal vacuole (pav) (from Reisser et al.: Protoplasma 119 [1984] 93). Magn.: a 500×, b 24 000×.

ates, may provide the cellulases for the digestion of the plant material.

The number of free-living protozoa containing intracellular bacteria is much higher than commonly assumed. In most cases, it is unknown whether these prokaryotes play any important role for their hosts. Whatever their function might be, almost every compartment of the cell is known to be invaded by bacteria (Fig. **332**). They can be found freely in the cytoplasm and in the karyoplasm, in different types of vacuoles, in the endoplasmic reticulum, in the nuclear envelope, in alveoli (especially in ciliates), and even, (although seldom), in the inner compartment of mitochondria. So far, they have not yet been detected inside plastids and dictyosomes.

A particular association with bacteria is known for hypermastigid flagellates. Besides intracellular bacteria, extracellular forms are also known which are tightly connected to the

plasma membrane of flagellates, as demonstrated for *Joenia* (Fig. **333**). Rod-like bacteria as well as spirochetes are freely suspended in the fluid of the paunch, but they are also attached to the surface of the flagellate. The metabolic pathways of this symbiotic association and the host insect are integrated (Fig. **334**).

In the trichomonad *Mixotricha paradoxa*, a rather peculiar situation is found. Spirochetes are regularly distributed over the cell surface alongside rod-like bacteria. The movement of these prokaryotes is coordinated by hydrodynamic interactions. Due to the highly viscous surrounding medium, metachronal waves are created, which are to some extent similar to what is observed, for instance, in ciliates. This coordinated movement of the ectobionts results in an enhanced motility of the host cell or, more probably, in a circulation of the surrounding medium and a more effective transport of nutrients.

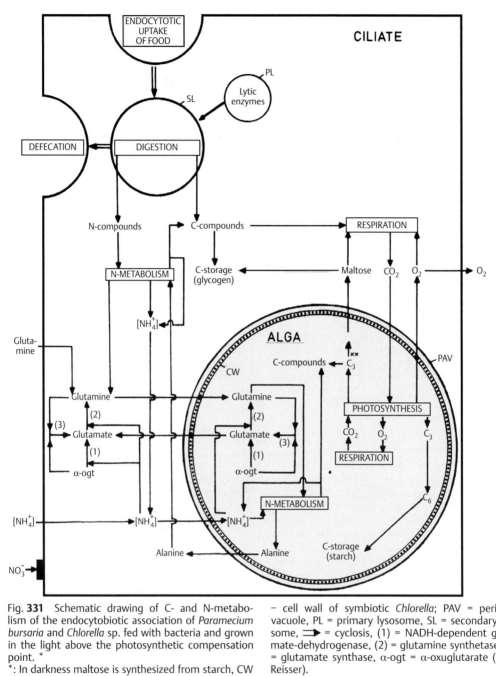

Fig. **331** Schematic drawing of C- and N-metabolism of the endocytobiotic association of *Paramecium bursaria* and *Chlorella* sp. fed with bacteria and grown in the light above the photosynthetic compensation point. *
*: In darkness maltose is synthesized from starch, CW – cell wall of symbiotic *Chlorella*; PAV = perialgal vacuole, PL = primary lysosome, SL = secondary lysosome, ⇒ = cyclosis, (1) = NADH-dependent glutamate-dehydrogenase, (2) = glutamine synthetase, (3) = glutamate synthase, α-ogt = α-oxuglutarate (after Reisser).

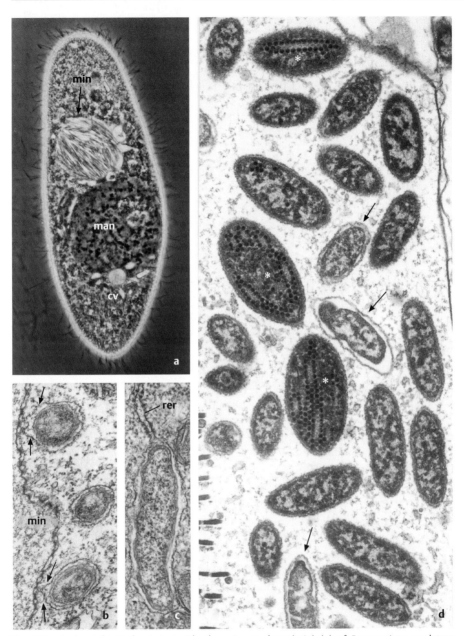

Fig. **332** Endocytobiotic bacteria inside the micronucleus (min) (**a**) of *Paramecium caudatum* (*Holospora elegans*), in the nuclear envelope of the micronucleus of *Paramecium caudatum* (arrows) (**b**), in the rough endoplasmic reticulum of *Entosiphon sulcatum* (rer) (**c**) and free or in vacuoles of *Trichodina pediculus*. Some bacteria are infected with viruses (*) (**d**). cv = contractile vacuole, man = macronucleus (a courtesy of H.-D. Görtz, Stuttgart. Magn: a 450×, b 12 000×, c 13 000×, d 10 000×.

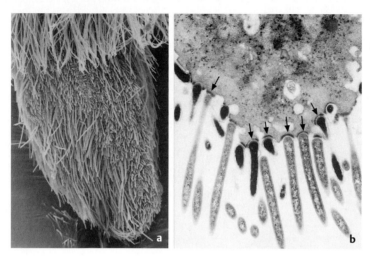

Fig. **333** Ectobiotic bacteria on the surface of the flagellate *Joenia* (**a**), fixed to the host cell membrane by special adhesion structures (arrows) (**b**) (from Radek and Hausmann: Acta Protozool. 31 [1992] 93). Magn.: a 1000×, b 8000×.

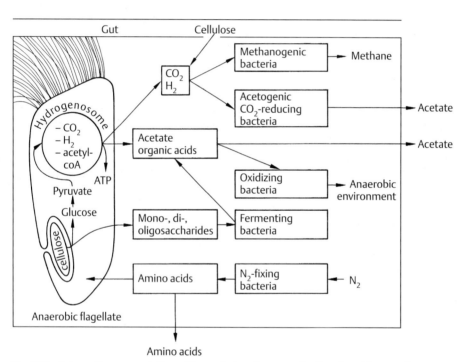

Fig. **334** Schematic representation of interrelations between the flagellated host and its endobionts in the termite hindgut (after Radek).

Biogeography of Protozoa

The majority of the free-living protozoa of temperate climate zones are probably cosmopolitan. However, in some cases it has been shown that certain species of a genus live in certain parts of the world. An an example, the worldwide distribution of 12 *Tetrahymena*-species is depicted in figure **335**. Interestingly, *Tetrahymena australis* and *T. capricornis* occur exclusively in Australia. Similarly the amoebae *Chaos carolinense* and *Ch. illinoisense* have not been found outside of North America. In most cases, however, no reliable data are available because

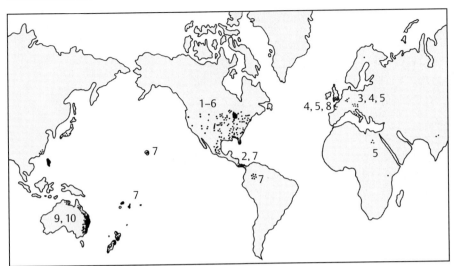

Fig. **335** Worldwide distribution of various biological species of the morphospecies complex *Tet-rahymena pyriformis*. 1 *T. thermophila*, 2 *T. americanis*, 3 *T. borealis*, 4 *T. cosmopolitanis*, 5 *T. pigmentosa*, 6 *T. canadensis* 7 *T. tropicalis*, 8 *T. hyperangularis*, 9 *T. australis*, 10 *T. capricornis* (after Elliot).

this aspect of protozoan ecology has not yet been studied in detail. Some species are apparently very rare, e. g., the naked amoeba *Thecamoeba verrucosa* or the ciliate *Folliculina boltoni* that is described as the only freshwater species of the 300 otherwise marine folliculinids.

Some protozoa live exclusively in tropical regions. This holds true for both freshwater and marine forms. For instance, the large foraminifers can only live in sunny, shallow waters of tropical shores. The area of distribution of parasitic or ectobiotic protozoa depends, however, mainly on the distribution of their vectors and hosts.

The reason for the world-wide distribution of certain protozoan species might be that the same or similar microhabitats are available, and that there is a wide range of mechanisms and possibilities for distribution. Due to their small size, many protozoa are transported over large distances by other organisms, e. g., in small drops of water or in humid material. Like other phototrophic and heterotrophic organisms, they may be transfered also by ships ballast water. Cysts might become widely distributed by the wind.

The increasing capacity for global transport of humans enhances the distribution of human protozoan parasites. Pathogenic agents, as well as their vectors, are transported into countries in which they never lived before or they become re-transported into regions from which they had been eradicated. For instance, around 1930, the mosquito *Anopheles gambia* was brought into Brazil. The consequence was a disastrous epidemic of malaria. Random sampling of air planes that landed on Hawaii between 1964 and 1968, revealed that 373 new insect species — including 65 mosquito species — were brought onto the island. In all probability, transfers such as these are the reason for a remarkable increases in cases of malaria during the last few decades in the USA and Northern European countries.

Man-made changes in the environment can also encourage the pathogenic habit in certain protozoa. A lethal meningitis due to the thermophilic amoeba *Naegleria fowleri* has been detected in temperate climates, caused by the organism living in artificially heated swimming pools or in circulation systems for cooling water.

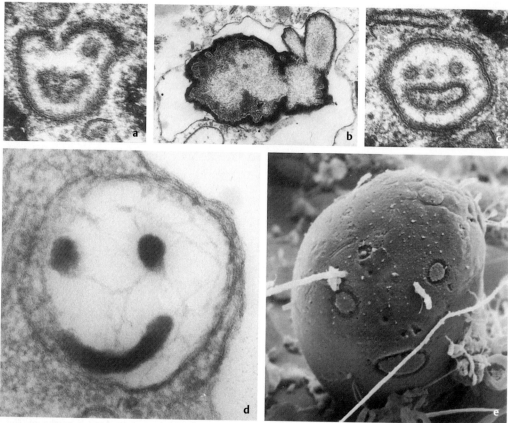

Fig. **336** Under certain circumstances, protozoa show not only metazoan (a, b) and human facets (c, d), but are even able to imprint human characters on organic matter (e): **a** parasomal sac of the ciliate *Pseudomicrothorax dubius*; **b** lithosome of the ciliate *Climacostomum;* **c** parasomal sac of *P. dubius*; **d** cortical vesicle of the heliozoon *Actinophrys sol*; d spore attacked by vampyrellid amoebae (b courtesy of D. Fischer-Defoy, Wiesbaden). Magn.: a 80 000×, b 20 000×, c 80 000×, d 55 000×, e 4000×.

Glossary

A

Aboral: With respect to a region of a cell opposite or away from the oral area.

Acanthopod: Fine-tapering pseudopod of *Acanthamoeba.*

Acrasin: Collective term for chemical attractants such as cAMP, produced by dictyostelid amoebae.

Acronematic flagellum: Trailing flagellum with a terminal mastigoneme or slender filament, as in the chrysomonads.

Actinopodium: Thin → pseudopodium, characteristic of heliozoans, acantharians, phaeodarians, and polycsytineans, stiffened by microtubules.

Activated sludge: Part of one type of the modern sewage treatment process in which sewage is violently aerated, and which supports large populations of ciliates.

Adhesive disc: Cup-shaped organelle at the aboral pole of peritrich ciliates *(e. g. Trichodina)* or diplomonads *(e. g. Giardia)* used for attachment to the surface of the host. Synonym: sucking disc.

Adoral: With respect to a region of the cell close to the oral apparatus.

Adoral zone of membranelles (**AZM**): A band of membranelles found in certain ciliates, consisting of numerous adoral membranelles arranged in parallel series at the left side of the → oral apparatus. Used mostly for feeding, but occasionally for locomotion.

Aflagellate: Flagella-less protozoa or a flagella-less stage in the life cycle. Synonyms: akont, amastigont.

Agamogony: Series of nuclear or cell divisions producing individuals that are neither gametes nor capable of forming gametes.

Agamont: Vegetative stage in the life cycle, characterized by asexual reproduction, such as → schizogony in Apicomplexa.

Agglutinated test: Test with a wall composed entirely of foreign particles, added during construction, being held together by an organic cement.

Aggregation: Stage in the life cycle of dictyostelids, in which hundreds or thousands of individual amoebae move to a common aggregation centre.

Akontobolocyst: Type of → extrusome.

Algivory: Mode of nutrition used by protozoa specialized for feeding more or less exclusively on algae.

Aloricate: Without a lorica.

Alternation of generations: Alternation of both vegetative and sexual stages and of haploid and diploid stages in the life cycle, as in Apicomplexa and foraminifers.

Alternation of hosts: Movement of a parasitic protozoon from the terminal host, in which the sexual stages occur, to an intermediate host, in which vegetative reproduction is performed, and vice versa.

Alveolus: Subplasmalemmal vesicle, sometimes containing plates (Apicomplexa, Dinoflagellata, Ciliophora).

Amastigote: Absence of flagella, or, as in Trypanosomatida, significant reduction of a flagellum to a remnant that is invisible at the light microscope level.

Amphiesmal vesicle: Subplasmalemmal vesicle (→ alveolus) of the dinoflagellates, often containing thecal plates.

Amphitrophy: Capability of gaining energy and nutrients by both → autotrophic and → heterotrophic means. Synonym: mixotrophy.

Amphizoic: The ability of some facultative parasitic protozoa to live both as a parasite and as a free-living bacterivore, as in *Acanthamoeba* or *Naegleria.*

Ampulla: Anterior invagination of the cell, especially in euglenoids.

Amylopectin: Storage polysaccharide of algae composed of α–1,4 glucoside linkages, with α–1,6 linked side chains.

Anabiosis: Restoration of active metabolism and growth from → cryptobiosis.

Anarchic field: Group of noncilioferous (barren) → kinetosomes giving rise to the infraciliary bases of the buccal organelles. Synonym: oral primordium.

Anastomosis: Fusion of pseudopodial projections to a network, e. g., in Labyrinthulea or Granuloreticulosa.

Anisogamy: Fusion of gametes that look identical but differ in overall size.

Anisokont: Equipped with two or more flagella, different in morphology and/or function.

Anlage: First visible, primordial part in the development of a distinct structure, e. g., of the macronucleus or the oral apparatus in ciliates. Synonym: primordium.

Aperture: An opening. Used in relation to tests or loricae to refer to the site of emergence of pseudopodia, flagella, or cells.

Apical complex: Complex organelle system of the Apicomplexa, consisting of conoidal rings, conoid, and polar ring, and serving for attachment to or for penetrating host cell surfaces.

Apochlorotic: Loss of photosynthetic pigments in organisms that once contained them or whose ancestors once contained them.

Argyrome: System of pellicular or cortical argentophilic structures or markings revealed by application of silver impregnation techniques.

Astropyle: Main aperture in the central capsule of Phaeodaria.

Athecate: Without shell or theca.

Aufwuchs: Community of benthic organisms living on the surface of organisms or other substrates.

Autogamy: Self-fertilization.

Autophagy: Digestion of cellular components.

Autotrophy: Trapping of energy from physical (e. g., light) or chemical sources and use of the energy to assemble organic macromolecules from inorganic compounds (especially CO_2).

Auxotrophy: Requirement for one or more organic growth factors or vitamins.

Axenic culture: The cultivation of conspecific organisms in the absence of any living organisms of other species.

Axoneme: Microtubular axis of the flagellum/cilium, normally in the typical 9×2+2 arrangement.

Axoplast: Microtubular organizing center (MTOC) at the axopodial base in actinopods.

Axopodium: Pseudopodial projection with an internal bundle of microtubules. Synonym: → actinopodium.

Axostyle: Axial organelle made of cross-linked microtubules, motile in trichomonads, nonmotile in parabasalians.

AZM: → Adoral zone of membranelles.

B

Bactivory: Feeding on bacteria. Synonym: microphagy.

Barren kinetosome: Basal body not always associated with a cilium. Synonyms: nonciliferous, aciliferous kinetosome.

Basal body: → Kinetosome.

Basal swelling: In some heterokont or euglenoid flagellates, the swelling at the base of the recurrent flagellum which is usually located in opposition to an intraplastidial or extraplastidial → eyespot (→ stigma).

Biflagellate: Equipped with two, mostly → heterokont flagella.

Binucleate: With two nuclei.

Bioluminiscence: Emission of visible light by living organisms such as certain dinoflagellates.

Birth pore: Opening or site of emergence of a larval form during the → budding process of reproduction in some suctorians.

Blepharoplast: Old term for → kinetosome.

Bloom: Dense growth of organisms, usually algae or cyanobacteria, typically short-lived (e. g., several days) and typically of one species.

Bothrosome: In Labyrinthulea, the interface between cytoplasmic content of the filopodial tracks and the innermost surface of the cell-like compartments. Synonyms: sagenogen, sagenogenetosome.

Bradyzoite: In the life cycle of *Sarcocystis, Toxoplasma* and other coccidians, a special merozoite stage with slow development.

Buccal apparatus: → Oral apparatus.

Budding: Especially in sessile ciliates but also in other protozoa, a special, modified binary or multiple kind of reproduction, mostly performed by exogenous gemmation of daughter cells, but also by endogenous formation of daughter cells. Synonym: gemmation.

C

Capsule: Intracellular skeletal or protective elements (central capsules in Polycystinea, polar capsules in Myxozoa).

Carnivory: Mode of nutrition referring to organisms that are heterotrophic and usually predators of nonprokaryotic heterotrophs.

Catenation: In ciliates and dinoflagellates, the formation of a chain of individuals by repeated incomplete fissions without separation. Catenoid colonies are formed e. g., by astomate ciliates.

Celestite: White mineral made of strontium sulfate comprising the spines of certain acantharians.

Central capsule: In Polycystinea and Phaeodarea, the structure lying between endoplasm and ectoplasm.

Centroplast: In centrohelid heliozoans, the single centrally located microtubules organizing center (MTOC) from which all axopods arise.

Chagas' disease: South American plague, caused by infection with *Trypanosoma cruzi*.

Chemoautotrophy: Type of → autotrophy using inorganic chemical sources of energy, e. g., oxidation of H_2S.

Chloroplast endoplasmic reticulum: Additional third membrane surrounding the plastids derived from eukaryotic symbionts. Mostly discernible by the presence of ribosomal aggregates.

Chrysolaminarin: → Leucosin.

Ciliary girdle: In peritrich ciliates, an aboral encircling band of cilia, the → telotroch.

Ciliary necklace: Intramembranous particles in ring-shaped conformation at the base of a cilium.

Ciliature: In ciliates, the whole complex of somatic and oral cilia.

Cilium: Synonym for flagellum, mostly used for ciliates. Whip-like projection with an axonemal complex (9×2+2) and a standard diameter of 0.25 μm. In general, cilia are less than 10 μm long, and are shorter than flagella.

Cingulum: Annular furrow (and the corresponding internal plates) between epicone and hypocone of armoured dinoflagellates.

Cirrus: Composite cluster of individual cilia formed by a → polykinetid, not covered by a common membrane, but acting as a functional unit in locomotion or food collection. Typical character of hypotrich ciliates.

Clathrocyst: → Extrusome.

Clavate cilia: Short, immotile cilia with a deviant axonemal pattern. They are present in the → scopula of certain peritrichs. Synonym: stereocilia.

Clonal culture: Culture of conspecific organisms derived by vegetative multiplication from a single cell.

Clone: Population of genetically identical cells with a single common ancestor, under natural conditions regularly occurring in trypanosomatids.

Cnidocyst: Type of → extrusome in dinoflagellates. Synonym: nematocyst.

Coccolith: Plate-like calcified scales formed in vesicles of the endoplasmic reticulum of the coccolithophorid prymnesiomonads (= haptomonads).

Coelopodium: Hollow → pseudopodium enclosing food, especially in radiolarians.

Collar: Cone-shaped, hollow circle of microvilli at the apex of choanoflagellates.

Conjugation: In ciliates, mating by partial and temporal sexual fusion of gamonts (= gamontogamy). In other cases, a type of spatial relation between gametes or structural complexes.

Conocyst: Type of → extrusome.

Conoid: In Apicomplexa, the cone-shaped structural unit made of helically running microtubules.

Contractile vacuole complex: Peripherally located expulsion vesicle (and the adhering microfilaments and microtubular ribbons) serving for the elimination of water acquired by osmosis.

Cortex: Complex outer layer of cells, in ciliates comprising the → pellicle and → infraciliature.

Costa: In trichomonads, the often contractile flagellar rootlet subtending the undulating membrane.

Costal strips: Siliceous strips which join to form the lorica of certain marine choanoflagellates.

Crenulate: With a regularly indented margin.

Cresta: Fibrillar, noncontractile structure below the basal portion of the trailing flagellum in Parabasalea.

Cruciform mitosis: In Plasmodiophorea, the cross-shaped arrangement of spindle apparatus with chromatids and the passively elongated nucleolus.

Cryptobiosis: Death-like condition brought on by starvation, desiccation, or freezing, reversible by → anabiosis.

Cyclosis: In ciliates, the rotationary streaming of endoplasm.

Cyrtos: In cyrtophorid ciliates, the basket-like microtubular (nematodesmal) apparatus surrounding the oral aperture. Synonym: nasse.

Cyst: Sexual or asexual stage in the life cycle characterized by the secretion of an extracellular envelope and serving for resting, multiplication, and digestion. Mostly formed in response to unfavorable conditions such as starvation or desiccation. As a cryptobiotic (dormant) propagule, responsible for the world-wide distribution of several species.

Cytopharyngeal apparatus (basket): In flagellates and ciliates, the cytopharynx and its associated microtubular or nematodesmal structures (forming the cyrtos or rhabdos in ciliates).

Cytopharynx: In ciliates, the distal part of the food ingestion apparatus; usually a channel formed by microtubules that draws newly formed food vacuoles away from the → cytostome and into the cell. Variations are also present in heterotrophic flagellates.

Cytoproct, Cytopyge: In ciliates, the permanently present slit-like cell anus usually located in the posterior part of the cell.

Cytostome: Cell mouth. Best used in reference to the area of the cell membrane covering the buccal cavity which forms and pinches off food vacuoles.

D

Defecation: Egestion (→ exocytosis) of of nondigestible material.

Detri(to)vory: Detritus eating.

Detritus: Fragments of dead plant and animal material before, during, and after breakdown by agents of decay. May incorporate inorganic matter, e. g., mud.

Deutomerite: In gregarines, the nucleus-containing posterior segment.

Diastole: Dilatation of the contractile vacuole.

Digestion: Degradation of ingested organic nutrients.

Dikinetid: In ciliates, the morphological complex of two neighboring (paired) kinetosomes with all the connected structures (cilia and rootlets).

Dimastigont: Paired kinetosomes.

Dinokaryon, Dinonucleus: In dinoflagellates, the nucleus with its unique features (uncoiled chromosomes, lack of histones and nucleosomes, persisting nuclear membranes).

Dinokont: The flagellation type in dinoflagellates, one transverse flagellum and one longitudinal flagellum.

Diplokaryon: Especially in Microspora, the pair formation of nuclei before → autogamy.

Diplozoic: In diplomonads, the doubling of all cell organelles as the result of an incomplete cell division.

Discobolocyst: Type of → extrusome.

E

Ectoplasm: Peripheral zone of the cytoplasm, in amoeboid cells or myxomycete plasmodia known as the gel tube, more rigid than the innermost endoplasm which with it is interchangeable in the ectoplasm → endoplasm (gel-sol) transformation at the → uroid region or at the end of a retracting pseudopod. In the hydraulic pressure theory, ectoplasmic actomyosins are said to be responsible for generation of the motive force.

Ecto(sym)biont: Organism living on the surface of another organism.

Egestion: → Defecation.

Ejectisome: → Extrusome.

Endo(sym)biont: Intracellular organism living inside a vacuole of the host cell or, in some cases, also in the cytoplasm.

Endocytobiose: Intracellular symbiosis.

Endocytosis: Uptake of fluid or particulate materials by a membrane invagination process (→ pinocytosis, → phagocytosis).

Endodyogeny: In *Sarcocystis* and related Apicomplexa, formation of two daughters cells (→ metrocysts, → merozoites) by → endogenous budding.

Endogenous budding, Endogenous cleavage: Formation of one or more cells inside the parent cell. The daughter cell is always located inside a parental vacuole and can be liberated to the outside by exocytosis (passing through a → birth pore). Common in sessile ciliates (chonotrichs, suctorians), and in Myxozoa and Paramyxea.

Endoplasm: More or less centrally located parts of the cytoplasm. Due to its more fluid character, it is often in motion (→ cyclosis, cytoplasmic streaming). Compare ectoplasm.

Endosome: Vesicle arising by endocytosis; but also a synonym for the nucleolus.

Epibiont: Ecological form of organism association, in which one cell lives on the surface of another (→ aufwuchs).

Epicone: The anterior part or hemisphere of dinoflagellates, separated from the → hypocone by the → cingulum (cingular plates, cingular furrow).

Epimerite: In septate gregarines, the most anterior part projecting from the protomerite and serving to anchor the protozoon to the host cell.

Epiplasm: Granular to filamentous layer beneath the plasmamembrane. In ciliates, a constituent of the → pellicle.

Euglenoid motion: Kind of distinctive cellular movement displayed by certain euglenids and produced by the relative shifting of pellicular protein plates. Synonym: metaboly.

Excystment: Emergence from a cyst by local softening of the cyst wall or by passing through a preformed annulus.

Exocytosis: Export of intravacuolar materials (or cellular formations) by the plasmalemma merging with the vacuolar membrane.

Extrusive organelle, Extrusome: A type of organelle, with its intravacuolar and dischargeable content, derived from typical dictyosomes and therefore present only in metakaryotic protozoans. The term comprises cnidocysts, discobolocysts, ejectisomes, kinetocysts, mucocysts, nematocysts, polar capsules, taeniocysts, toxicysts, and trichocysts.

Eyespot: Intra- or extraplastidial pigmented area in a number of flagellates, sometimes having a special spatial relationship with the basal swellings of the recurrent flagellum, as in chrysomonads. Synonym: stigma.

F

Falx: In opalinids, a special region of the cortex from which kineties arise.

Feeding veil: A delicate protoplasmic cone protruding from the upper sulcal region and acting as a means of prey entrapment in some dinoflagellates. Synonym: pallium.

Filopodium: Filiform pseudopodium with a central axis of 7 nm filaments (possibly actin). Commonly found in several groups of amoeboid organized cells.

Filter feeding: Type of feeding in which suspended particles are concentrated and ingested. Needs cilia or flagella to direct a current of water to the cell and a filter device to concentrate the particles before enclosure within a food vacuole.

Flagellum: → Cilium. In contrast to cilia, the flagella are normally longer than 10 μm and can have → mastigonemes.

Flagellar pocket: → Reservoir.

Flagellar hairs: → Mastigoneme. Synonym: flimmer.

Flagellar rootlet: Cross-banded fibrillar structures extending from the kinetosome into the cytoplasm of a number of flagellates. In contrast to ciliary rootlets, they do not comprise microtubules.

Flagellar swelling: Thickening near the flagellar base of euglenoids.

Flimmer: → Mastigoneme.

Frustule: Outer siliceous shell of a diatom.

Funis: Ribbon of microtubules facilitating the parallel alignment of the recurrent flagellum with the posterior end of the cell in diplomonads.

Fusule: Aperture in the central capsule of polycystineans through which the axopodial microtubular bundles pass.

G

Gamete: Sexually active cell capable of uniting with a compatible cell to form a zygote.

Gametogamy: Fusion of → gametes.

Gamogony: The sexual phase in the life cycle of Apicomplexa and Foraminiferea.

Gamone: Soluble mating type substance active in inducing conjugation.

Gamont: Male, female, or hermaphrodite cell producing haploid reproductive units (gametes or nuclei).

Gamontogamy: Mating of gamonts.

Gemmation: → Budding.

Germling: Offspring of a spore or cyst.

Girdle lamella: In chrysomonads, the circular band of thylakoids beneath the inner plastidial membrane.

Glycosome: Organelle in kinetoplastids which contains enzymes of the glycolytic metabolic pathway.

Glyoxysome: Microbody harboring the enzymes of glyoxylate metabolism.

Granellae: Crystals of barium sulfate found in large numbers in the cytoplasm of xenophyophoreans.

Granuloreticulopodium: Pseudopodium with internal cables of microtubules and with bidirectional movement of particles. They are capable of anastomosing with each other to form pseudopodial networks. Pseudopodium type of Granuloreticulosa and other rhizopods.

Gullet: Buccal cavity at the apex of cryptomonads from where the two flagella arise. Possible remnant of an ancrestral oral apparatus.

H

Haplosporosome: Membrane-bound organelle with internal membranous constituents, in haplosporidians and possibly paramyxeans.

Haptocyst: Type of → extrusome.

Haptonema: Autapomorphic organelle of prymnesiomonads. Formed as microtubular appendage with internal cisterna of the endoplasmic reticulum and serving as a holdfast, or for crawling or food uptake.

Herbivory: Eating of algae.

Heterodynamic flagella: Flagella that have independent patterns of beating; not correlated with each other.

Heterokaryotic: With nuclear dimorphisms.

Heterokont: Biflagellated, with one smooth flagellum and the other covered with → mastigonemes.

Heterotrich: Equipped with different kinds of cilia, e. g., somatic monokinetids and oral membranelles, as in heterotrich ciliates.

Heterotrophy: Mode of nutrition in which the consumer relies for energy and nutrients upon organic compounds created by other organisms (e. g., algae).

Histophagy: Kind of heterotrophy in which tissues of prey or host organisms are ingested.

Holdfast organelle: Attachment structure by which cells are fixed to a surface.

Homodynamic flagella: Flagella of the same cell, with the same pattern of beating.

Homokaryotic: The presence of only one type of identical nuclei.

Hyaloplasm: Light microscopically organelle-free zone of cytoplasm, as in lamellipodia or veils.

Hydrogenosome: Membrane-bound organelle in anaerobic protozoans such as parabasalians and rumen ciliates. Site of generation of H_2.

Hypocone: Posterior hemisphere of dinoflagellates; → epicone.

I

Idiosome: Structure produced by the organism. Refers to the elements that comprise (or adhere to) the tests of some amoebae.

Infraciliature: Assemblage of all kinetosomes and their associated subpellicular structures.

Ingestion apparatus: → Oral apparatus.

Inner membrane complex: Flattened vesicular complex underneath the cell membrane of Apicomplexa (alveolar structure).

Interkinetal fission: Fission in a plane parallel to the ciliary or flagellar rows (kineties), e. g., in Opalinida.

Isogamy: Fusion of gametes which look identical.

Isokont: With flagella identical in size and function.

K

Kappa-particles: Bacterial endobionts in *Paramecium.*

kDNA: DNA of the → kinetoplast, visible by Feulgen reaction, organized in macro- and minicircles.

Karyomastigont system: → Mastigont system which incorporates the nucleus by additional microtubules or microtubular rootlets.

Kinetid: Basal apparatus of cilia and flagella, consisting of a least one kinetosome with all associated rootlets (monokinetid), but also of two (dikinetid) or more kinetosomes (polykinetid). Somatic and oral kinetids can be separated, depending on their location.

Kinetocyst: Type of → extrusome.

Kinetodesmal fiber: Cross-striated rootlet fibril, as one part of the ciliate kinetid, originating from a kinetosome to the right side and running mostly in the anterior direction.

Kinetodesma: Bundle of overlapping kinetodesmal fibers.

Kinetoplast: In Kinetoplastida, a modified region of the single mitochondrion containing kDNA. Located mostly nearby the two kinetosomes and in the past often confused with these.

Kinetosome: Basal body. Basal termination of cilia and flagella. Cylindrical organelle, 0.25 μm in diameter and 1.5–4 μm in length, consisting of 9 peripheral microtubular triplets in the 9×3+0 arrangement. The two inner microtubules (A-, B-tubule) extend into the flagellar shaft when present. Serving as MTOC also for cortical microtubular rootlets.

Kinety: Mostly longitudinally running row of somatic kinetids (→ mono-, → dikinetids).

L

Lamellipodium: Flat lobate pseudopodial process of amoeboid cells.

Lamina corticalis: Dense fibrillar or filamentous layer beneath the pellicle of various ciliates, marking the ecto–endoplasmic boundary.

Leishmaniasis: Infection by *Leishmania* species causing destruction of macrophages of vertebrates.

Leucosin: Reserve polysaccharide composed of β-1,4-linked polymers of glucose. Synonym: chrysolaminarin.

Lieberkühn's organelle: Lenticular, refractile structure invariably and exclusively found beneath the pellicle close to the left side of the buccal cavity of ophryoglenid ciliates.

Lithosome: Membrane-bound cytoplasmic inclusion, found in a number of ciliates and acantharians, comprised of some organic and inorganic material laid down in concentric layers.

Lobopodium: Lobate or finger-shaped pseudopodium of amoeboid cells.

Longitudinal fission: Fission occurring between the kineties, or interkinetal, mostly longitudinal.

Lorica: Extracellular envelope or protective covering, usually formed as a test, envelope, case, basket, or shell, usually open at one end.

M

Macrogamete: Large female gamete.

Macronucleus: Somatic nucleus in protozoa with nuclear dimorphism (Ciliophora, Foraminifera).

Macrostome: In *Tetrahymena* and other ciliates, cells with an enlarged → oral apparatus.

Mastigoneme: Hair-like projection of the flagellar shaft, often detailed or tripartite. Synonym: flimmer.

Mastigont (system): In flagellated taxa, the kinetid with associated microtubular organelles such as basal bodies or axostyles. When the nucleus is part of the system, it is called → karyomastigont.

Mating type: In sexual processes, a strain of conspecific individuals not able to undergo fusion with each other but only with members of a complementary mating type. Complementary mating types form a biospecies (→ syngen).

Maupas' body: Highly birefringent vacuolar structure in cryptomonads, possibly representing autophagosomes.

Membranelle: Compound structure comprised of numerous cilia, and associated with the oral apparatus of a ciliate.

Merogony: Stage of life cycle of Apicomplexa with multiple fissions (→ schizogonies) leading to the formation of meronts.

Meront: → Merogony.

Merozoite: Multiple fission product of meronts. In the life cycle, merozoites are precursors of gamonts.

Mesokaryon: Older designation of the dinokaryon of most dinoflagellates reflecting their putative position between prokaryotes and metakaryotes. In the modern view, seen as an autapomorphic, not a primitive character.

Metaboly: Euglenoid motion.

Metachronal waves, **Metachrony:** Synchronous waves of the flagellar or ciliary beat, connected with phase shifts between the corresponding kineties. Not autochthonous, but the results of hydrodynamic coupling.

Metrocyte: Large, round proliferative cell found in the last-generation meronts of the apicomplexan genera *Sarcocystis* and *Frenkelia*.

Microgamete: Small, male gamete.

Micronemes: Strongly branched tubular system in the apical complex of Apicomplexa, possibly connected with the → rhoptries and involved in penetration of the host cell.

Micronucleus: The generative mucleus in protozoa with nuclear dimorphism (e. g., Ciliophora).

Microphagous: Denoting ingestion of particulate matter in the pico- to nanosize range (0.2–20 μm).

Micropore: Lateral invagination of the cell membrane in sporozoites of gregarines and coccidians, and the place of endocytosis.

Microstome: In *Tetrahymena* and relatives, a small cell with an inconspicuous small oral apparatus.

Mixotrophy: Ability to live phototrophically or heterotrophically or by both simultaneously.

Monad or **Monadoid:** The flagellated stage in the life cycle of protozoa and algae.

Monokinetid: In ciliates, the derived formation of unpaired individual kinetids.

Monopodial: With only a single pseudopodium.

Morphogenesis: Morphological transformation accompanying growth and differentiation in ciliates, with maintenance of the patterned array of cytoarchitectural substructures.

MTOC: Microtubules organizing centers such as kinetosomes, centrioles, or other structurally less organized zones from which microtubules arise.

Mucocyst: Type of → extrusome.

Müller's body: Vacuole containing barium concretions, described from various karyorelictid ciliates.

Myonemes: In ciliates, intracellular filamentous ribbons, sometimes with contractile properties. Derived from kinetosomes and part of the cortex.

Myophrisks: Myonemens at the spiculae of Acantharea.

Myxameba: Amoeba of the single-cell stage of acellular slime molds (Myxogastra).

N

Nasse: → Cyrtos.

Nemadesmata, Nematodesmata: Rod-like microtubular bundles arising from the distal side of neighboring kinetosomes and cross linked, often in a hexagonal pattern.

Nematocyst: Nonhomologous extrusome type in Dinoflagellata and Myxozoa characterized by the presence of a coiled, extrusible tube. Synonym: cnidocyst.

Nuclear dualism: Simultaneous presence of both somatic (macro)nuclei controlling the vegetative cellular functions and generative (micro)nuclei responsible for horizontal and vertical gene transfer. They occur permanently in Ciliophora and transitionally during the developmental cycle of distinct foraminifers.

Nucleomorph: In cryptomonads, the residual nucleus of an endosymbiotic eukaryotic partner.

O

Ocellus: In some dinoflagellates, a modified region of the chloroplast resembling in ultrastructure a light-perceiving organelle.

Omnivory: Eating animal-like food as well as plant-like food.

Oocyst: In the life cycle of Apicomplexa, the outer cystic envelope containing the oocyte and, later, also sporocysts.

Oogamy: Sexual dimorphism, especially the fertilization of a nonmotile female macrogamete by a motile male microgamete.

Ookinete: Motile zygote of haemosporidian Apicomplexa.

Opisthe: The posterior half of a dividing ciliate, normally receiving the cytoproct.

Oral apparatus: In flagellates and ciliates, the whole complex of → kinetids and → infraciliature forming the oral aperture. It normally comprises the buccal cavity, filamentous systems, → cytopharynx and → cytostome.

Oral kinetid: → Kinetids forming the ciliature and → infraciliature of the oral apparatus.

Oral primordium: → Anarchic field.

Oral region: The anterior (or ventral) region of the cell where the oral apparatus is located.

Oral trapez: Organelle found in some ciliates, used to break open the cell wall of the prey and to ingest its content.

Osmotrophy: Nutrition in which soluble compounds are taken up by the organisms, either by pinocytosis or by transmembrane transport.

P

Pallium: → Feeding veil.

Palmella stage: Resting state of nonflagellated, phototrophic unicells embedded in an amorphous gelatinous matrix.

Panmixis: Random mating.

Pansporoblast: In the life cycle of Myxozoa, the initial multinucleate stage from which the differentiation of spores starts.

Parabasal apparatus: In Parabasalea, rod-shaped or tubular dictyosomal aggregates, adhering to striated parabasal filaments, arising in the vicinity of the flagellar kinetosomes and running posteriorly through the cell.

Paraflagellar rod: → Paraxial rod.

Paramylon: Starch-like polysaccharide made of β-1,3-glucose serving as a nutritional storage product in Euglenoida.

Parasitophorous vacuole: Vacuole harboring an intracullular parasite.

Parasomal sac: In Ciliophora, an invagination of the plasmalemma immediately beside a cilium, possibly homologous with the → micropore of Apicomplexa. Like them, they run between the → alveoli into the cytoplasm and function as sites of endocytosis and possibly also exocytosis.

Paraxial rod: Filamentous bundle running parallel to the axoneme of the flagellar shaft of Kinetoplastida and Euglenozoa, causing the flagellum to appear relatively thick.

Paroral: Near the oral apparatus.

Paroral membrane: Single or multiple row of cilia (membranelle) running along the right side of the peristome of ciliates.

Pedicel: Type of → holdfast organelle.

Pedogamy: Sexual fusion of daughter nuclei deriving from meiotic divisions of the same nucleus. This type of autogamy is common in some heliozoa.

Peduncle: Cytoplasmic extension used as a stalk or means of affixing an organism.

Pellicle: Functional-morphological unit formed by the plasmalemma plus the underlying filamentous structures, → alveoli, and → epiplasm.

Pelta: In trichomonads and some other Parabasalia, the crescent-shaped microtubular structure associated with the anterior protion of the → axostyle.

Peniculus: Membranelle type of the left oral ciliature of ciliates such as *Paramecium*.

Perialgal vacuole: Vacuole enclosing an algal symbiont.

Pericyte: A cell surrounding another cell. Brought about by endogenous budding (e. g., in Myxozoa) or self-phagocytosis (as in Ascetospora).

Perilemma: Extracellular trilamellar layer of organic materials on the surface of the plasmalemma of some ciliates (e. g., oligotrichs, hypotrichs).

Periplast: In cryptomonads, intracellular layer of proteinaceous plates forming a more or less rigid skeleton.

Peristome: Special term for the buccal cavity in some ciliates.

Perkinetal fission: Fission plane perpendicular to the orientation of kineties, as found in ciliates.

Phaeodium: In Phaeodaria, a pigment-rich deposit around the astropyle, possibly involved in silica metabolism.

Phagocytosis: Ingestion of particulate matter.

Phoresy: Organism deriving benefit from a transporting host.

Phoront: In the life cycle especially of hypostome and apostome ciliates, a stage living on the surface of a host.

Photoautotrophy: → Autotrophy dependent on light energy.

Photoauxotrophy: Photoautotrophy, but requiring vitamins.

Phyllae: Ribbons of microtubules, in the oral region of some ciliates.

Pinocytosis: Process of ingesting fluids or mucus.

Plasmodium: Multinucleate cell. In contrast to a syncytium, the multinucleation is brought about by a number of karyokineses which are not followed by cytokinesis.

Plasmotomy: Type of asexual (multiple) fission of multinucleated protozoa, not necessarily combined with karyokinesis.

Plastid endoplasmic reticulum: → Chloroplast endoplasmic reticulum.

Pleuronematic flagellum: Flagellum with one or more rows of → mastigonemes.

Polar filament: In Myxozoa, extrusible filamentous structure used for anchoring the spore to the host cell surface. Compare with → polar tube.

Polar ring: In Apicomplexa, ring-like structure at the apical region of → sporozoites and → merozoites.

Polar tube: In Microspora, tubular structure formed by extrusion, through which the infectious amoebula is injected into the host cell.

Polaroplast: Lamellated structural complex in the spores of Microspora, involved in the formation of the polar filament.

Polyenergid: Equipped with several identical (homokaryotic) nuclei or corresponding amounts of DNA.

Polykinetid: Group of kinetids forming a cirrus or a membranelle.

Postciliary (microtubular) ribbons: In the cortex of ciliates, microtubular ribbons originating from the → kinetosome and running to the posterior part of the cell.

Postciliodesmata: Postciliary ribbons composed of stacks of microtubules.

Posterosome: Vacuolar component in the posterior half of spores of Microspora, possibly involved in the extrusion mechanism of the polar filament.

Primite: Anterior gamont of a → syzygial tandem association of gregarines, posterior gamont: → satellite.

Primordium: → Anlage.

Proboscis: Elephant trunk-like extension of the anterior end of certain flagellates and ciliates.

Promastigote: In trypanosomatids, cells with an anteriorly directed flagellum.

Protomerite: In gregarines (Apicomplexa), the anucleate first body segment bearing the epimerite as an appendix and the deutomerite as a nucleated compartment.

Pseudopodium: Transient cell projection serving for locomotion and/or food uptake. In sarcodinal organized cells, different types occur: → axopodium, → filopodium, → lamellipodium, → lobopodium, etc.

Pseudostome: Opening of the shell or → lorica of testate amoebae through which pseudopodia are extended.

Pusule: In both freshwater and marine dinoflagellates, a vacuolar system underneath the plasma membrane, including invaginations of the plasmalemma, possibly involved in osmoregulation.

Pyrenoid: Proteinaceous center for starch accumulation lying inside some types of chloroplasts.

Q

Quadrulus: Buccal ciliary organelle with an infraciliary base four → kinetosomes in width and many in length, known especially for *Paramecium*.

R

Radiolarite: Microcrystalline quartz in fossil formations; product of the skeletons of radiolarians.

R-body: Virus-like body in the → kappa-particles of *Paramecium* and related genera.

Recurrent flagellum: Flagellum that curves posteriorly from its site of insertion, to trail along and/or behind the body.

Red tide: Mass development of dinoflagellates or other protozoa, leading to a reddish coloration in marine waters.

Replication (reorganization) band: Lightly staining cross band of a macronucleus which migrates either from the midpoint to the ends, or ends up at the center of the nucleus. The area concerned is involved in DNA replication and histone synthesis.

Reservoir: Anterior invagination or → vestibulum of Euglenida. Not necessarily involved in phagocytosis, as with → gullet or the → cytopharyngeal apparatus. Synonym: flagellar pocket.

Resting cyst: Multilayered cyst in the life cycle of several protozoa containing condensed cytoplasm. Serves to withstand periods of frost and desiccation.

Reticulopodium: Descriptive term used for a net-like pseudopodium with anastomoses of → filopodia.

Rhabdos: In ciliates, tubular cytopharyngeal ingestion apparatus made of → nematodesmata and → transverse microtubules.

Rhizoplast: Cross-banded structure extending from the kinetosome to the nucleus or other cell components.

Rhizopodium: Fine pseudopodium with a branching or reticulate pattern.

Rhoptry: In → sporozoites and → merozoites of Apicomplexa, two (or up to eight) bottle-shaped organelles arising at the apex and running to the center of the cell. Possibly containing lytic enzymes used for the penetration of the host cell membrane.

Rohr: In Plasmodiophora, the tubular structure of the appressorial apparatus.

Rootlet: Portion of the → kinetid, deriving from the → kinetosome.

Rosette: In apostome ciliates, an ultrastructurally complex septate structure of unknown function.

Rostellum: Small → rostrum.

Rostrum: Apical end of a protozoan cell which has the appearance of a beak or shows a distinctive protuberance of some kind.

S

Sagenogen, Sagenogenetosome: → Bothrosome.

Sapropel: Sites that are very rich in decaying organic matter and that lack oxygen. Usually it refers to sediments.

Satellite: The posterior gamont in a → syzygial tandem association of gregarines: anterior cell: → primite.

Scale: Platelets of inorganic or organic materials deposited on the plasmalemma of the cell body and sometimes also on flagella.

Schizogony: Asexual multiple fission stage in the life cycle of Apicomplexa (and other protozoa).

Schizont: Multinucleate cell as precursor of merozoites or sporozoites.

Scintillons: Luminescent particles inside luminescing dinoflagellates.

Sclerotium: Desiccation-resistant resting stage of acellular slime molds (Myxogastra) formed in response to dehydration.

Scopula: Organelle complex from which stalks, → holdfast organelles, and → sucking disks of peritrich ciliates arise.

Shuttle streaming: Type of protoplasmic streaming in Myxogastra, characterized by the rhythmic alternation of streaming direction (time interval about 2 min.).

Siphon: Tubular ingestion apparatus, e. g., in the euglenoid *Entosiphon*.

Somatic ciliature: In contrast to the oral ciliature, the ciliature of the cell body used for locomotory purposes.

Somatonemes: External tubular hairs adhering to the cell surface of the proteromonads; homologous with the → mastigonemes of Heterokonta.

Sorocarp: Designation for the nonhomologous fruiting body (stalk with several terminal cysts) formed in the life cycles of the ciliate *Sorogena* or dictyostelids.

Spasmoneme: Special kind of → myoneme found in the stalk of peritrich ciliates.

Spicules: Delicate, pointed structures lying external to the body: like → spines, but invariably excreted.

Spindle trichocyst: Type of → extrusome.

Spines: Pointed structure; either part of the cell or a structure secreted by the cell.

Spongiome: System of secretory tubules in the vicinity of the → contractile vacuole complex.

Spore: Mostly, desiccation-resistante propagule and infection-inducing stage in the life cycle of several parasitic protozoa.

Sporoblasts: Precursor cell (mostly a zygote) which forms spores.

Sporocyst: Cyst formed inside the → oocyst of Apicomplexa and containing sporozoites.

Sporogony: In the life cycle of Apicomplexa, the mitotic fission stage leading to the formation of spores.

Sporoplasm: Infective amoeboid content of the spore (Myxozoa, Paramyxea) or the amoebula of Microspora.

Sporozoite: In Apicomplexa, the motile infective agent released from the spore.

Stachel: In Plasmodiophorea, a spine-shaped organelle used for penetration of host cell wall.

Stalk: Type of → holdfast organelle.

Stephanocont: Having a ring or crown of flagella.

Stereocilia: →Clavate cilia.

Stigma: Extra- or intraplastidial accumulation of reddish carotene lipid globuli with putative function in light perception. Synonym: eyespot.

Stomatocyst: Type of cyst with a siliceous wall and a single plugged opening.

Stomatogenesis: Formation of the new oral apparatus in dividing ciliates.

Subkinetal microtubules: One or more ribbons of microtubules running beneath the proximal ends of → kinetosomes in certain ciliates.

Sucking disc: → Adhesive disc.

Sulcus: Longitudinal groove of the posterior half of armored dinoflagellates in which the proximal part of the posterior flagellum runs.

Suture: Furrow or seam between adjacent cell domains. In ciliates, a small line without ordered kineties.

Swarmer: Flagellated or nonflagellated motile stage of an otherwise stationary organism.

Symphoriont: Cell adhering to the surface of an animal host and therefore transported passively.

Syncilium: Group of closely packed cilia forming a special tuft; characteristic for entodiniomorphids.

Syncytium: Multinucleate and mostly large or giant cell. In contrast to → plasmodium, the multinucleation is brought about by the fusion of formerly individual cells.

Syngamy: → Gametogamy.

Syngen: Complex of two or more sexually compatible mating types, e. g., in *Paramecium* or *Tetrahymena*.

Systole: Contraction of the → contractile vacuole.

Syzygy: In gregarines, the lateral or catenate presexual association of male and female gamonts.

T

Tachyzoite: In Toxoplasma or related coccidia, guickly developing merozoite.

Taeniobolocyst: Type of → extrusome.

Tectin: Complex of protein and mucopolysaccharides occuring in some tests.

Telotroch: Circular girdle of cilia in otherwise naked regions of peritrich ciliates, formed only in swarming cells.

Tentacle: Long cell projection, mostly stiffened by microtubular bundles (nematodesmata in suctorians). Also known from *Noctiluca* as a nonhomologous structure.

Test: General term for shell or lorica. Used especially for the covering of testate amoebae.

Theca: General designation for all types of extracellular coverings such as shells, tests, valves, frustules, etc.

Theront: Stage in the life cycle of certain ciliates, in which the organism typically does not feed but moves quickly (hunter).

Tinsel: → Mastigoneme.

Tomite: Free-swimming, but nonfeeding stage in the life cycle of histophagous ciliates (hypostome and apostome Ciliophora), fission product of a → tomont.

Tomont: Dividing stage of hypostome and apostome ciliates.

Toxicyst: Type of → extrusome.

Transition zone: Region between the kinetosome and the flagellar base.

Transitional helix: In Heterokonta, a helical structure (spiral body) at the → transition zone of the flagellar shaft, possibly made of ribonuclear proteins.

Transverse fission: → Perikinetal fission.

Transverse microtubules: In ciliates, ribbon of 3–5 microtubules arising tangenitally from a kinetosome at its left anterior side. Longer extensions are involved in the formation of some cytopharyngeal structures called → rhabdos.

Trichite: → Nematodesmos.

Trichocyst: Type of → extrusome. Synonym: akontobolocyst.

Trophont: Adult trophic and growing stage in the life cycle of some apostome and hypostome ciliates from wich → tomonts succeed.

Trophozoite: General designation of cells in a developmental stage characterized principally by food uptake.

Trypomastigote: Developmental stage in the life cycle of trypanosomatids characterized by the formation of an undulating membrane.

U

Undulating membrane: In ciliates, a paroral membrane made of inidividual cilia; in flagellated cells, a velum-like extension between the flagellum and lateral cell surface.

Undulipodium: Bütschli's term for a putative common ancestor of pseudopodia and flagella.

Ur-: German prefix indicating ancestral nature as in Ur-Thiere = protozoa.

Uroid: The mostly folded or wrinkled posterior part of a moving amoeboid cell, the site of permanent pinocytosis.

V

Valve: Greater part or one half of a shell or test. In Myxozoa, the outer covering (shell valves) of the mature spore.

Valvogenic cells: In Myxozoa, the cells of the → pansporoblast producing the shell → valves of the mature spores.

Velum: Lamellipodial flat cellular projection, especially in heterotrophic dinoflagellates, serving for locomotion and food uptake.

Vestibulum: Invagination (buccal cavity) of the ventral or apical cell surface, in some ciliates leading to the → cytostome-cytopharyngeal complex.

W

Watchglass organelle: → Lieberkühn's organelle in *Ophryoglena* and related ciliates.

X

Xenosomes: Foreign bodies such as sand grains or diatom frustules used for the construction of tests or loricae (mostly in foraminifers, testate amoebae and some heterotrich ciliates). Also collective term for endosymbionts such as zoochlorellae, zooxanthellae, or cyanellae etc.

Z

Zoochlorellae: Members of the genus *Chlorella* or other green algae living as intracellular symbionts in several protozoa and metazoa.

Zoogloea: Slime produced by bacteria.

Zoospore: Flagellated swarmer of otherwise immotile organisms produced by an asexual multiplication.

Zooxanthellae: Term used mainly for dinoflagellates (named after the genus *Zooxanthella*) but also some chrysomonad or cryptomonad genera living as intracellular symbionts in a number of heterotrophic organisms (Acantharea, Polycystinea, Foraminiferea, Cnidaria etc.).

Bibliography

Protozoological Journals and Periodicals

Acta Protozoologica, founded 1963, published by the Nencki Institute of Experimental Biology, Warszawa, Poland

Archiv für Protistenkunde, founded 1901, published by Gustav Fischer Verlag, Jena, Germany

Journal of Protozoology, founded 1947, 1993 renamed Journal of Eukaryotic Microbiology, published by the Society of Protozoologists, printed by Allen Press, Lawrence, USA

Protistologica, founded 1965, published by CNRS, Paris, France 1987 renamed European Journal of Protistology, published by Gustav Fischer Verlag, Stuttgart, Germany

Progress in Protistology, founded in 1986, edited by J. O. Corliss, Maryland, and D. J. Patterson, Bristol/Sydney, published by Biopress, Bristol, England

Progress in Protozoology, Proceedings of the International Congress of Protozoology

I: Prague, Czechoslovakia, 1961, Ludvík, J., J. Lom, J. Vávra (eds), Czechoslovak Academy of Sciences, Prague 1963

II: London, England, 1965, Neal, R. A. (ed), Excerpta Medica Foundation, Amsterdam 1965

III: Leningrad, USSR, 1969, Strelkov, A. A., K. M. Sukhanova, I. B. Raikov (eds), Nauka, Leningrad 1969

IV: Clermont-Ferrand, France, 1973, Puytorac, J. de, J. Grain (eds), Paul Couty, Clermont-Ferrand 1974

V: New York, USA, 1977, Hutner, S. H., L. K. Bleyman (eds), Pace University, New York 1978

VI: Warszawa, Poland 1981, Dryl, S., S. L. Kazubski, L. Kuznicki, J. Ploszaj (eds), Acta Protozoologica, Special Congress Volume, part I + II, Warszawa 1982 + 1984

VII: Nairobi, Kenya, 1985, Cardiner, P. R., L. H. Otieno (eds), Insect Sci. Applic., Special Congress Volume, Oxford 1986

VIII: Tsukuba, Japan, 1989, Nozawa, Y. (ed), Zoological Science, Supplement, Daigaku Letterpress, Hiroshima 1990

IX: Berlin, Germany, 1993, Hausmann, K., N. Hülsmann (eds), Gustav Fischer Verlag, Stuttgart 1994

History

Bradbury, S.: The microscope. Past and present. Pergamon, Oxford 1968

Corliss, J. O.: A salute to fifty-four great microscopists of the past: a pictorial footnote to the history of protozoology. Trans Amer. Microsc. Soc. 97 (1978) 419; 98 (1979) 26

Corliss, J. O.: Historically important events, discoveries, and works in protozoology from the mid-17th to the mid-20th century. Rev. Soc. Mex. Hist. Nat. 42 (1991) 45

Dobell, C.: Antony van Leeuwenhoek and his "little animals". Bale, Sons & Danielsson, London 1932

Ford, B. J.: The Leuwenhoek legacy. Biopress, Bristol 1991

General Textbooks

Anderson, O. R.: Comparative protozoology. Ecology, physiology, life history. Springer Verlag, Berlin 1987

Bütschli, O.: Protozoa. In: Bronn, H. G. (Hrsg.): Klassen und Ordnungen des Tierreichs. Winter, Heidelberg 1880–1889

Chen, T. T.: Research in protozoology, vol. I–IV. Pergamon, New York 1967–1972

Cooms, G. H., M. J. North (eds): Bichemical protozoology. Taylor & Francis, New York 1991

Doflein, F., E. Reichenow: Lehrbuch der Protozoenkunde, 6. Aufl., Gustav Fischer Verlag, Stuttgart, 1949–1953

Dogiel, V. A. (revised by J. I. Poljanski, E. M. Chejsin): General protozoology, 2nd ed. Claredon, Oxford 1965

Farmer, J. N.: The protozoa. Introduction to protozoology. Mosby, St. Louis 1980

Grassé, P.-P.: Protozoaires. In: Traité de Zoologie, vol. I, 1+2, and II, 1+2. Masson, Paris 1952/1953, 1984, 1994

Grell, K. G.: Protozoology, 3rd ed. Springer Verlag, Berlin 1973

Harrison, F. W., J. O. Corliss (eds): Protozoa. In: F. W. Harrison (ed): Microscopic anatomy of invertebrates, vol. 1. Wiley-Liss, New York 1991

Hyman, L. H.: The invertebrates, vol. 1: Protozoa through Ctenophora. McGraw Hill, New York 1940

Kidder, G. W.: Protozoa. In: I. M. Florkin, B. T. Scheer (eds): Chemical zoology, vol. I. Academic Press, London 1967

Kudo, R. R.: Protozoology 6th ed. Thomas, Springfield 1971

Lee, J. J., S. H. Hunter, E. C. Bovee (eds): Illustrated guide to the protozoa. Allen Press, Lawrence 1985

Levandowsky, M., S. H. Hutner: Biochemistry and physiology of protozoa, vol. I–V. Academic Press, London 1979–1981

Mackinnon, D. L., R. S. J. Hawes: An introduction to the study of protozoa. Clarendon, Oxford 1961

Margulis, L., J. O. Corliss, M. Melkonian, D. J. Chapman (eds): Handbook of protoctista. Jones & Barlett, Boston 1990

Margulis, L., H. I. McKhann, L. Olendzenski (eds): Illustrated glossary of protoctista. Jones & Barlett, Boston 1993

Mehlhorn, H. A. Ruthmann: Allgemeine Protozoologie. Gustav Fischer Verlag, Jena 1992

Pitelka, D. R.: Electron-microscopic structure of protozoa. Pergamon, Oxford 1963

Puytorac, de P., J. Grain, J.-P. Mignot: Precis des protistologie. Boubée, Paris 1987

Rondanelli, E. G., M. Scaglia: Atlas of human protozoa. Masson, Milano 1993

Sleigh, M.: Protozoa and other protists, 2nd ed. Arnold, London 1989 (first published as "The biology of protozoa," 1973)

Monographs on the Four Principal Organization Types of Protozoa

Flagellates

Bold, H. C., M. J. Wynne: Introduction to the algae. Structure and reproduction. Prentice-Hall, Engelwood Cliffs 1978

Bourrelly, P.: Les algues d'eau douce, tome I–III. Boubée, Paris 1966–1970

Cox, E. R. (ed): Phytoflagellates. Elsevier Science Publishers, Amsterdam 1980

Dodge, J. D.: The fine structure of algal cells. Academic Press, London 1973

Green, J. C., B. S. C. Leadbeater (eds): The haptophyte algae. Oxford University Press, Oxford 1994

Hoare, C. A.: The trypanosomes of mammals. Blackwell Scientific Publisher, Oxford 1972

Hoek, C. van den, D. G. Mann, H. M. Jahns: Algae. Cambridge University Press, Cambridge 1995

Lumsden, W. H. R., D. A. Evans: Biology of the Kinetoplastida, vol. I + II. Academic Press, London 1976 + 1979

Moestrup, Ø.: Flagellar structure in algae: A review, with new observations particularly on the Chrysophyceae, Phaeophyceae (Fucophyceae), Euglenophyceae, and *Reckertia*. Phycologia 21 (1982) 427

Patterson, D. J., J. Larsen: The biology of free-living heterotrophic flagellates. Clarendon Press, Oxford 1991

Pickett-Heaps, J. D.: Green algae. Structure, reproduction and evolution in selected genera. Sinauer, Sunderland 1975

Pringsheim, E. G.: Farblose Algen: Ein Beitrag zur Evolutionsforschung. Gustav Fischer Verlag, Jena 1963

Round, F. E.: The ecology of algae. Cambridge University, Cambridge 1983

Spector, D. L.: Dinoflagellates. Academic Press, New York 1984

Taylor, F. J. R. (ed): The biology of dinoflagellates. Blackwell Scientific Publications, Oxford 1987

Wiessner, W., D. G. Robinson, R. C. Starr (eds): Algal development. Molecular and cellular aspects. Springer Verlag, Berlin 1987

Sarcodines

Anderson, O. R.: Radiolaria. Springer Verlag, Berlin 1983

Boltovsky, E., R. Wright: Recent Foraminifera. Junk, The Hague 1976

Hedley, R. H., C. G. Adams: Foraminifera, vol. I + II. Academic Press, London 1974–1978

Hemleben, C. H., M. Spindler, O. R. Anderson: Modern planktonic Foraminifera. Springer Verlag, Berlin 1989

Jones, R. W.: The Challenger Foraminifera. Oxford University Press, Oxford 1994

Lee, J. J., O. R. Anderson: Biology of Foraminifera. Academic Press, London 1991

Loeblich, A. R., H. Tappan: Foraminiferal genera and their classification, part I + II. Chapman & Hall, New York 1988

Martinez, A. J.: Free-living amebas: natural history, prevention, diagnosis, pathology, and treatment of disease. CRC Press, Boca Raton 1985

Murray, J. W.: Ecology and palaeoecology of benthic Foraminifera. Wiley & Sons, New York 1991

Page, F. C., F. J. Siemensma: Nackte Rhizopoden und Heliozoea. In: Matthes, D. (Hrsg.): Protozoenfauna, Vol. 2. Gustav Fischer Verlag, Stuttart 1991

Rondanelli, E. G. (ed): Amphizoic amoebae. Human Pathology. Piccin, Padova 1987

Sporozoans

Davies, S. F. M., L. P. Joyner, S. B. Kendall: Coccidiosis. Oliver & Boyd, London 1963

Garnham, P. C. C.: Malaria parasites and other haemosporidia. Blackwell, Oxford 1966

Hammond, D. M., P. L. Long: The Coccidia. University Park, Baltimore 1973

Kreier, J. P.: Malaria, vol. I–III. Academic Press, London 1980

Levine, N. D.: Protozoan parasites of domestic animals and of man, 2nd ed. Burgess, Minneapolis 1973

Levine, N. D.: The protozoan phylum Apicomplexa, vol. I + II. CRC Press, Boca Raton 1988

Long, P. L.: The biology of Coccidia. Arnold, London 1982

Scholtyseck, E.: Fine structure of parasitic protozoa. Springer Verlag, Berlin 1979

Ciliates

Corliss, J. O.: The ciliated protozoa. Characterization, classification and guide to the literature, 2nd ed. Pergamon, Oxford 1979

Curds, C. R.: British and other freshwater ciliated protozoa, Part I. Ciliophora: Kinetofragminophora. In: Kermack, D. M., R. S. K. Barnes (eds): Synopsis of the British fauna. Cambridge University, Cambridge 1982

Curds, C. R., M. A. Gates, D. McL. Roberts: British and other freshwater ciliated protozoa, Part II. Ciliophora: Oligohymenophora and Polyhymenophora. In: Kermack, D. M., R. S. K. Barnes (eds): Synopsis of the British fauna. Cambridge University, Cambridge 1983

Dragesco, J.: Cilies libres du Cameroun. Ann. Fac. Sci. Yaounde, Yaounde 1970

Dragesco, J., A. Dragesco-Kernéis: Ciliés mesopsammiques littoraux. Systematique, morphologie, ecologie. Trav. Stat. Biol. Roscoff (N. S.) 12 (1960) 1

Dragesco, J., A. Dragesco-Kernéis: Ciliés libres de l'Afrique intertropicale. Editions de l'ORSTOM, Paris 1986

Fenchel, T., B. J. Finlay: The biology of free-living anaerobic ciliates. Europ. J. Protistol. 26 (1991) 201

Hausmann, K., P. C. Bradbury (eds): Ciliates: cells as organisms. Gustav Fischer, Stuttgart 1996

Jones, A. R.: The ciliates. Hutchinson, London 1974

Lynn, D. H.: The organization and evolution of microtubular organelles in ciliated protozoa. Biol. Rev. 56 (1981) 243

Matthes, D., W. Guhl, G. Haider: Suctoria und Urceolariidae (Peritricha). In: Matthes, D. (Hrsg.): Protozoenfauna, vol. 7, 1. Gustav Fischer Verlag, Stuttgart 1988

Monographs on Selected Groups and Species

Acanthamoeba
Byers, T. J.: Growth, reproduction and differentiation in *Acanthamoeba*. Int. Rev. Cytol. 61 (1979) 283

Amoeba
Jeon, K. W. (ed): The biology of *Amoeba*. Academic Press, London 1973

Blepharisma
Giese, A. C.: *Blepharisma*. The biology of a ligth-sensitive protozoon. Stanford University, Stanford 1973

Chlamydomonas
Harris, E. H.: The *Chlamydomonas* sourcebook: a comprehensive guide to biology and laboratory use. Academic Press, London 1989

Chrysomonadida
Duff, K. E., B. A. Zeeb, J. P. Smol: Atlas of chrysophycean cysts. Kluwer Academic Publishers, Dordrecht 1995

Colpodea
Foissner, W.: Class Colpodea (Ciliophora). In: Matthes, D. (Hrsg.): Protozoenfauna, Vol. 4/1. Gustav Fischer Verlag, Stuttgart 1993

Dictyostela
Raper, K. B.: The dictyostelids. Princeton University Press, Princeton 1984

Entamoeba
Ravdin, J. (ed): Amebiasis: human infection by *Entamoeba* histolytica. Churchill, Livingstone 1988

Euglena
Buetow, D. E.: The biology of *Euglena*, vol. I–IV. Academic Press, London 1968–1989

Gojdics, M.: The genus *Euglena*. University of Wisconsin, Madison 1953

Leedale, G. F.: Euglenoid flagellates. Prentice-Hall, Englewood Cliffs 1967

Leedale, G. F.: The euglenoids. Oxford University Press, Oxford 1971

Wolken, J. J.: *Euglena*. An experimental organism for biochemical and biophysical studies, 2nd ed. Appleton, New York 1967

Folliculinidae
Hadži, J.: Studien über Follikuliniden. Dela Slov. Akad. Znan. Umet. Hist. Nat. Med. 4 (1951) 1

Giardia
Meyer, E. A. (ed): Giardiasis. Elsevier Science Publishers, Amsterdam 1990

Thompson, R. C. A., J. A. Reynoldson, A. J. Lymbery (eds) *Giardia*: from molecules to diseases. CAB International, Oxon 1994

Mallomonas

Lavau, S., R. Wetherbee: Structure and development of the scale case of *Mallomonas adamas* (Synurophyceae). In: Wetherbee, R., R. A. Andersen, J. D. Pickett-Heaps (eds): The protistan cell surface. Springer Verlag, Wien 1994

Siver P. A.: The biology of *Mallomonas*. Morphology, taxonomy and ecology. Kluwer Academic Publishers, Dordrecht 1991

Microspora

Bulla, L. A., T. C. Cheng: Comparative Pathobiology, vol. 1: Biology of the Microsporidia, vol. II: Systematics of the Microsporidia. Plenum, London 1976, 1977

Canning, E. U., J. Lom: The Microsporidia of vertebrates. Academic Press, London 1986

Sprague, W.: Systematics of the Microsporidia. Plenum, New York 1977

Weiser, J.: Die Mikrosporidien als Parasiten der Insekten. In: Zwölfer, W. (Hrsg.): Monographien zur Angewandten Entomologie, Nr. 17. Paul Parey, Hamburg 1961

Mycetozoans

Olive, L. S.: The myzetozoans. Academic Press, London 1975

Naegleria

John, D. T.: Primary amebic meningoencephalitis and the biology of *Naegleria fowleri*. Ann. Rev. Microbiol. 36 (1982) 101

Opalina

Metcalf, M. M.: The opalinid ciliate infusorians. U. S. Nat. Mus. Bull. 120 (1923) 1

Metcalf, M. M.: Further studies on the opalinid ciliate infusorians and their hosts. Proc. U. S. Nat. Mus. 87 (1940) 465

Ophryoglena

Canella, M. F., I. Rocchi-Canella: Biologia des *Ophryoglena* (ciliés hyménostomes histophages). Ann. Univ. Ferrara 3 (1976) 1

Paramecium

Ehret, C. F., G. de Haller: Origin, development, and maturation of organelles and organelle systems of the cell surface in *Paramecium*. J. Ultrastruct. Res., Suppl. 6 (1963) 1

Görtz, H.-D.: *Paramecium*. Springer Verlag, Berlin 1988

Janisch, R.: Biomembranes in the life and regeneration of *Paramecium*. Purkyne University, Brno 1987

Jurand, A., G. G. Selman: The anatomy of *Paramecium aurelia*. Macmillan St. Marin's Press, New York 1969

Kalmus, H.: *Paramecium*. Das Pantoffeltierchen. Gustav Fischer Verlag, Jena 1931

van Wagtendonk, W. J.: *Paramecium*. A current survey. Elsevier Science Publishers, Amsterdam 1974

Wichterman, R.: The biology of *Paramecium*, 2nd ed. Plenum, New York 1986

Pneumocystis

Hopkin, J. M.: *Pneumocystis carinii*. Oxford University Press, Oxford 1991

Walzer, P. D.: *Pneumocystis carinii* pneumonia, 2nd ed. Marcel Dekker, New York 1993

Stentor

Tartar, V.: The biology of *Stentor*. Pergamon, Oxford 1961

Tetrahymena

Elliott, A.: The biology of *Tetrahymena*. Dowden, Hutchinson & Ross, Stroudsburg 1973

Hill, D. L.: The biochemistry and physiology of *Tetrahymena*. Academic Press, London 1972

Toxoplasma

Dubey, J. P., C. P. Beattie: Toxoplasmosis of animals and man. CRC Press, Boca Raton 1988

Trichomonadida

Kulda, J., J. Cerkasov (eds): Trichomonads and trichomoniasis. Acta Universitatis Carolinae – Biologica 30 (1986) 177

Trypanosoma

Englund, P. T., S. L. Hajduk, J. C. Marini: The molecular biology of trypanosomes. Ann. Rev. Biochem. 51 (1982) 695

Suza, W. de: Cell biology of *Trypanosoma cruzi*. Int. Rev. Cytol. 86 (1984) 197

Urceolariidae

Haider, G.: Monographie der Familie Urceolariidae (Ciliata, Peritricha, Mobilia) mit besonderer Berücksichtigung der im süddeutschen Raum vorkommenden Arten. Parasitol. Schr. Reihe 17 (1964) 1

General Systematics

Cavalier-Smith, T.: Eukaryote kingdoms: seven or nine? BioSystems 14 (1981) 461

Cavalier-Smith, T.: Kingdom Protozoa and its 18 phyla. Microbiol. Rev. (1993) 953

Corliss, J. O.: An interim utilitarian ("user-friendly") hierarchical classification and characterization of the protists. Acta Protozool. 33 (1994) 1

Grell, K. G.: Unterreich Protozoa, Einzeller oder Urtiere. In: Gruner, H.-E. (Hrsg.): Lehrbuch der Speziellen Zoologie, 5. Aufl. Gustav Fischer Verlag, Stuttgart 1993

Honigberg, B. M., E. C. Bovee, J. O. Corliss, M. Gojdics, R. P. Hall, R. R. Kudo, N. D. Levine, A. R. jr. Loeblich, J. Weiser, H. D. Wenrich: A revised classification of the phylum Protozoa. J. Protozool. 11 (1964) 7

Levine, N. D., J. O. Corliss, F. E. G. Cox, G. Deroux, J. Grain, B. M. Honigberg, G. F. Leedale, A. R. Loeblich III, J. Lom, D. Lynn, E. G. Merinfeld, F. G. Page, G. Poljanski, V. Sprague, J. Vavra, F. G. Wallace: A newly revised classification of the protozoa. J. Protozool. 27 (1980) 37

Evolution

Fenchel, T., B. J. Finlay: The evolution of life without oxygen. Amer. Sci. 82 (1994) 22
Gibbs, S. R.: The chloroplast endoplasmatic reticulum: structure, function, and evolutionary significance. Int. Rev. Cytol. 72 (1981) 49
Gray, M. W., W. F. Doolittle: Has the endosymbiont hypothesis been proven? Microbiol. Rev. 46 (1982) 1
Hülsmann, N., K. Hausmann: Towards a new perspective in protozoan evolution. Europ. J. Protistol. 30 (1994) 365
Jeon, K. W.: Integration of bacterial endosymbionts in amoebae. Int. Rev. Cytol. (suppl.) 14 (1983) 29
Lee, R. E., P. Kugrens: Relationship between the flagellates and the ciliates. Microbiol. Rev. 56 (1992) 529
Mignot, J.-P.: Les coenobes chez les Volvocales: un example du passage des unicellulaires aux pluricellulaires. Ann. Biol. 24 (1985) 1
Ragan, M. A., D. J. Chapman: A biochemical phylogeny of the protists. Academic Press, London 1978
Schlegel, M.: Protist evolution and phylogeny as discerned from small subunit ribosomal RNA sequence comparisons. Europ. J. Protistol. 27 (1991) 207
Sitte, P.: Symbiogenetic evolution of complex cells and complex plastids. Europ. J. Protistol. 29 (1993) 131
Sogin, M. L., H. J. Elwood, J. H. Gunderson: Evolutionary diversity of eukaryotic small-subunit rRNA genes. Proc. Natl. Acad. Sci. USA 83 (1986) 1383
Taylor, F. J. R.: On dinoflagellate evolution. BioSystems 13 (1980) 65
Taylor, F. J. R., D. J. Blackbourn, J. Blackbourn: The readwater ciliate Mesodinium rubrum and its "incomplete symbionts". A review including new ultrastructural observations. J. Fish. Res. Bd. Canada 28 (1971) 391

Morphology and Physiology

Skeletal Structures

Amos, L. A., W. B. Amos: Molecules of the cytoskeleton. Macmillan Education, London 1991
Cachon, J. P., M. Cachon: Les systèmes axopodiaux. Ann. Biol. 13 (1974) 523

Corliss, J. O., S. C. Esser: Comments on the role of the cyst in the life cycle and survival of free-living protozoa. Trans. Amer. Microsc. Soc. 93 (1974) 578
Grain, J.: The cytoskeleton in protists: nature, structure, and functions. Int. Rev. Cytol. 104 (1986) 153
Melkonian, M., R. A. Andersen, E. Schnepf: The cytoskleton of flagellate and ciliate protists. Springer Verlag, Wien 1991
Morrill, L. C., A. R. Loeblich III: Ultrastructure of the dinoflagellate amphiesma. Int. Rev. Cytol. 82 (1983) 151
Mulisch, M.: Chitin in protistan organisms. Distribution, synthesis, and deposition. Europ. J. Protistol. 29 (1993) 1
Wetherbee, R., R. A. Anderson, J. D. Pickett-Heaps (eds): The protistan cell surface. Springer Verlag, Wien 1994

Holdfast Organelles

Hovasse, R., J.-P. Mignot, L. Joyon, J. Baudoin: Etude comparée des dispositifs servants à la fixation chez les protistes. Ann. Biol. 11 (1972) 1

Extrusomes

Haacke-Bell, B., R. Hohenberger-Bregger, H. Plattner: Trichocysts of Paramecium: secretory organelles in search of their function. Europ. J. Protistol. 25 (1990) 289
Hausmann, K.: Extrusive organelles in protists. Int. Rev. Cytol. 52 (1978) 197
Hovasse, R.: Trichocystes, corps trichocystoides, cnidocystes et colloblastes. Protoplasmatologia III F (1965) 1
Hovasse, R., J.-P. Mignot: Trichocystes et organites analogues chez les protistes. Ann. Biol. 14 (1975) 397
Krüger, F.: Die Trichocysten der Ciliaten im Dunkelfeldbild. Zoologica 91 (1936) 1
Kugrens, P., R. E. Lee, J. O. Corliss: Ultrastructure, biogenesis, and functions of extrusive organelles in selected non-ciliate protists. In: Wetherbee, R., R. A. Andersen, J. D. Pickett-Heaps (eds): The protistan cell surface. Springer Verlag, Wien 1994

Contractile Vacuoles

Dodge, J. D.: The ultrastructure of the dinoflagellate pusule: a unique osmoregulatory organelle. Protoplasma 75 (1972) 285
Hausmann, K., D. J. Patterson: Contractile vacuole complexes in algae. In: Wiesner, W., D. Robinson, R. C. Starr (eds): Compartments in algal cells and their interaction. Springer Verlag, Berlin 1984

Heuser, J., Q. Zhu, M. Clarke: Proton pumps populate the contractile vacuoles of *Dictyostelium* amoebae. J. Cell Biol. 121 (1993) 1311

Kitching, J. A.: Contractile vacuoles. Biol. Rev. 13 (1938) 403

Kitching, J. A.: Contractile vacuoles of protozoa. Protoplasmatologia III D (1956) 1

Patterson, D. J.: Contractile vacuoles and associated structures: their organization and function. Biol. Rev. 55 (1980) 1

Selected Organelles

Fenchel, T., B. J. Finlay: The structure and function of Müller vesicles in loxodid ciliates. J. Protozool. 33 (1986) 69

Hannert, V., P. A. M. Michels: Structure, function and biogenesis of glycosomes in Kinetoplastida. J. Bioenerg. Biomembr. 26 (1994) 205

Lynn, D. H., S. Frombach, M. S. Ewing, K. M. Kocan: The organelle of Lieberkühn as a synapomorphy for the Ophryoglenina (Ciliophora: Hymenostomataida). Trans. Am. Microsc. Soc. 110 (1991) 1

Müller, M.: The hydrogenosome. J. Gen. Microbiol. 139 (1993) 2879

Opperdoes, F. R., P. A. M. Michels: The glycosomes of the Kinetoplastida. Biochimie 75 (1993) 231

Plattner, H. (ed): Membrane traffic in protozoa, part I + II. JAI Press, Greenwich 1993

Motility

Allen, R. D., N. Kamiya (eds): Primitive motile systems in cell biology. Academic Press, New York 1964

Amos, W. B., J. G. Duckett (eds): Prokaryotic and eukaryotic flagella. Symposium no. 35 of the Society of Experimental Biology. Cambridge University Press, Cambridge 1982

Bloodgood, R. A. (ed.): Ciliary and flagellar membranes. Plenum Press, New York 1990

Cachon, J., M. Cachon: Movement by non-actin filament mechanisms. BioSystems 14 (1981) 313

Febvre-Chevalier, C., J. Febvre: Buoyancy and swimming in marine planktonic protists. In: Maddock, L., Q. Bone, J. M. V. Rayner (eds): Mechanics and physiology of animal swimming. Cambridge University Press, Cambridge 1994

Fukui, Y.: Toward a new concept or cell motility: cytoskeletal dynamics in amoeboid movement and cell division. Int. Rev. Cytol. 144 (1993) 85

Grębecki, A.: Cortical flow in free-living amoebae. Int. Rev. Cytol. 148 (1994)

Komnick, H., W. Stockem, K.-E. Wohlfarth-Bottermann: Cell motility: mechanisms in protoplasmic streaming and amoeboid movement. Int. Rev. Cytol. (1973) 169

Melkonian, M. (ed): Algal cell motility. Current Phycology, vol. III. Chapman and Hall, New York 1992

Parducz, B.: Ciliary movement and coordination in ciliates. Int. Rev. Cytol. 21 (1967) 91

Preston, T. M., C. A. King, J. S. Hyams: The cytoskeleton and cell motility. Chapman and Hall, New York 1990

Sleigh, M. A.: Cilia and flagella. Academic Press, London 1974

Stebbings, H., J. S. Hyams: Cell motility. Longman, London 1979

Stockem, W., W. Kłopocka: Ameboid movement and related phenomena. Int. Rev. Cytol. 112 (1988) 137

Stockem, W., K. Brix: Analysis of microfilament organization and contractile activities in Physarum. Int. Rev. Cytol. 149 (1994) 145

Taylor, D. L., J. S. Condeelis: Cytoplasmic structure and contractility in amoeboid cells. Int. Rev. Cytol. 56 (1979) 57

Ingestion, Digestion, Defecation

Allen, R. D.: Membranes of ciliates: ultrastructure, biochemistry and fusion. In: Poste, G., L. Nicolson (eds): Membrane fusion. Elsevier Science Publishers, Amsterdam 1978

Allen, R. D.: *Paramecium* phagosome membrane: from oral region to cytoproct and back again. J. Protozool. 31 (1984) 1

Chapman-Andresen, C.: Studies on pinocytosis in amoebae. C. R. Lab. Carlsberg 33 (1962) 73

Dutta, G. P.: Recent advances in the cytochemistry and ultrastructure of cytoplasmic inclusions in Mastigophora and Opalinata (Protozoa). Int. Rev. Cytol. 36 (1973) 93

Dutta, G. P.: Recent advances in cytochemistry and ultrastructure of cytoplasmic inclusion in Ciliophora (Protozoa). Int. Rev. Cytol. 39 (1974) 285

Kitching, J. A.: Food vacuoles. Protoplasmatologia III D (1965) 1

Mast, S. O: The food-vacuole in *Paramecium*. Biol. Bull. 92 (1947) 31

Nilsson, J. R.: Physiological and structural studies on *Tetrahymena pyriformis* GL. C.R. Lab. Carlsberg 40 (1976) 215

Nisbet, B.: Nutrition and feeding strategies in protozoa. Croom Helm, London 1984

Radek, R., K. Hausmann: Endocytosis, digestion, and defecation in flagellates. Acta Protozoologica 33 (1994) 127

Schnepf, E., M. Elbrächter: Nutritional strategies in dinoflagellates. A review with emphasis on cell biological aspects. Europ. J. Protistol. 28 (1992) 3

Sikora, J.: Cytoplasmic streaming in *Paramecium*. Protoplasma 109 (1981) 57

Stockem, W., K. E. Wohlfarth-Bottermann: Pinocytosis (endocytosis). In: Lima de Faria, A. (ed): Handbook of molecular cytology. Elsevier Science Publishers, Amsterdam 1969

Nuclei and Sexuality

Beale, G. H.: The genetics of *Paramecium aurelia*. Monographs of experimental biology. Cambridge University Press, Cambridge 1954

Bělař, K.: Der Formwechsel der Protistenkerne. Gustav Fischer Verlag, Jena 1926

Bell, G.: Sex and death in protozoa. The history of an obsession. Cambridge University Press, Cambridge 1989

Dini, F., D. Nyberg: Sex in ciliates: In: Gwynfryn Jones, J. (ed): Advances in Microbioal Ecology, Plenum Press, New York 1993

Heywood, P., P. T. Magee: Meiosis in protists. Bact. Rev. 40 (1976) 190

Miyake, A.: Cell communication, cell union, and initiation of meiosis in ciliate conjugation. Curr. Top. Dev. Biol. 12 (1978) 37

Ng, S. F.: Developmental heterochrony in ciliated protozoa: overlap of asexual and sexual cycles during conjugation. Biol. Rev. 65 (1990) 19

Raikov, 1. B.: The protozoan nucleus. Morphology and evolution. Springer Verlag, Wien 1982

Smith-Sonneborn, J.: Genetics and aging in protozoa. Int. Rev. Cytol. 73 (1981) 319

Morphogenesis and Reproduction

Aufderheide, K. J., J. Frankel, N. E. Williams: Formation and positioning of surface-related structures in protozoa. Microbiol. Rev. 44 (1980) 252

Frankel, J. (ed): Pattern formation: ciliate studies and models. Oxford University Press, New York 1989

Fulton, C.: Cell differentiation in *Naegleria gruberi*. Ann. Rev. Microbiol. 31 (1977) 597

Grain, J., J. Bohatier: La régénération chez les protozoaires ciliés. Ann. Biol. 16 (1977) 194

Grimes, G. W., K. J. Aufderheide: Cellular aspects of pattern formation: the problem of assembly. Monographs in developmental biology, vol. 22. Karger, Basel 1991

Grimstone, A. V.: Fine structure and morphogenesis in protozoa. Biol. Rev. 36 (1961) 97

Jeon, K. W., J. F. Danielli: Microsurgical studies with large free-living amebas. Int. Rev. Cytol. 56 (1979) 57

Jerka-Dziadosz, M.: Control of pattern formation in ciliated protozoa. In: Nover, L., F. Lynen, K. Mothes (eds): Cell compartimentation and metabolic channeling, Gustav Fischer Verlag, Jena 1980

Lwoff, A.: Problems of morphogenesis in ciliates. Wiley, New York 1950

Nanney, D. L.: Experimental ciliatology: an introduction to genetic and developmental analysis in ciliates. Wiley, New York 1980

Molecular Biology

Blackburn, E. H.: Structure and function of telomeres. Nature 350 (1991) 569

Blum, B., N. Bakalara, L. Simpson: A model for RNA editing in kinetoplastid mitochondria: "Guide" RNA molecules transcribed from maxicircle DNA provide edited information. Cell 60 (1990) 189

Borst, P., J. H. Gommers-Ampt, M. J. L. Ligtenberg, G. Rudenko, R. Kieft, M. C. Taylor, P. A. Blundell, F. van Leeuwen: Control of antigenic variation in African trypanosomes. Cold Spring Harb. Symp. Quant Biol. 58 (1993) 105

Cech, T. R.: RNA as an enzyme. Sci. Amer. 11 (1986) 76

Cross, G. A. M.: Cellular and genetic aspects of antigenic variation in trypanosomes. Annu. Rev. Immunol. 8 (1990) 83

Feagin, J. E.: The extrachromosomal DNAs of apicomplexan parasites. Annu. Rev. Microbiol. 48 (1994) 81

Gall, J. G. (ed.): The molecular biology of ciliated protozoa. Academic Press, New York 1986

Howard, R. J., B. L. Pasloske: Target antigens for asexual malaria vaccine development. Parasitology Today 9 (1993) 369

Kemp, D. J., R. L. Coppel, R. F. Anders: Repetitive proteins and genes of malaria. Annu. Rev. Microbiol. 41 (1987) 181

Pays, E., L. Vanhamme, M. Berberof: Genetic controls for the expression of surface antigens in African trypanosomes. Annu. Rev. Microbiol. 48 (1994) 25

Prescott, D. M.: The DNA of ciliated protozoa. Microbiol. Rev. 58 (1994) 233

Steinbrück, G.: Recent advances in the study of ciliate genes. Europ. J. Protistol. 26 (1990) 2

Behavior

Carlile, M. J.: Primitive sensory and communication systems. The taxes and tropisms of micro-organisms and cells. Academic Press, London 1975

Colombetti, G.: New trends in photobiology: photomotile responses in ciliated protozoa. J. Photochem. Photobiol. B 15 (1990) 243

Jennings, H. S.: Behaviour of lower organisms. Columbia University Press, New York 1906

Kung, C., Y. Saimi: The physiological basis of taxes in *Paramecium*. Ann. Rev. Physiol. 44 (1982) 519

Machemer, H.: Motor control of cilia. In: Görtz, H. D. (ed): *Paramecium*. Springer Verlag, Berlin 1988

Machemer, H.: Cellular behaviour modulated by ions: electrophysiological implications. J. Protozool. 36 (1989) 563

Machemer, H.: Ciliate sensory physiology. In: Elsner, N., H. Beer (eds): Proceedings of the 22nd Göttingen Neurobiology Conference, vol. I. Thieme Verlag, Stuttgart 1994

Machemer, H., R. Bräucker: Gravireception and graviresponses in ciliates. Acta Protozool. 31 (1992) 185

Machemer, H., J. W. Deitmer: Mechanoreception in ciliates. In: Autrum, H., D. Ottoson, E. R. Perl, R. F. Schmidt, H. Shima, W. D. Willis (eds): Progress in Sensory Physiology, vol. 5. Springer Verlag, Heidelberg 1985

Melkonian, M., H. Robenek: The eyespot apparatus of flagellated green algae: a critical review. Progr. Phycol. Res. 3 (1984) 193

Nakaoka, Y., T. Kurotani, H. Itoh: Ionic mechanism of thermoreception in *Paramecium*. J. Exp. Biol. 127 (1987) 95

Nultsch, W., D.-P. Häder: Photomovement of motile microorganisms. Photochem. Photobiol. 29 (1979) 423

Schöne, H.: Orientierung im Raum. Formen und Mechanismen der Lenkung des Verhaltens im Raum bei Tier und Mensch. Wissenschaftliche Verlagsgesellschaft, Stuttgart 1980

Song, P.-S., K. L. Poff: Photomovement. In: Smith, K. C. (ed): The science of photobiology, 2nd ed., Plenum, New York 1989

Tominaga, T., Y. Naitoh: Comparison between thermoreceptor and mechanoreceptor currents in *Paramecium caudatum*. J. Exp. Biol. 189 (1994) 117

Van Houten, J.: Chemosensory transduction in eukaryotic microorganisms. Ann. Rev. Physiol. 54 (1992) 639

Ecology

Free-living Protozoa

Arndt, H.: A critical review of the importance of rhizopods (naked and testate amoebae) and actinopods (heliozoa) in lake plankton. Marine Microbial Food Webs 7 (1993) 3

Baden, D. G.: Marine food-borne dinoflagellate toxins. Int. Rev. Cytol. 82 (1983) 99

Boney, A. D.: Phytoplankton, 2nd ed. Arnold, New York 1989

Capriulo, G. M. (ed): Ecology of marine protozoa. Oxford University Press, New York (1990)

Cairns, J.: Freshwater protozoan comunities. In: Bull, A. T., J. H. Slater (eds): Microbial interactions and communities, vol. 1. Academic Press, London 1982

Curds, C. R.: The ecology and role of protozoa in aerobic sewage treatment processes. Ann. Rev. Microbiol. 36 (1982) 27

Curds, C. R.: Protozoa in the water industry. Cambridge University Press, Cambridge 1992

Darbyshire, J. E. (ed): Soil protozoa. CAB International, Oxon 1994

Fenchel, T.: The ecology of marine microbenthos, I–IV. Ophelia 4 (1967) 121; 5 (1968) 73; 5 (1968) 123; 6 (1969) 1

Fenchel, T.: Ecology of protozoa: the biology of free-living phagotrophic protists. Springer Verlag, Berlin 1987

Fenchel, T., B. J. Finlay: Communities and evolution in anoxic worlds. In: May, R. M., P. Harvey (eds): Oxford series in ecology and evolution. Oxford University Press, Oxford 1995

Finlay, B. J.: Physiological ecology of free-living protozoa. Adv. Microb. Ecol. 11 (1990) 1

Finlay, B. J., C. Ochsenbein-Gattlen: Ecology of free-living protozoa. Rep. Freshw. Biol. Ass. 17 (1982) 1

Finlay, B. J., T. Fenchel: Physiological ecology of the ciliated protozoon *Loxodes*. Rep. Freshw. Biol. Ass. 54 (1986) 73

Fukuyo, Y., H. Takano, M. Chihara, K. Matsuoka: Red tide organisms in Japan. Uchida Rokakuho, Tokyo 1990

Granéli, E., B. Sundström, D. M. Anderson: Toxic marine phytoplankton. Elsevier Science Publishers, Amsterdam 1990

Hallegraeff, G. M.: Aquaculturists' guide to harmful Australian microalgae. SCIRO, Hobart 1991

Laybourn-Parry, J.: Protozoan plankton ecology. Chapman & Hall, New York 1992

Nilsson, J. R.: *Tetrahymena* in cytotoxicology: with special reference to effects of heavy metals and selected drugs. Europ. J. Protistol. 25 (1989) 2

Sieburth, J. McN: Sea microbes. Oxford University, New York 1979

Parasites

Aikawa, M., C. R. Sterling: Intracellular parasitic protozoa. Academic Press, London 1974

Baker, J. R.: Parasitic protozoa. Hutchinson, London 1969

Dollet, M.: Plant diseases caused by flagellate protozoa *(Phytomonas)*. Ann. Rev. Phytopathol. 22 (1984) 115

Gardiner, C. H., R. Fayer, J. P. Dubey: An atlas of protozoan parasites and animal tissues. US Dept. of Agriculture, Agriculture Handbook No. 651, Washington 1988

Kreier, J. P.: Parasitic protozoa, 8 volumes, 2nd ed Academic Press, London 1991–1994

Kreier, J. P., J. R. Baker: Parasitic protozoa. Allen & Unwin, Winchester 1987

Mehlhorn: Parasitology in focus. Springer Verlag, Berlin 1988

Lom, J., I. Dykova: Protozoan parasites of fishes. Elsevier Science Publishers, Amsterdam 1992

Rondanelli, E. G., G. Carosi, G. Filice: Human pathogenic protozoa. Piccin, Padova 1987

Symbionts

Breznak, J. A.: Intestinal microbiota of termites and other xylophagous insects. Ann. Rev. Microbiol. 36 (1982) 323

Finlay, B. J., T. Fenchel: Methanogens and other bacteria as symbionts of free-living anaerobic ciliates. Symbiosis 14 (1992) 375

Görtz, H.-D.: Endonuclear symbionts in ciliates. Int. Rev. Cytol. 14 (suppl.) (1983) 145

Heckmann, K.: Endosymbionts of *Euplotes*. Int. Rev. Cytol. 14 (suppl.) (1983) 111

Honigberg, B. M.: Protozoa associated with termites and their role in digestion: In: Krishna, K., F. M. Weesner (eds): Termites, vol. II. Academic Press, London 1970

Jennings, D. H., D. L. Lee: Symbiosis. Symp. Soc. Exp. Biol. 29 (1975) 175

Lee, J. J.: Perspective on algal endosymbionts in larger Foraminifera. Int. Rev. Cytol. 14 (suppl.) (1983) 49

Ogimoto, K., S. Imai: Atlas of rumen microbiology. Japan Sci. Soc., Tokyo 1981

Preer, J. R., J. L. B. Preer, A. Jurand: Kappa and other endosymbionts in *Paramecium aurelia*. Bact. Rev. 38 (1974) 113

Reisser, W.: Enigmatic chlorophycean algae forming symbiotic associations with ciliates. In: Seckbach, J. (ed): Evolutionary pathways and enigmatic algae: *Cyanidium caldarium* (Rhodophyta) and related cells. Kluwer Academic Publishers, Dordrecht 1994

Reisser, W., W. Wiessner: Autotrophic eukaryotic freshwater symbionts. In: Linskens, H.-F., J. Heslop-Harrison (eds): Cellular interactions. Encyclop. Plant Physiol. (New Series), vol. 17, Springer Verlag, Berlin 1984

Soldo, A. T.: The biology of the xenosome, an intracellular symbiont. Int. Rev. Cytol. 14 (suppl.) (1983) 79

Taylor, D. L.: Chloroplasts as symbiotic organelles. Int. Rev. Cytol. 27 (1970) 29

Williams, A. G., G. S. Coleman: The rumen protozoa. Springer Verlag, Berlin 1992

Identification Keys and Guides

Canter-Lund, H., J. W. G. Lund: Freshwater algae. Their microscopic world explored. Biopress, Bristol 1995

Cash, J., G. H. Walles: The British freshwater rhizopoda and heliozoa, vol. I–V. Roy. Soc., London 1905–1921

Dodge, J. D.: Atlas of dinoflagellates. Ferrand Press, London 1985

Ettl, H., J. Gerloff, H. Heynig, D. Mollenhauer (Hrsg.): Süßwasserflora von Mitteleuropa, 20 Bände. Gustav Fischer Verlag, Stuttgart 1978 – im Druck

Foissner, W., H. Blatterer, H. Berger und F. Kohmann: Taxonomische und ökologische Revision der Ciliaten des Saprobiensystems.
Band 1: Cyrtophorida, Oligotrichida, Hypotrichida, Colpodea.
Band 2: Peritrichia, Heterotrichida, Odontostomatida.
Band 3: Hymenostomata, Prostomatida, Nassulida.
Band 4: Gymnostomata, Suctoria, *Loxodes*.
Informationsberichte des Bayerischen Landesamtes für Wasserwirtschaft, München 1991–1995

Finlay, B. J., A. Rogerson, A. J. Cowling: A beginner's guide to the collection, isolation, cultivation and identification of freshwater protozoa. CCAP, Ambleside 1988

Hausmann, K., D. J. Patterson: Taschenatlas der Einzeller. Protisten. Arten und mikroskopische Anatomie, 2. Aufl. Franckh'sche Verlagshandlung, Stuttgart 1987

Kahl, A.: Urtiere, oder Protozoa. I. Wimpertiere oder Ciliata (Infusoria). In: Dahl, F. (Hrsg.): Die Tierwelt Deutschlands. Gustav Fischer Verlag, Jena 1930–1935

Murray, J. W.: An atlas of British recent foraminiferids. Heinemann, London 1971

Ogden, C. G., R. H. Hedley: An atlas of freshwater testate amoebae. Oxford University, Oxford 1980

Page, F. C.: An illustrated key to freshwater and soil amoebae with notes on cultivation and ecology. Freshwater Biological Association, Sci. Publ. No. 34, Wilson, Kendal 1976

Page, F. C.: Marine Gymnaboebae. Institute of Terrestrial Ecology, printed by Lavenham Press, Lavenham 1983

Page, F. C.: A new key to freshwater and soil gymnamoebae with instructions for culture. Freshwater Biological Association, Ambleside 1988

Patterson, D.: Free-living freshwater protozoa. A color guide. CRC Press, Boca Raton 1992

Streble, H., D. Krauter: Das Leben im Wassertropfen. Mikroflora und Mikrofauna des Süßwassers, 8. Aufl. Franckh'sche Verlagshandlung, Stuttgart 1988

Zhukov, B. F., S. A. Karpov: Freshwater choanoflagellates. Nauka, Leningrad 1985

Cultivation of Protozoa

Lee, J. J., A. T. Soldo: Protocols in protozoology. Society of Protozoologists, Allen Press, Lawrence 1992

Teaching Protozoology

Röttger, R. (Hrsg.): Praktikum der Protozoologie. Gustav Fischer, Stuttgart 1995

Index

The numbers in **bold type** refer to pages with illustrations.